Lecture Notes in Electrical Engineering

Volume 278

For further volumes:
http://www.springer.com/series/7818

Aly A. Farag · Jian Yang · Feng Jiao
Editors

Proceedings of the 3rd International Conference on Multimedia Technology (ICMT 2013)

Editors
Aly A. Farag
Electrical and Computer Engineering
University of Louisville
Kentucky, KY
USA

Feng Jiao
Nanjing University of Information Science and Technology
Nanjing
People's Republic of China

Jian Yang
Department of Electronic Engineering
Tsinghua University
Beijing
People's Republic of China

ISSN 1876-1100
ISSN 1876-1119 (electronic)
ISBN 978-3-662-51375-0
ISBN 978-3-642-41407-7 (eBook)
DOI 10.1007/978-3-642-41407-7
Springer Heidelberg New York Dordrecht London

Softcover reprint of the hardcover 1st edition 2014

Printed on acid-free paper

Springer is part of Springer Science+Business Media (www.springer.com)

Organizing Committee

General Chair

Prof. Aly A. Farag	Professor of University of Louisville, USA
Prof. Rangding Wang	Executive Dean, Ningbo Institutes of Advanced Technology, China

Technical Program Committee Chair

Prof. Jian Yang	Tsinghua University, China
Prof. Naixue Xiong, Ph.D.	School of Computer Science, Colorado Technical University, USA

Publication Chair

Jiao Feng, Ph.D.	Nanjing University of Information Science and Technology, China

Technical Program Committee

Wenyu Liu	Huazhong University of Science and Technology, China
Mingyan Jiang	Shandong University, China
Bo Sun	Beijing Normal University, China
Aiguo Song	Southeast University, China
Canhui Cai	Huaqiao University, China
Yougang Yang	Northwest A&F University, China
Chenglin Zhao	Shaoyang University, China
Chuanhe Huang	Wuhan University, China

Xianglin Huang	Communication University of China, China
Guiping Liao	Hunan Agricultural University, China
Junying Gan	Wuyi University, China
Feng Liang	Zhejiang Wanli University, China
Xiangyang Xue	Fudan University, China
Wei Xu	Jilin Agricultural Science and Technology College, China
Xun Wang	Zhejiang Gongshang University, China
Daifeng Zha	Jiujiang University, China
Fernando Ferri	IRPPS, CNR, Roma
Shashikant Patil	SVKM'S Nmims University, India
Yong BIAN	Canada Center for Remote Sensing, Natural Resources Canada, Canada
Xiaofei ZHANG	Nanjing University of Aeronautics and Astronautics, China
LV Teng	Army Officer Academy, China
Chuanrong ZHANG	University of Connecticut, USA
VPS Naidu	National Aerospace Laboratories, India
Francesco Zirilli	Universita di Roma La Sapienza, Italy
LI Bingzhao	Beijing Institute of Technology, China
Sukumar Senthilkumar	Chon Buk National University, Korea
Rajkumar Kannan	Bishop Heber College, India
Bormin Huang	University of Wisconsin-Madison, USA
Jian WANG	Cyber Physical System R&D Center, The Third Research Institute of Ministry of Public Security, China
Aniruddha Bhattacharjya	Amrita Vishwa Vidyappeetham University, India

Contents

Chapter 1
Spherical Mirror Estimation Using Phase Retrieval Wavefront Sensor Technology

Xinxue Ma, Jianli Wang, Bin Wang and Tianyu Lv

Abstract In order to verify the estimated wavefront ability of the phase retrieval wavefront sensor (PRWS), a measured spherical mirror of experiment platform was set up with the method of PRWS. PRWS technology is based on the focal plane image information wavefront solver in the focal plane wavefront measured technology, whose principle is sampling a number of the given defocus images; get the wavefront phase information by solving the optical system wavefront with Fourier optical diffractive theory and mathematics optimization. In order to validate the veracity of PRWS, both the PRWS measurement results and ZYGO interferometer measurement results were compared, experimental results demonstrate that agreement is obtained among the errors distribution, PV value and RMS value of ZYGO interferometer, so PRWS technology can effectively estimate the aberrations of spherical mirror.

Keywords Phase retrieval · Wavefront sensor · Spherical mirror · Zernike polynomial · Aberration

1.1 Introduction

In the machining processing of large-scale optics mirror in situ real-time estimation and alignment with the use of the optical system during dynamic measurement of wave aberrations is difficult to accomplish for the present tradition optical

---

X. Ma · J. Wang · B. Wang (✉) · T. Lv
Changchun Institute of Optics, Fine Mechanics and Physics, Chinese Academy of Sciences, Changchun 130033, China
e-mail: eatingbeen@hotmail.com

X. Ma
University of Chinese Academy of Sciences, Beijing 100039, China

A. A. Farag et al. (eds.), *Proceedings of the 3rd International Conference on Multimedia Technology (ICMT 2013)*, Lecture Notes in Electrical Engineering 278,
DOI: 10.1007/978-3-642-41407-7_1,

inspection equipment [1–3]. In order to control the optical quality of the telescope, we need a simple and with high accuracy method. This article proposed phase retrieval wavefront sensor (PRWS) technology [4–8] based on the focal plane image information wavefront solver in the focal plane estimated wavefront technology, whose principle is given by sampling a number of the defocus images [9–14]; solve the optical system wavefront by Fourier optical methods. System hardware is simple, free from the environment (especially the vibration) influence of optical components and systems for dynamic estimation, real-time display of measurements [7, 15], which has good application prospects in the field of the optical processing, system alignment, active optics, adapt optics etc.

In order to verify the estimated wavefront ability of PRWS, this paper set up a measured spherical mirror of experiment platform with the method of PRWS based on phase retrieval (PR) theory research and experimental verification. This paper compared PRWS measurement results with ZYGO interferometer [16–22] measurement results, experimental results demonstrate that agreement is obtained among the errors distribution, PV value and RMS value of ZYGO interferometer, so using PRWS technology can effectively test the spherical mirror aberration, which illustrates the feasibility and accuracy of PRWS measurement methods.

This paper is organized as follows: the theory of PRWS is presented in Sect. 1.2, the design of experiment in Sect. 1.3 and the summary in Sect. 1.4.

1.2 The Principles of PRWS

PR system is the wavefront detector of a focal plane waves; a laser spot light on the object plane is a target designated from the focal plane image acquisition, use the acquired image, the defocus of the corresponding image, known pupil size and shape to reverse solve the aberration of the optical system [23]. The structure of the PR system is shown in Fig. 1.1.

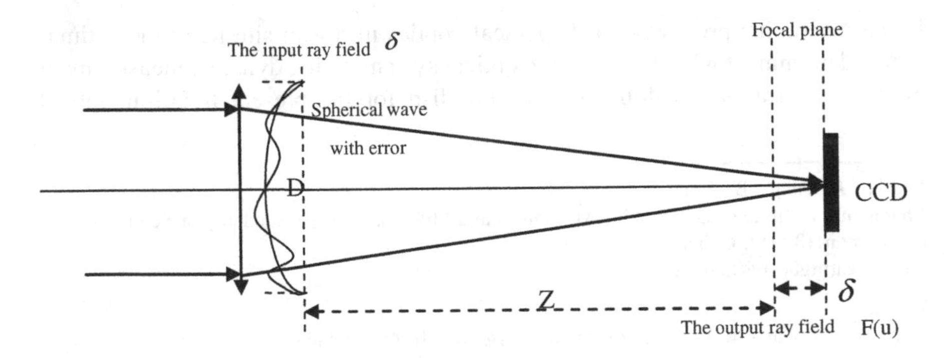

Fig. 1.1 Schematic of optical path of PR

Assuming that the aperture of a measured optical system is D, the focal length is Z, the center wavelength of the laser light source is λ, whose pupil constraint function is $|f(x)|$, the generalized pupil function for focus plane is

$$f(x) \;=\; |f(x)| \;\exp\,[i\theta(x)], \tag{1.1}$$

where θ is wavefront distortion and can be obtained with Zernike polynomial fitting: $\theta(x) = \sum_n \alpha_n Z_n(x)$, the real number α_n represents the first nth terms of polynomial coefficients, Z_n indicates the first nth terms of Zernike polynomials basement.

For linear optical system, when the generalized pupil $f(x)$ whose defocus is δ in the plane, the impulse response function $F(u)$ is

$$F(u) = |F(u)| \;\exp\,[i\psi(u)] = \mathcal{F}\{f(x)\;\exp\,[\varepsilon(x,\delta)]\}, \tag{1.2}$$

where x is the coordinates of the pupil domain, u is the coordinates of the image domain, both of them are two-dimensional vector field coordinates. ψ is the phase part of the impulse response, $\mathcal{F}$ is two-dimensional Fourier transform, $\varepsilon(x,\ \delta)$ is wavefront aberration caused by defocus δ in the position x.

For a PR system, $|f(x)|$ of Eq. (1.1) is the priori conditions of a known optical system, corresponds to the size and shape of the pupil. $|F(u)|^2$ is the image collected by CCD where the defocus is δ. Therefore, the purpose that we estimate wavefront by PR is to get α_n by the above known quantity. So formal description of the problem for: $|f(x)|$, $\delta_1, |F_1(u)|^2$, $\delta_2, |F_2(u)|^2, \ldots, \delta_M$, $|F_M(u)|^2$are known. Image acquisition distance from the focal plane at $\delta_1,\ \delta_2, \ldots, \delta_M$, respectively, is $|F_1(u)|^2$, $|F_2(u)|^2, \ldots,\ |F_M(u)|^2$.

The objective function and the partial derivative of PR objective function with respect to α_n, respectively, is Eqs. (1.3) and (1.4)

$$B_k = E_{Fk}^2 = N^{-2}\sum_{m=1}^{M}\sum_{u}\left[\left|G_{m,\,k}(u)\right| - |F(u)|\right]^2, \tag{1.3}$$

$$\partial_{\alpha_n} B_k \;=\; -2\sum_{m}\sum_{x}|f(x)|\left|g'_{m,k}(x)\right|\;\sin[\theta'_{m,k}(x) - \theta_{m,k}(x)]\;Z_n(x) \tag{1.4}$$

With the objective function (1.3) and its impact on the Zernike coefficient derivative (1.4), we can use the mathematical optimization algorithm to solve various Zernike wavefront coefficient values, here we use L-BFGS algorithm that the phase diversity (PD) [24–33] experiment has been able to solve. The following is solving steps:

Step 0 Selected starting point $\alpha^0 \in \mathcal{R}^n$ and the initial symmetric positive definite matrix $H_0 \in \mathcal{R}^{n\times n}$. Set the number of search accuracy ε bigger than zero and limited memory times m. Compute the gradient $\partial_\alpha B(\alpha^0)$, and order k is zero.

Step 1 If $\|\partial_\alpha B(\alpha^k)\| \leq \varepsilon$, the algorithm will terminate, the optimal solution is α^k,which is the wavefront Zernike coefficient. Otherwise, order d^k is $d^k = -H_k \partial_\alpha B(\alpha^k)$.

Step 2 Adopt the strategy of non-precise linear search, according to the Eqs. (1.3) and (1.4) determining step c_k, update $\alpha^{k+1}$is $\alpha^{k+1} = \alpha^k + c_k d^k$, and according to Eq. (1.4) for calculating the gradient values $\partial_\alpha B(\alpha^{k+1})$.

Step 3 Use the initial value H_0 or the intermediate information structure $H_k^{(0)}$, repeat using Eq. (1.5) for $m + 1$ times amending and get H_{k+1},

$$H_{k+1} = (I - \frac{s_k y_k^T}{s_k^T y_k}) H_k^{(0)} (I - \frac{y_k s_k^T}{s_k^T y_k}) + \frac{s_k s_k^T}{s_k^T y_k}, \tag{1.5}$$

where, $s_k = \alpha^{k+1} - \alpha^k$, $y_k = \partial_\alpha B(\alpha^{k+1}) - \partial_\alpha B(\alpha^k)$.

Step 4 Let $k = k + 1$, turn to Step 1. Where $\alpha = [\alpha_1, \ldots, \alpha_n]'$, α^k represents the value of α for k times iteration. In the L-BFGS algorithm, we only need to store $m + 1$ vector group $\{s_i, y_i\}_{i=k-m}^k$, and calculate the next iteration of the inverse of the Hessian matrix approximation. In practical calculation, usually select the appropriate m to control the amount of storage based on the problem of the size and machine performance. Generally the value of m ranges from 3 to 20, in this paper m is 5.

1.3 The Design of the Experiments

1.3.1 Experimental Theory and Components

The schematic diagram of PRWS is shown in Fig. 1.2. Gaussian beam emitted from the laser through the pinhole into a spherical wave, passes through the lens two into a parallel light, the light projects in the prism through the prism is divided into two parts, a part need not be considered, another part of the parallel reflects after a light through the lens 1 converge after the measured mirror, the reflected beam with phase information (i.e., aberration), again divided into two groups by a beam splitter, one part backtrack, the other part through the converging lens 3 converged at CCD camera, which is placed on a movable platform, move along the optical axis and the angle of the camera posture fine-tuning to get the focus before and after receiving a different amount of defocus images, used in realizing the estimated wavefront based on the PR. We can obtain aberration of the measured spherical mirror with the PR algorithm.

The focal length of measured spherical mirror is 0.2 m, the center wavelength is 632.5 nm, focal length of *3* in the experimental system is 0.15 m, the exit pupil caliber is 0.012 m, and the depth of focus is about 0.286 mm. In the experiment,

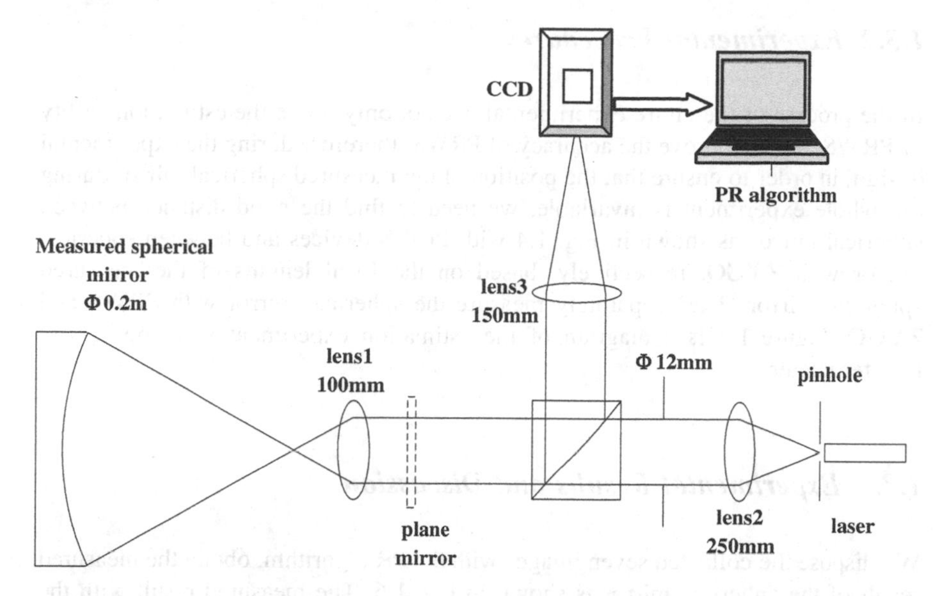

Fig. 1.2 Schematic diagram of PRWS

the defocus we select is 0, ±1, ±1.5, ±2 mm. Camera pixel size is 6.45 μm, each defocus position, respectively, intercept 128*128 pixel size of target region, the exposure time is 20 ms, the accuracy of mobile platform is ±5 μm. The experimental optical path is shown in Fig. 1.3.

Fig. 1.3 The experimental system of PRWS

1.3.2 Experimental Procedures

In the process of the entire experimental, we not only prove the estimation ability of PRWS, but also prove the accuracy of PRWS, therefore during the experimental design, in order to ensure that the position of the measured spherical mirror during the whole experiment is invariable, we need to find the good distance between spherical mirror as shown in Fig. 1.4 with PRWS devices and between spherical mirror with ZYGO, respectively, based on the focal lengths of the measured spherical mirror. Then separately measure the spherical mirror with PRWS and ZYGO. Figure 1.5 is a diagram of the estimation experiment with the ZYGO interferometer.

1.3.3 Experimental Results and Discussion

We dispose the collected seven images with the PR algorithm, obtain the measured result of the spherical mirror is shown in Fig. 1.6. The measured result with the ZYGO interferometer is shown in Fig. 1.7.

In order to illustrate the accuracy and viability of the PRWS better, we rotated the spherical mirror a definite degree, and then estimate the spherical mirror with steps one to nine, the obtained measurement results are shown in Figs. 1.8 and 1.9.

In order to illustrate the accuracy and repeatability of PR, we rotated the spherical mirror some degree, and then estimate the spherical mirror with steps one to nine, the obtained measurement results are shown in Fig. 1.10.

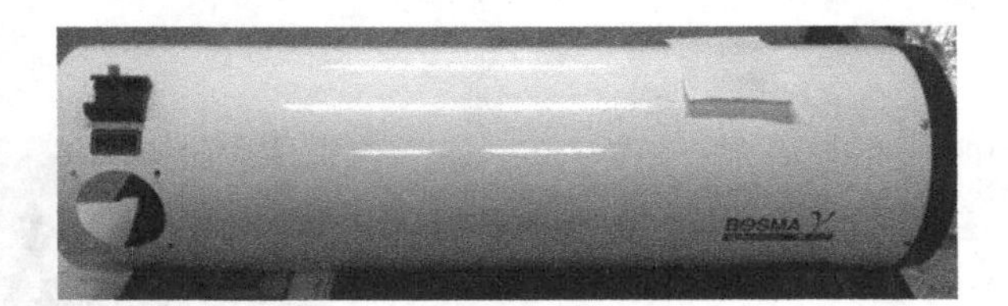

Fig. 1.4 Tested telescope

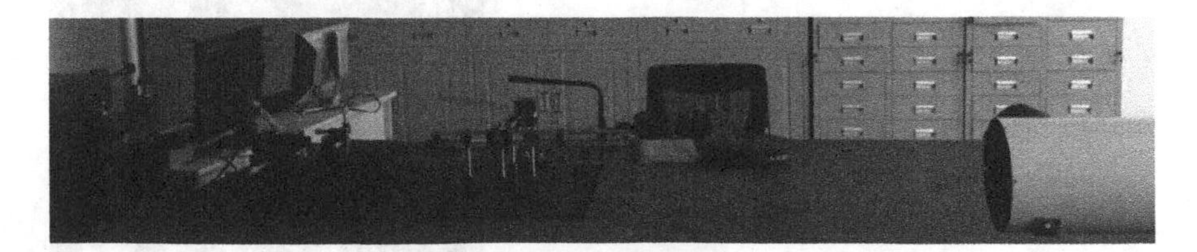

Fig. 1.5 The experimental system of ZYGO interferometer measurement

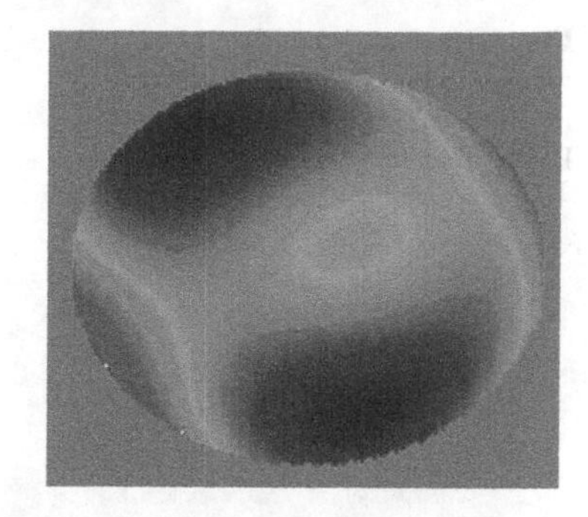

Fig. 1.6 Result of PRWS, RMS = 0.272λ, PV = 1.608λ

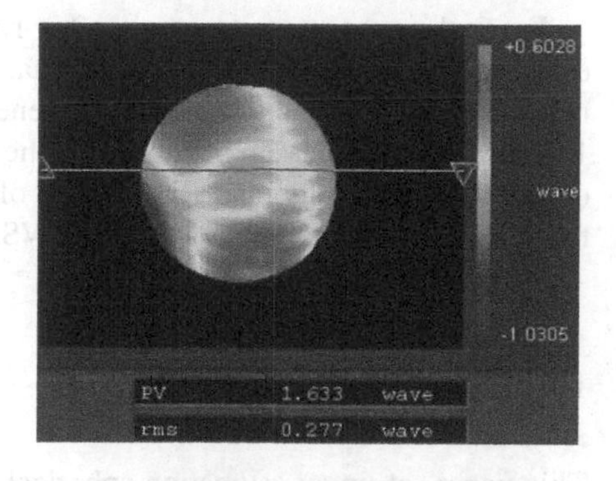

Fig. 1.7 Result of ZYGO interferometer, RMS = 0.277λ, PV = 1.633λ

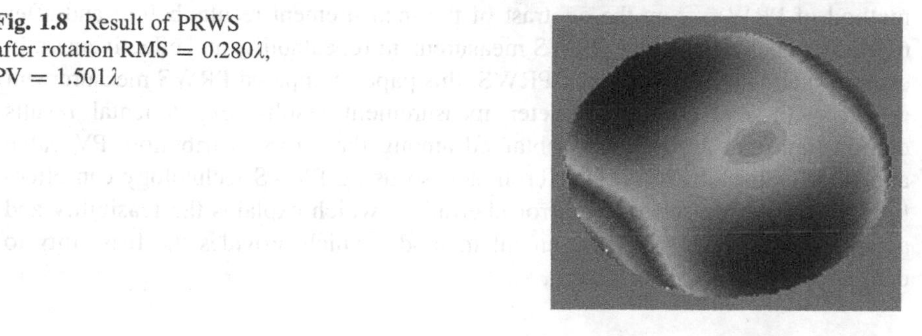

Fig. 1.8 Result of PRWS after rotation RMS = 0.280λ, PV = 1.501λ

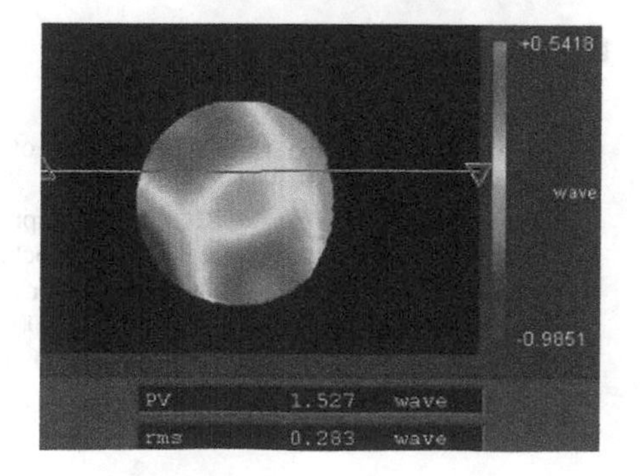

Fig. 1.9 Result of ZYGO interferometer after rotation, RMS = 0.283λ, PV = 1.527λ

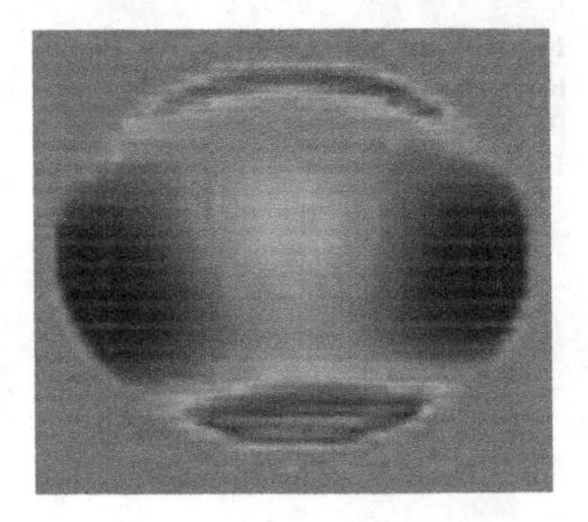

Fig. 1.10 Result of PR measurement after rotation, RMS = 0.271λ, PV = 1.659λ

From the measurement result after rotating, we can see that the results of estimated wavefront before and after rotation are agreement, and verify the PRWS measurement repeatability and effectiveness. Seeing from Figs. 1.6, 1.7, 1.8, and 1.9, for the same measurement mirror, the agreement is obtained among the errors distribution, PV value and RMS value of ZYGO interferometer, which explains the feasibility and accuracy of the PRWS measurement methods.

1.4 Conclusions

This paper set up an estimation spherical mirror of experiment platform with the method of PRWS, from the contrast of the measurement results before and after rotation, which verified the PRWS measurement repeatability and effectiveness. In order to validate the veracity of PRWS, this paper compared PRWS measurement results with ZYGO interferometer measurement results, experimental results demonstrate that agreement is obtained among the errors distribution, PV value and RMS value of ZYGO interferometer, so using PRWS technology can effectively estimate the spherical mirror aberration, which explains the feasibility and accuracy of the PRWS measurement methods, which provides the feasibility to data support for our later search.

References

1. Brown B, Aaron M (2001) The politics of nature. In: Smith J (ed) The rise of modern genomics, 3rd edn. Wiley, New York
2. Brady Gregory R, Fienup JR (2004) Improved optical metrology using phase retrieval. Optical fabrication and testing, vol 10. Rochester, NY, 2004, p 1–3
3. Yang H, Gong C (2011) Phase retrieval for a kind of wavefront sensor based on pupil phase diversity. Acta Optica Sinica 31(11):1112002
4. Osten W (2008) Some answers to new challenges in optical metrology. Proc SPIE 7155: 715503-1–715503-16

5. Ohara CM, Faust JA, Lowman AE et al (2004) Phase retrieval camera optical testing of the advanced mirror system demonstrator. SPIE 5487:1744–1756
6. Gerchberg RW, Saxton WO (1972) A practical algorithm for the determination of phase from image and diffraction phase pictures. Optic 35(2):237–246
7. Fienup JR (1982) Phase retrieval algorithms: a comparison. Appl Opt 21(15):2758–2769
8. Fienup JR, Marron JC, Schulz TJ et al (1993) Hubble space telescope characterized by using phase-retrieval algorithms. Appl Opt 32(10):1747–1767
9. Dean Bruce H, Aronstein David L, Smith JS et al (2006) Phase retrieval algorithm for JWST flight and testbed telescope. Proc SPIE 6265:1–17
10. Fei Li, Changhui Rao (2011) Study on phase diversity wavefront sensor. Acta Optica Sinica 31(8):0804001
11. Liang S, Yang J, Xue B (2010) A new phase diversity wave-front error sensing method based on genetic algorithm. Acta Optica Sinica 30(4):1015–1019
12. Jiang P, Ma H, Zou Y et al (2011) Study of aberration correction in light path of adaptive optical system. Acta Optica Sinica 31(12):1214002
13. Devaney AJ, Childlaw R (1978) On the uniqueness question in the problem of phase retrieval from intensity measurement. JOSA A 68(10):1352–1354
14. Han B, Xiao W, Pan F et al (2012) Optimization of space sampling distance of phase retrieval algorithm for in-line digital holography. Laser Optoelectron Prog 49:120903
15. Fu F, Bin Z (2011) Recovery of high frequency phase of laser beam with wavefront distortio. Chin J Lasers 38(4):0402009
16. Brady Gregory R, Fienup JR (2005) Phase retrieval as an optical metrology tool. In: Optical fabrication and testing. Topical meeting of the optical society of America, SPIE Technical Digest 2005, TD03, pp 139–141
17. Millerd JE, Wyant JC (2005) Simultaneous phase-shifting fizeau interferometer.US Patent 20050046864
18. Deck Leslie (1996) Vibration-resistant phase-shifting interferometry. Appl Opt 35(34):6655–6662
19. Burge JH, Wyant JC (1995) Applications of computer-generated holograms for interferometric measurement of large aspheric optics. Proc SPIE 2576:258–269
20. Reichelt S, Pruss C, Tiziani HJ (2003) Absolute interferometric test of aspheres by use of twin computer- generated holograms. Appl Opt 42(22):4468–4479
21. Sommargren GE, Phillion DW, Campbell EW (1999) Sub-nanometer interferometry for aspheric mirror fabrication. In:The 9th international conference on production engineering, Osaka, Japan, 1999
22. Reichelt S,Tiziani HJ (2003) Twin-CGHs for absolute calibration in wavefront testing interferometry. Opt Commun 220:23–32
23. Ma X, Wang J, Wang B (2012) Study on phase retrieval algorithm. Laser Infrared 42(2):217–221
24. Wang JL, Wang ZY, Wang B et al (2011) Image restoration by phase-diversity speckle. Opt Precis Eng 19(5):1165–1170
25. Wang B, Wang ZY, Wang JL et al (2011) Phase-diverse speckle imaging with two cameras. Opt Precis Eng 19(6):1384–1390
26. Zhao JY, Chen ZF, Wang B et al (2012) Improvement of phase diversity object function's parallelity. J Opt Precis Eng 20(2):431–438
27. Wang B, Ma XX et al (2013) Calibration of no-common path aberration in AO system using multi-channel phase-diversity wave-front sensing. Opt Precis Eng 21(7):1683–1692
28. Wang ZY, Wang B, Wu YH et al (2012) Calibration of non-common path static aberrations by using phase diversity technology. Acta Optica Sinica 32(7):0701007
29. Zhao JY, Ma XX et al (2012) Image restoration based on real time wave-front information. Opt Precis Eng 20 (6):1350–1356
30. Byrd RH, Lu P, Nocedal J (1995) A limited-memory algorithm for bound-constrained optimization. SIAM J Sci Stat Comput 16(5):1190–1208

31. Mahdi H, Stojan R et al (2012) Memory-enhanced noiseless cross-phase modulation. Light Sci Appl 40
32. Lingling H, Xianzhong C et al (2013) Helicity dependent directional surface plasmon polariton excitation using a metasurface with interfacial phase discontinuity. Light Sci Appl 2:e70
33. Dai D, Bauters J, Bowers JE (2012) Passive technologies for future large-scale photonic integrated circuits on silicon: polarization handling, light non-reciprocity and loss reduction. Light Sci Appl 1:e1

Chapter 2
Combined Utility and Adaptive Residence Time-Based Network Selection for 4G Wireless Networks

J. Shankar, C. Amali and B. Ramachandran

Abstract Next generation wireless networks are expected to include heterogeneous wireless networks to offer a diverse range of multimedia services to mobile users. Due to the heterogeneity and the diversity of access networks, various user applications with different Quality of Service (QoS) requirements pose new challenges on multi-interface Mobile Terminal (MT) in designing optimal network selection algorithm for guaranteeing seamless QoS support to the users. Thus, Vertical Hand Off (VHO) is necessary to provide uninterrupted services to mobile users anywhere and anytime in 4G networks. In this paper, Service continuity with guaranteed QoS and minimum VHO rate is considered as a challenging issue. In order to achieve this, discovered networks are evaluated periodically according to the velocity and direction of Mobile Terminal. Network evaluation is carried out through the integration of utility function calculation and cell residence time estimation. The proposed scheme avoids unnecessary VHO by selecting the network based on the capabilities of MT, QoS requirements of ongoing service and characteristics of networks. Thus, it eliminates the probability of HO failure by considering all possible critical factors and adaptive Residence Time (RT) threshold in 4G wireless networks. The results are compared against the performance of utility-based handoff and residence time-based handoff schemes. The simulation results show that the proposed scheme reduces the number of unnecessary handoffs which in turn minimizes the VHO rate and probability of HO failure in 4G wireless networks.

Keywords Vertical · Handoff · Heterogeneous network · Utility function · Adaptive residence time · QoS

J. Shankar (✉) · C. Amali · B. Ramachandran
SRM University, Chennai, India
e-mail: shankarjayaraj@hotmail.co.in

C. Amali
e-mail: amali.vec@gmail.com

B. Ramachandran
e-mail: ramachandran.b@ktr.srmuniv.ac.in

A. A. Farag et al. (eds.), *Proceedings of the 3rd International Conference on Multimedia Technology (ICMT 2013)*, Lecture Notes in Electrical Engineering 278,
DOI: 10.1007/978-3-642-41407-7_2,

2.1 Introduction

The future wireless and mobile environments are likely to have users to access multiple networks at the same time. In fourth generation (4G) wireless networks, users are able to roam freely from one type of wireless access network to another while preserving the main characteristics of their connections. Therefore, there is a need to have mechanisms to decide which network is the most suitable for each user at each moment for every application that the user requires. Vertical Hand Off (VHO) is the capability to switch on going connections from one Radio Access Network (RAN) to another. The trend is to utilize high bandwidth wireless local area network (WLAN) resource for mobile users in hotspots and switch to Cellular Networks (CN) when the coverage of WLAN is not available. The strength of 4G systems is to integrate existing and newly developed wireless systems instead of putting efforts in developing new radio interfaces and technologies to provide seamless mobility and better service quality for 4G users. In this paper, the coexistence of UMTS, WLAN, and WiMAX access networks are considered as a heterogeneous wireless network which is illustrated in Fig. 2.4. Hence, it is expected that mobile users could enjoy seamless mobility and ubiquitous service access in an always best connected mode, the development of an appropriate interworking solution for these heterogeneous wireless networks is crucial.

The criteria involved in VHO decision are very important to achieve uninterrupted mobility scenarios in taking decisions for switching to the target network from both application requirements and mobile terminal capabilities. The decision making process of handoff may be centralized or decentralized (i.e., the handoff decision may be made at the network or Mobile Terminal (MT)). From the decision process point of view, one can find at least three different kinds of handoff decisions. They are, (1) Network assisted handoff (2) Mobile assisted handoff and (3) Mobile controlled handoff. Out of these three, mobile controlled and network assisted handoff are combined to get Mobile Controlled Network Assisted (MCNA) handoff, because only the MT has the knowledge about the networks available in the coverage area.

If the mobile terminal velocity and moving pattern are irregular, more unnecessary handoff can occur. In our proposed algorithm, these two factors are considered as important parameters to select the target network in heterogeneous wireless networks. When the MT is moving with high velocity, it is necessary to find out how much time the MT will stay in the target network. If the estimated residence time is less than the predefined threshold, handoff to such a network is not beneficial as, it requires more number of handoff to complete the ongoing service.

The remainder of this paper is organized as follows: The related and existing works with their shortcomings are discussed in Sect. 2.2. The system model of the proposed scheme to improve the network selection mechanism is explained in Sect. 2.3. Section 2.4 presents the different modules of the proposed algorithm. The simulation environment and the results are discussed in Sect. 2.5. Finally, the paper is concluded with Sect. 2.6.

2.2 Related Work

Recently, various network selection algorithms based on Multiple Attribute Decision Making (MADM) approach have been developed to improve the VHO decision in heterogeneous wireless networks. An Analytic Hierarchy Process (AHP)-based network selection algorithm for UMTS and WLAN is presented in [1]. AHP and Grey Relational Analysis (GRA)-based network selection mechanism for UMTS and WLAN is presented in [2] for next generation networks. In [3], a fuzzy logic-based MADM problem is formulated in which multiple parameters are considered to perform VHO decision in heterogeneous networks. Fuzzy logic is used to represent the imprecise information of different parameters of the networks and the preferences of the user. The method for discovering the reachable wireless networks is proposed in [4] and it is the first step for VHO. After discovering the reachable candidate networks, the mobile terminal decides whether to perform handoff or not.

In [5], a Position Aware Vertical Handoff decision algorithm (PAHO) has been proposed which considers the MS's position, its moving pattern and speed and also the coverage area of the current BS to select the target network for performing VHO. In [6], a target network is selected by calculating a utility function which considers parameters such as available bandwidth, SINR, traffic load, and MS speed. In [7], a movement aware VHO algorithm is proposed to exploit the MT velocity and movement pattern for eliminating ping pong effects and reducing unnecessary handovers. The handoff decision is made based on the comparison of cost functions of different access networks [8]. In most existing cost function-based network selection algorithm, a set of network and user parameters are chosen as the cost factors and fixed weights are assigned in designing the cost function. As Next Generation Wireless Networks (NGWN) is expected to support users with different profiles and service applications with different Quality of Service (QoS) requirements, cost function with fixed weights cannot efficiently reflect the QoS requirements on communication service, resulting in low efficiency in network selection. Thus, in [9] a Modified Weight Function-based Network Selection Algorithm (MWF-NSA) that considers user preference and application profile has been proposed in deciding the weight functions of the networks. In [10], traveling distance prediction is used to perform handoff necessity estimation and also to avoid unnecessary handoffs from cellular network to WLAN.

In this paper, the discovered networks are preselected based on the velocity of MT. Utility function is calculated for evaluating the preselected networks. The network which has maximum utility function is considered for residence time estimation. Thus, our proposed work aims to incorporate the characteristics of networks, QoS requirements of different traffics, MT velocity, and moving pattern to select the best suitable network to perform VHO in heterogeneous wireless networks.

2.3 System Model of the Proposed Algorithm

The performance of the proposed algorithm is compared with utility-based and residence time-based VHO schemes. In utility-based VHO scheme, optimized utility function is used to select the target network by introducing trade-off between user satisfaction and network efficiency. Whereas, residence time-based VHO decision scheme estimates the traveling time of mobile terminal in the candidate networks using MT velocity and direction. In this scheme, time threshold value should be high for high velocity of MT in order to increase the resource utilization and also to provide uninterrupted services to mobile users.

In the proposed scheme, the above two approaches are combined and named as Combined Utility and Adaptive Residence Time (CU-ART)-based network selection algorithm which select the best network based on the capabilities of MT, network conditions, and QoS requirements of running applications. An appropriate system model is constructed to give the sequence of processes involved in the proposed algorithm. This model is then analyzed and simulated using utility function and mathematical techniques.

Figure 2.1 represents the system model of proposed CU-ART algorithm. New networks are discovered based on Received Signal Strength (RSS) measurements in a given coverage area. These newly discovered networks are filtered through a network preference block based on velocity of MT. Utility block finds the utility value of the networks in the filtered lists. The network which has maximum utility function is considered for residence time estimation. Residence time is the time for which the mobile terminal stays in the particular network. Longer the residence time lesser will be the frequency of handoffs. The network with residence time greater than the adaptive Residence time threshold is selected as a target network. Thus, the proposed system model not only satisfies the QoS requirements of mobile users but also improves the system performance by reducing the unnecessary VHO.

2.4 Working of CU-ART Network Selection Algorithm

The three modules of the proposed algorithm are, (1) Discovery and Preselection of networks (2) Utility function calculation and (3) Residence time estimation.

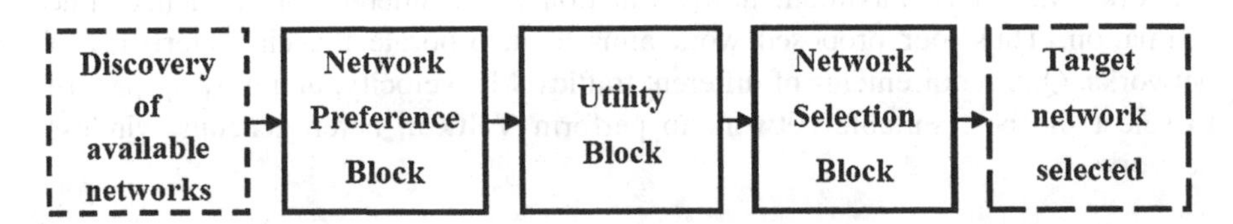

Fig. 2.1 System model

2.4.1 Discovery and Preselection of Networks

The networks available in the given coverage area are discovered by measuring the RSS from BS. The networks which have RSS greater than the threshold value are considered for preselection process. This module obtains the information about velocity and moving direction of MT from GPS system. Then, the discovered networks are preselected based on the speed that supports. Networks with low coverage cannot provide services to high speed mobile users. As a result, large number of handover is required to maintain the ongoing service. WLAN is capable of providing low mobility and high bandwidth services. But WiMAX and UMTS support applications which require large coverage area and medium bandwidth. By selecting the networks according to the MT velocity, number of VHO required to complete the ongoing connection can be minimized.

Figure 2.2 illustrates the working of proposed CU-ART-based network selection algorithm. The network preselection is done with respect to velocity of MT. Average pedestrian speed (i.e., 1.4 m/s) is considered to filter the discovered networks. Utility function is calculated only for filtered networks to reduce computation time which in turn minimizes the HO delay. The network with maximum utility and residence time greater than the adaptive residence time threshold is selected as a target network. It is explained further in the following subsections.

2.4.2 Utility Function Calculation

The Utility module reads the network parameters such as bandwidth, connection delay, and cost whenever networks are discovered. To support user applications with different levels of QoS requirements, 3GPP has defined four traffic classes such as Conversational class, Streaming class, Interactive class, and Background class. The characteristics of different traffics must be taken into account in the design of network selection algorithm in order to provide better performance to the users in the integrated networks. Appropriate weight factor should be assigned to each metric to account for its importance in providing QoS requirements to particular application. The importance levels of High, Medium, and Low for a particular parameter ‘i’ is defined as i_H, i_M, and i_L, respectively. They are in the order $0 < i_L < i_M < i_H < 1$. The values assumed are $i_L = 0.3$, $i_M = 0.6$ and $i_H = 0.9$. The weight factors of the importance levels are W_H, W_M, and W_L, respectively, where $W_H + W_M + W_L = 1$.

$$W_L = \frac{i_L}{(i_H + i_M + i_L)}, \quad W_M = \frac{i_M}{(i_H + i_M + i_L)}, \quad W_H = \frac{i_H}{(i_H + i_M + i_L)}$$

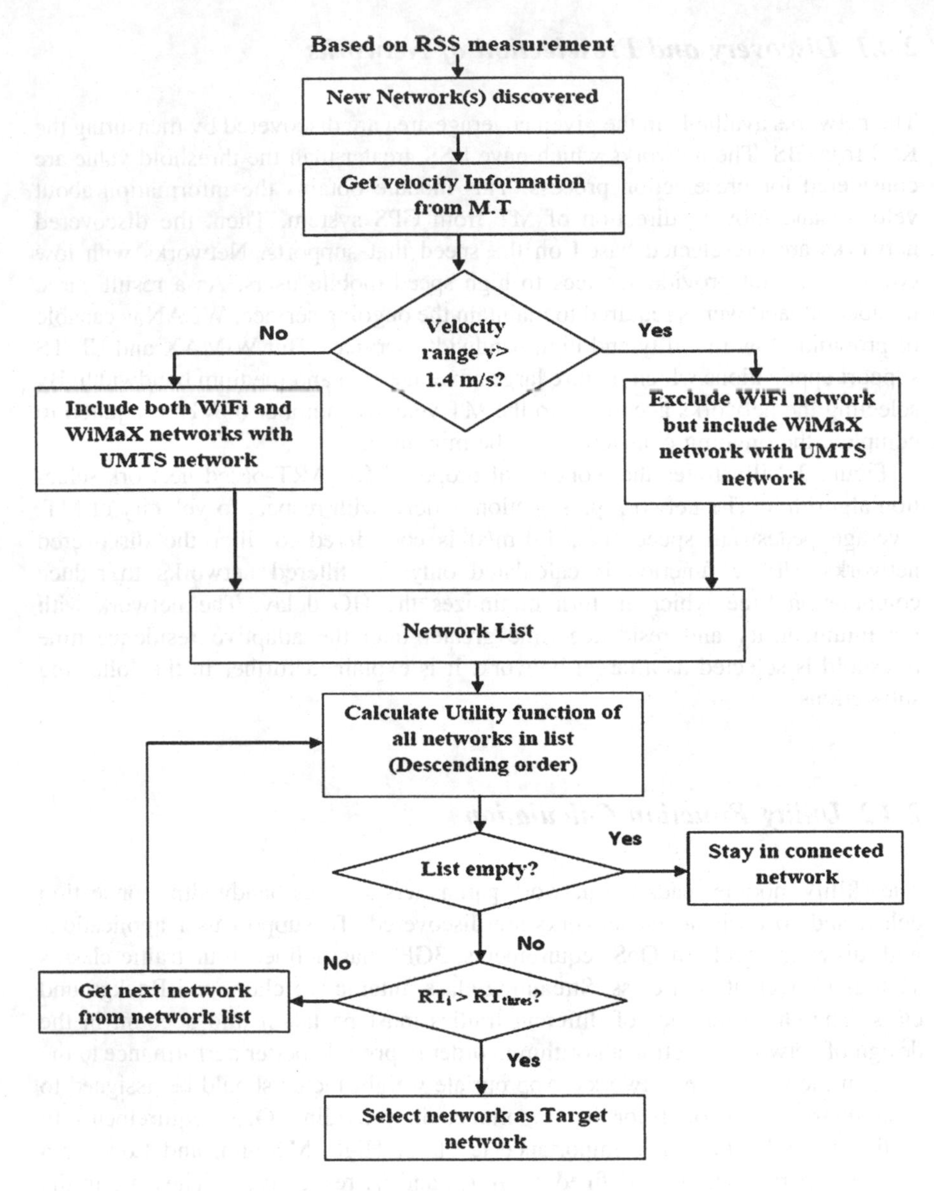

Fig. 2.2 Flowchart of CU-ART algorithm

W_H, W_M, and W_L are high medium and low weights respectively. Then, utility function is calculated for the integration of Wi-Fi, UMTS, and WiMAX based on the network conditions and QoS requirements of ongoing service.

The utility function is calculated using the following formulae given in [9].

$$\text{NEF}_j = W_D(1 - D_j) + W_B B_j + W_C(1 - C_j)$$

Table 2.1 Utility values of networks

Application types	Utility values					
	For velocity < 1.4 m/s			For velocity >=1.4 m/s		
	Wi-Fi	UMTS	WiMAX	Wi-Fi	UMTS	WiMAX
Video call	0.388893	0.333322	0.277785	0	0.54544	0.45456
Voice call	0.388887	0.333322	0.277791	0	0.545435	0.454565
FTP	0.4	0.33332	0.26668	0	0.555533	0.444467
Video Stream	0.388887	0.333322	0.277791	0	0.545435	0.454565
Browsing	0.388887	0.333322	0.277791	0	0.545435	0.454565
Messaging	0.388887	0.333322	0.277791	0	0.545435	0.454565
Navigation Application	0.388893	0.333322	0.277785	0	0.54544	0.45456
Gaming	0.375	0.333325	0.291675	0	0.53332	0.46668

where j corresponds to Wi-Fi, UMTS, and WiMAX and W_D, W_B, and W_C are weights for delay, bandwidth, and cost respectively. They are calculated according to the QoS requirements of applications for each network which are shown in Table 2.1. Application types that are considered in this algorithm are (1) Video call (2) Voice call (3) FTP (4) Video Stream (5) Browsing (6) Messaging (7) Navigation Application and (8) Gaming. The numerical results of utility function module for different applications are shown in Table 2.1 based on the velocity of MT. From the table, it is inferred that for MT moving with velocity greater than 1.4 m/s (average pedestrian speed), utility function is available only for UMTS and WiMAX to serve the mobile users with minimum VHO frequency.

2.4.3 Residence Time Estimation

In order to minimize the probability of HO failures and unnecessary HO, residence time is estimated for the networks available in the given direction of MT. The network with maximum residence time is selected to serve the mobile users. Figure 2.3 illustrates the residence time calculation using trigonometric method. P_1 and P_2 are arbitrary points on the boundary through which MT enters into the network from which the value of θ is determined. The value of θ is uniformly distributed in $[0, 2\pi]$.

2.4.3.1 Residence Time (t)

Residence time of MT is estimated by using the parameters such as direction (θ), velocity (v), and the radius of network coverage (R).

In Fig. 2.3, considering right angle triangle OAP_1.

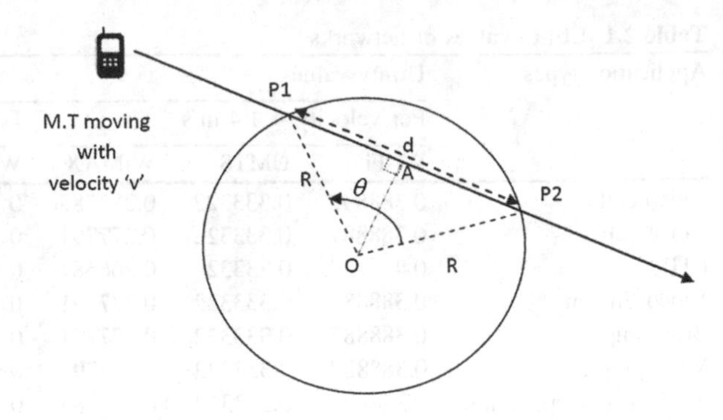

Fig. 2.3 Scenario for residence time estimation

$$\sin {}^{\theta}\!/_{2} = \frac{{}^{d}\!/_{2}}{R} \quad \text{Squaring on both sides gives}$$

$$\sin^2 {}^{\theta}\!/_{2} = \frac{d^2}{4R^2} \text{ leads to } \frac{(1-\cos\theta)}{2} = \frac{d^2}{4R^2}$$

where d is the distance between P_1 and P_2 and θ is the angle between points P_1 and P_2 with respect to base station.

From $d = v_* t$, the residence time of MT can be expressed as:

$$t = \sqrt{\frac{2R^2(1-\cos\theta)}{v^2}} \tag{2.1}$$

2.4.3.2 Adaptive Residence Time Threshold (T)

Adaptive residence time threshold is needed for high speed users to reduce the HO failure probability and HO frequency. From Eq. 2.1, it is known that the estimated RT is a function of θ, v, and R. In the given scenario, more number of trajectories can be generated by considering different values of θ for the evaluation of discovered networks. For $\theta = 0°$, the point P_1 and P_2 coincide and path of MT just graces through the boundary. For this case, the estimated RT is zero. For $\theta = 180°$, the path of the MT goes through the diameter of network coverage offering maximum residence time.

Using the theorem stated in [10], the pdf of estimated RT is given by:

$$f(T) = \begin{cases} \frac{2}{\pi\sqrt{4R^2 - v^2T^2}}; & 0 \le T < \frac{2R}{v} \\ 0; & \text{Otherwise} \end{cases} \tag{2.2}$$

The Eq. 2.2 is evaluated to obtain the pdf for different velocities.

$$f_{\text{mean}}(T) = \text{mean}(f(T)) \quad \forall \quad 1 < \text{T} < \text{T}_{\text{max}} \tag{2.3}$$

T_{max} for Wi-Fi, UMTS, and WiMAX is 12 s, 160 s, 400 s, respectively. It is calculated based on the typical network radius of 150 m, 2000 m, and 5000 m, respectively.

Mean of pdf is considered from worst case estimated residence time of 1 s to maximum limit (T_{max}).

From Eq. 2.2, the adaptive residence time threshold is given by:

$$T = \frac{1}{v}\sqrt{\left(4R^2 - \frac{4}{\pi^2 (f(T))^2}\right)} \tag{2.4}$$

From Eqs. 2.3 and 2.4,

$$T = \frac{1}{v}\sqrt{\left(4R^2 - \frac{4}{\pi^2 (f_{\text{mean}}(T))^2}\right)} \tag{2.5}$$

The adaptive RT threshold for a particular velocity is calculated using Eq. 2.5. Where, v is velocity of MT, R is radius of network, $f_{mean}(T)$ is the mean pdf value of RT evaluated under $\frac{2R}{v}$.

2.4.3.3 Probability of Handoff (P_{HO})

Handoff occurs when estimated RT of MT is greater than the adaptive residence time threshold T. The probability of handoff to occur can be evaluated from the cdf of estimated RT which is obtained by integrating the pdf in Eq. 2.2.

$$\begin{aligned} P_{\text{HO}} &= F(T) = P(t \geq T) \\ P_{\text{HO}} &= \int_{T}^{\infty} f(T)dT \\ &= \begin{cases} 1 - \frac{2}{\pi}\arccos\left(\frac{vT}{2R}\right) & ; \quad 0 \leq T \leq \frac{2R}{v} \\ 0 & ; \quad \text{otherwise} \end{cases} \end{aligned} \tag{2.6}$$

The probability of handoff occurrence is given by Eq. 2.6 for the networks available in the direction of MT moving with a velocity 'v'.

2.5 Simulation Results and Discussion

The three modules of proposed CU-ART algorithm are simulated using Matlab. The network coverage radius for Wi-Fi, UMTS, and WiMAX is assumed as 150, 2000, and 5000 m, respectively. The heterogeneous scenario used for simulation is

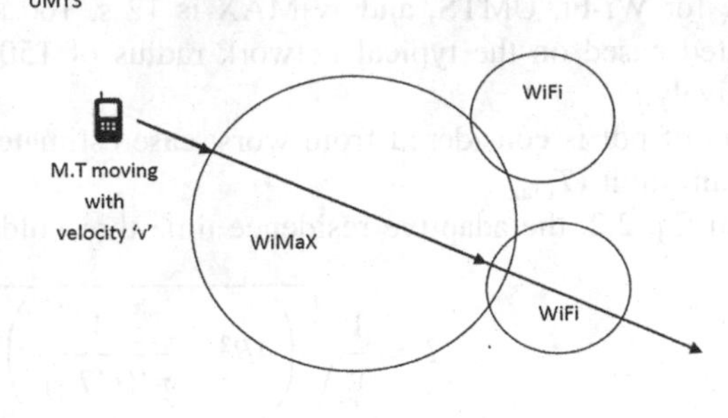

Fig. 2.4 Heterogeneous scenario

shown in Fig. 2.4. In this scenario, it is assumed that UMTS network covers the entire area because of widely deployed 3G network.

2.5.1 Residence Time

Residence time of MT across the Wi-Fi network for different random trajectories (θ) obtained through simulation is shown in Fig. 2.5 with respect to the velocity of the MT. Five different values of θ are considered (i.e., $\theta = 0°, 30°, 60°, 90°, 180°$) for simulation. The estimated residence time of MT across the Wi-Fi network is shown in Fig. 2.5 which shows the effect of velocity and moving direction on residence time estimation. From simulation, it is also found that the estimated residence time values for UMTS and WiMAX are 400 and 2000 s respectively for a particular velocity of 5 m/s with $\theta = 60°$.

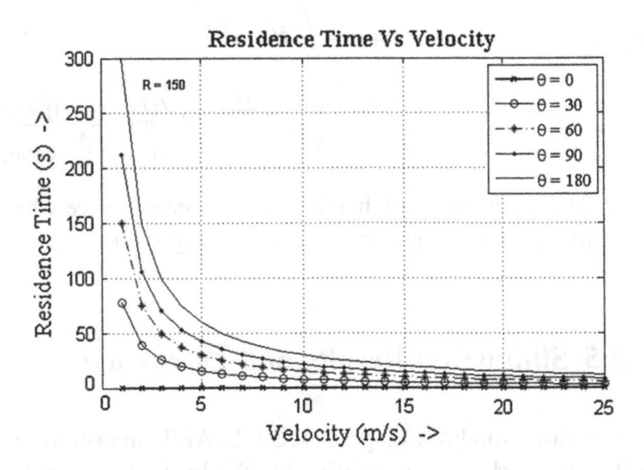

Fig. 2.5 Residence time for Wi-Fi network

2.5.2 Adaptive Residence Time Threshold

Adaptive residence time threshold for Wi-Fi, UMTS, and WiMAX have been simulated for MT velocity from 1 to 25 m/s and is given in Table 2.2. From the Table 2.2, it is inferred that when velocity increases, adaptive residence time threshold also increases to select the network which offers maximum residence time and also to initiate the HO as early as possible for high speed users. Thus, the probability of HO failure and unnecessary HO is maintained within the predesigned limits.

The threshold value of RT varies from 7.682 to 12 s, 93.12 to 160 s, and 231.7 to 238.2 s based on the velocity of MT for Wi-Fi, UMTS, and WiMAX networks respectively to minimize the HO failure probability and VHO rate in heterogeneous networks. The adaptive RT threshold allows the MT to make use of selected target network as maximum as possible. Thus, it improves the resource utilization efficiency and also improves the system performance by reducing unbeneficial VHO. The adaptive residence time threshold values for different networks obtained through simulation are listed in Table 2.2.

2.5.3 Number of Handoffs

Simulation is performed to obtain the probability of HO for the networks available in the direction of MT ($\theta = 60°$) for velocity from 1 to 25 m/s. The number of HO is obtained from the probability of HO occurrence during the movement of MT over a particular trajectory. The proposed algorithm is simulated at 300 points along the path through which MT traverse across the networks. The mean pdf is calculated for a particular velocity ranging from 1 to 25 m/s over 300 points. For simulation purpose, the proposed algorithm is repeated periodically to obtain the number of VHO accurately. Finally, the number of VHO is obtained by multiplying the mean pdf with number of iterations.

The total number of handoff occurred for Utility-based VHO algorithm [9], Adaptive RT-based VHO algorithm [10] and the proposed CU-ART-based network selection algorithms are shown in Fig. 2.6. In utility-based VHO algorithm,

Table 2.2 Adaptive RT threshold for different networks

Velocity (m/s)	Adaptive residence time threshold (s)		
	Wi-Fi	UMTS	WiMAX
5	7.3905	93.1878	231.6057
10	7.4904	94.4146	232.3163
15	7.6920	96.8482	233.5453
20	8.1079	101.6637	235.3693
25	12	160	237.9195

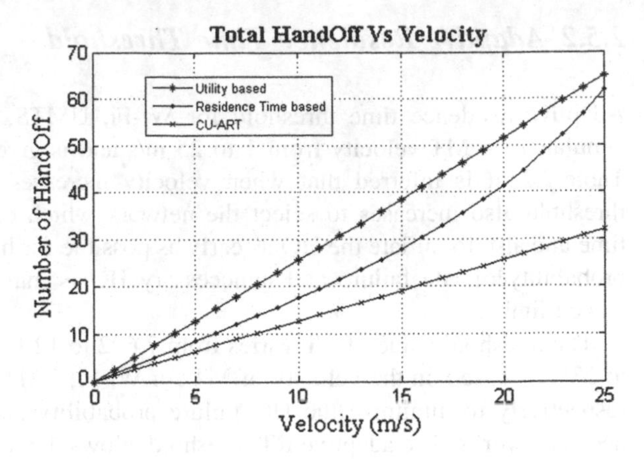

Fig. 2.6 Comparison of algorithms with number of handoffs

residence time of available networks is not considered for target network selection. Therefore, number of HOs required to complete the ongoing service is more compared to other two approaches. Whereas, in adaptive RT-based VHO algorithm, the probability of HO failure occurs only when the target network does not support the QoS requirements of mobile users.

Thus, the number of HO is less compared to utility-based approach. But, the proposed CU-ART network selection algorithm selects the network with maximum utility function and estimated residence time as a best suitable network to provide uninterrupted services to mobile user. Also, the networks which support the MT velocity only considered for evaluation process. Thus, probability of HO occurrence during the movement of MT across the heterogeneous wireless networks can be minimized.

2.6 Conclusion and Further Work

In this paper, CU-ART-based network selection algorithm was proposed to select the best suitable network based on MT velocity, direction, utility function, and estimated residence time. Adaptive residence time threshold is also calculated for MT velocity varying from 5 to 25 m/s. It is observed that adaptive residence time threshold is high for high speed users to maintain service continuity with guaranteed QoS and minimal VHO rate. Through simulation, it is proved that the proposed CU-ART network selection algorithm minimizes the number of handoff to more than 50 % of utility and residence time estimation based VHO schemes taken separately for high speed users. Thus, the probability of HO failure is eliminated by considering both static and dynamic HO decision parameters and adaptive RT threshold in the proposed CU-ART algorithm.

In our future work, the CU-ART-based network selection algorithm will be analyzed based on other performance metrics like throughput and HO delay. The cooperation techniques between different radio access technologies will also be considered to achieve minimal VHO probability, QoS, and power saving in heterogeneous wireless networks.

References

1. Song Q, Jamilipour A (2005) Network selection in an integrated wireless LAN and UMTS Environment using mathematical modeling and computing techniques. IEEE Trans Wirel Commun 12(3):42–48
2. Fu J, Wu J, Zhang J, Ping L, Li Z (2010) Novel AHP and GRA based handover decision mechanism in heterogeneous wireless networks. Proc CICLing 2:213–220
3. Kunarak S, Suleesathira R (2010) Predictive RSS with fuzzy logic based vertical handoff algorithm in heterogeneous wireless networks. In: IEEE International Symposium on Communications and Information Technologies (ISCIT), Tokyo, 26–29 Oct 2010, pp 1235–1240
4. Chen WT, Liu JC, Huang HK (2004)An adaptive scheme for vertical handoff in wireless overlay networks. In: IEEE proceeding of the 10th international conference on parallel and distributed system, 7–9 July 2004, pp 541–548
5. He F, Wang F (2008) Position aware vertical handoff decision algorithm in heterogeneous wireless networks. In: 4th international conference on wireless communications, networking and mobile computing, WiCOM '08, Dalian, 12–14 Oct 2008, pp 1–5
6. Lee D, Han Y, Hwang J (2006) QoS based vertical handoff decision algorithm in heterogeneous systems. In: Proceeding of 17th annual IEEE international symposium on personal, indoor, and mobile radio communications, Helsinki, 11–14 Sept 2006, pp 1–5
7. Lee W, Kim E, Kim J, Lee C (2007) Movement aware vertical hand-off of WLAN and mobile WiMAX for seamless ubiquitous access. IEEE Trans Consum Electron 53(4):1268–1275
8. Lee SY, Sriram K, Kim K (2009) Vertical handoff algorithms for providing optimized performance in heterogeneous wireless networks. IEEE Trans Veh Tech 58(2):865–881
9. Amali C, Ramachandran B (2012) Modified weight function based network selection algorithm for 4G wireless networks. In: ACM International Conference on Advances in Computing, Communication and Informatics (ICACCI 2012) Chennai, India, 3rd–5th Aug 2012, pp 292–299
10. Hassane AI, Li R, Zeng F (2012) Handover necessity estimation for 4G heterogeneous networks. Int J Inf Sci Tech (IJIST) 2(1):1–13

Chapter 3
Three-Dimensional Compressed Sensing Imaging Using Phase-Shift Laser Range Finding

Bingbing Guo, Ping Wei and Jun Ke

Abstract In this paper, we propose a novel algorithm for three-dimensional compressed sensing (CS) imaging using phase-shift range finding. This system focuses on the application of high-resolution terrain measurement from a medium or short-range. Both the SNR and the imaging speed are improved due to the system structure consist of single photo-detector and no scanning components. Simulation experiments to reconstruct terrain data show that this algorithm can achieve an average MSE (mean square error) less than 0.05 and a ranging accuracy in centimeter magnitude. In our experiments, CS is utilized in a laser range finding system, and much fewer measurements are needed to reconstruct depth maps. Compared with other 3D imaging approaches, our algorithm is more robust in the case of noise and complex target properties.

Keywords Compressed sensing · Phase-shift · Three-dimensional imaging · Single photo-detector

3.1 Introduction

Laser range finding is now widely used in range detection and three-dimensional imaging areas. Compared to other range detecting methods, laser radar can achieve better depth resolution, farther detecting range, and be more robust against perturbation.

P. Wei (✉)
Beijing Institute of Technology, No. 5, Zhongguancun Road,
Haidian, Beijing, China
e-mail: pwei@bit.edu.cn

B. Guo · P. Wei · J. Ke
Key Laboratory of Photo-electronic Imaging Technology and System, Ministry of Education of China Optoelectronic School, Beijing Institute of Technology, Beijing, China

A. A. Farag et al. (eds.), *Proceedings of the 3rd International Conference on Multimedia Technology (ICMT 2013)*, Lecture Notes in Electrical Engineering 278,
DOI: 10.1007/978-3-642-41407-7_3,

Today, there are two popular kinds of laser radar framework. One includes mechanical scanning device, and the other uses a 2D array of sensing pixels. Scanning laser radar obtains depth map through scanning the target point by point, which makes a heavy and complex system, and consumes more time to imaging. In order to overcome the disadvantages of scanning laser radar, 2D sensor array is widely used to replace a scanning device. However, the SNR of nonscanning laser radar is relatively low because the long distance transmitting will reduce the power of the incident signal on each pixel of the sensor. Lincoln Laboratory of MIT has developed a 32×32-pixel APD array for nonscanning laser radar to improve the SNR in 2000 [1]. Nevertheless, the spatial resolution is limited to 32×32 by the APD array which is difficult to enlarge on account of the semiconductor development.

In case of the limitations of the above laser radars, we introduce optical multichannel method to get rid of 2D array sensor. Multichannel means each time we make a measurement, signals from several points of the target are added and sensed by one photo-detector. In this way, single photo-detector [4] could be applied to nonscanning laser radar and the SNR will be increased significantly. Spatial Light Modulator (SLM) is necessary in multichannel approach. For example, DMD can be used as SLM in laser radar detecting system. In general, full measurements should be applied to acquire a complete depth map through DMD devices, so the DMDs have to change 1,024 times, thus making 1,024 measurements if the spatial resolution of the map is 32×32. That process must cost a lot of time. Compressed sensing (CS) provides a powerful technique to estimate a signal from measurements much fewer than the full measurements, so CS framework is used in our system to achieve a high ranging speed supported by single photo-detector.

The prototype that makes use of CS theory is named as compressive laser range finder, which becomes studied by many researchers recently. In 2012, Rochester University and MIT cooperated to develop a model of CS laser radar based on pulse laser [3]. The model used DMD to modulate the reflected signals and then focused them to a single APD. The result turned out feasible when experiments were conducted at 2.1 m from the imaging device at low light levels.

In this paper, we propose a novel method using phase-shift principle in laser range finder to improve the accuracy of depth ranging further. Simulation results show that reconstruction MSE is less than 0.05 and the ranging accuracy is in centimeter magnitude. Section 3.2 is a brief summary of CS theory; Sect. 3.3 describes the general project and the algorithm model; Sect. 3.4 discusses the reconstruction results of simulations; Sect. 3.5 makes conclusion of the entire work.

3.2 CS Theory

Research of CS laser radar is based on the new theory of CS proposed by Candes, Romberg, and Tao. CS theory provides a new way to acquire and process signal, and it is different from traditional Shannon-Nyquist sampling theory [2]. As long as the primitive signal satisfies the sparsity condition, CS approach could sample the signal in a rate far lower than Nyquist approach and reconstruct primitive signal at high possibility as well. Lower sampling rate will release the burden of sampling, transmitting, and storing the large quantity of signal greatly. Most heavy work is taken by reconstruction algorithm which is performed by effective computer software.

There are three important steps for CS theory. The first step is sparse transformation. Considered that signal x ($x \in R^{N\times 1}$) is sparse in certain orthogonal basis $\{\psi_i\}_{i=1}^{N}$, we denote $\Psi = \{\psi_i\}_{i=1}^{N}$ as the sparsity matrix. Thus, we can get sparse transformation result f:

$$x = \Psi^{-1} f \tag{3.1}$$

The second step is to make observations for x. Denote $\Phi(M \times N(M \ll N))$ as the observation matrix which is incoherent with Ψ, the outcome is y:

$$y = \Phi x \tag{3.2}$$

Third, by substituting x in Eq. 3.2 with Eq. 3.1 we have:

$$y = \Phi \Psi^{-1} f \tag{3.3}$$

The process of recovering x from y is 'impossible' for M is far less than N, but it is possible to recover f from y because f is sparse that it contains great number of zero or zero-close coefficients. It has been proved that if there exists 2 K columns linearly independent from each other in matrix $\Phi\Psi^{-1}$, at least one K-sparse vector f satisfies Eq. 3.3, in this case, we can recover x through solving a nonlinear optimization problem. Till now some reconstruction algorithm such as BP (Basis Pursuit) and OMP (Orthogonal Matching Pursuit) is stable and effective to a CS problem.

3.3 System Description

As described in the Introduction Section, we aim at improving ranging accuracy and SNR at the same time. Taking the current research progress of compressive range finding into account, we transmit continuous wave and lay the ranging principle on phase-shift method. Phase-shift method provides accuracy and SNR of high quality when in a medium or short-range compared to direct time-delay method based on pulse laser source. Next, we will describe the details of the overall architecture and algorithm model of the system.

The overall architecture for our system is depicted in Fig. 3.1.

At first, we use continuous cosine signal to modulate the laser diode to transmit continuous wave. After being reflected by the scene, back wave is imaged on DMD by lens system. The DMD functions as SLM with several chosen patterns. For each pattern the reflected light is focused to the APD by DMD. At last, phase-shift and related depth are measured and analyzed in the processor.

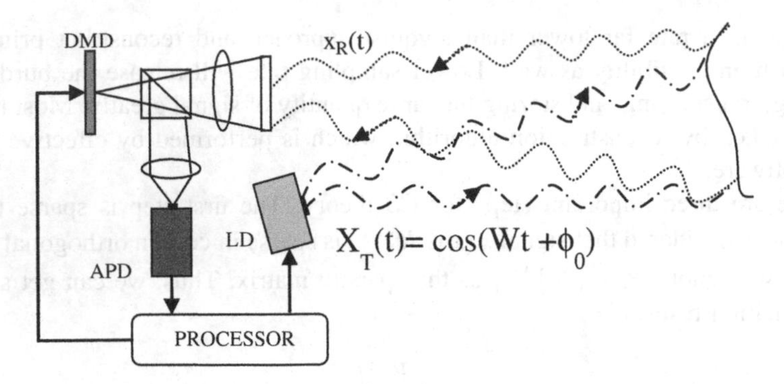

Fig. 3.1 The proposed architecture consists of a laser diode, a DMD device, one single APD, a set of imaging lens, and a processor

The transmitted signal $x_T(t)$ is in the form of cosine wave, with a frequency f and an initial phase φ_O:

$$x_T(t) = \cos(2\pi f t + \varphi_0) \tag{3.4}$$

The phase of reflected signal changes for a phase-shift is produced by time-delay caused by transmission. Phase-shift over 2π ends up with ambiguity, so there is a maximum distance that can be measured accurately decided by 2π:

$$d_{\max} = \frac{c}{2f}(c : \text{speed of light}) \tag{3.5}$$

For example, when $f = 1\,\text{MHz}$, $d_{\max} = 150\,\text{m}$; $f = 500\,\text{kHz}$, $d_{\max} = 300\,\text{m}$, in the application of 3D terrain measurement while landing a tank, a frequency between 500 kHz–1 MHz is appropriate.

Algorithm model is discussed below. For the convenience of discussion, we start modeling from a single point. For the whole terrain target, the reflected light can be regarded as a superposition of light from multiple points.

3.3.1 Regular Response Without DMD Modulation

By substituting $w = 2\pi f$ in Eq. 3.4, we have:

$$x_T(t) = \cos(wt + \varphi_0) \tag{3.6}$$

Assuming that the spatial resolution of the 2D image is $N \times N$, then a point of index (i, j) could represent all points in the target. Let t_{rij} denote the time delay of transmission and neglect attenuation, we get the function of the receiving signal:

$$x_{Rij}(t) = \cos\left[w\left(t - t_{rij}\right) + \varphi_0\right] = \cos[wt + (\varphi_0 - wt_{rij})] \tag{3.7}$$

So the phase-shift of the signal is $\Delta\varphi_{ij} = \varphi_0 - (\varphi_0 - wt_{rij}) = wt_{rij}$. Depending on 'sum to product' formula, Eq. 3.7 can be transformed to

$$\begin{aligned} x_{Rij}(t) &= \cos\left[(wt + \varphi_0) - \Delta\varphi_{ij}\right] \\ &= \cos(wt + \varphi_0)\cos\Delta\varphi_{ij} + \sin(wt + \varphi_0)\sin\Delta\varphi_{ij} \end{aligned} \tag{3.8}$$

When applying Eq. 3.8 to all the $N \times N$ points, we can get the total reflected signal by adding every reflected signal together:

$$\begin{aligned} x_R(t) &= \sum_{i=1}^{N}\sum_{j=1}^{N} x_{Rij}(t) \\ &= \sum_{i=1}^{N}\sum_{j=1}^{N} [\cos(wt + \varphi_0)\cos\Delta\varphi_{ij} + \sin(wt + \varphi_0)\sin\Delta\varphi_{ij}] \\ &= \cos(wt + \varphi_0)\left(\sum_{i=1}^{N}\sum_{j=1}^{N} \cos\Delta\varphi_{ij}\right) + \sin(wt + \varphi_0)\left(\sum_{i=1}^{N}\sum_{j=1}^{N} \sin\Delta\varphi_{ij}\right) \end{aligned} \tag{3.9}$$

3.3.2 Response to Binary DMD Patterns

The DMD pixels discretize the FOV (field of view) into $N \times N$ pixels, since the reflected light is modulated by the observation matrix Φ in Eq. 3.2, DMD should change its pattern based on Φ. Define the values of Φ as $C_{ij}(i \le N, j \le N)$, then Eq. 3.9 is altered to

$$\begin{aligned} x_R(t) = \sum_{i=1}^{N}\sum_{j=1}^{N} C_{ij}x_{Rij}(t) &= \cos(wt + \varphi_0)\left(\sum_{i=1}^{N}\sum_{j=1}^{N} C_{ij}\cos\Delta\varphi_{ij}\right) \\ &+ \sin(wt + \varphi_0)\left(\sum_{i=1}^{N}\sum_{j=1}^{N} C_{ij}\sin\Delta\varphi_{ij}\right) \end{aligned} \tag{3.10}$$

For notational simplicity, let

$$a = \sum_{i=1}^{N}\sum_{j=1}^{N} C_{ij}\cos\Delta\varphi_{ij};\ b = \sum_{i=1}^{N}\sum_{j=1}^{N} C_{ij}\sin\Delta\varphi_{ij}$$

With this notation, we simplify Eq. 3.10 as

$$x_R(t) = a\cos(wt + \varphi_0) + b\sin(wt + \varphi_0) \tag{3.11}$$

Combining sine and cosine term of the same frequency, we can obtain the polar form of Eq. 3.11, in which M is amplitude of $x_R(t)$ while $\Delta\psi$ denotes phase-shift of $x_R(t)$:

$$x_R(t) = M\cos[(wt + \varphi_0) - \Delta\psi] \tag{3.12}$$

$$M = \sqrt{a^2 + b^2} \tag{3.13}$$

$$\tan\Delta\psi = -{}^{b}/_{a} \tag{3.14}$$

Equation 3.11 proves that no matter how the scene shapes, the total reflected signal is linear combination of two basis: $\sin(wt + \varphi_0)$ and $\cos(wt + \varphi_0)$. Equation 3.12 demonstrates that the frequency and signal form do not change after transmission. Compare Eqs. 3.11 and 3.12 we get

$$\sum_{i=1}^{N}\sum_{j=1}^{N} C_{ij}\cos\Delta\varphi_{ij} = M\cos\Delta\psi = a; \sum_{i=1}^{N}\sum_{j=1}^{N} C_{ij}\sin\Delta\varphi_{ij} = M\sin\Delta\psi = b \tag{3.15}$$

3.3.3 Observations and Depth Map Recovery

The full collection of binary DMD values is denoted $\{C_{ij}^{k}(i \le N, j \le N), 1 \le k \le P\}$, so P observations are made to get a depth map. Each time we receive a signal, we obtain the values of a and b, the way to determine a or b is a novel access to phase-detection of the back wave named Synchronous Demodulation. After we get the values of a and b, M and $\Delta\psi$ can be calculated too.

With P observations and phase-detection by synchronous demodulation, now we have

$$A = [a^1, a^2, \ldots, a^k \ldots, a^P]^T; B = [b^1, b^2, \ldots, b^k \ldots, b^P]^T \tag{3.16}$$

Define vector Dc and Ds in which include the sin and cosine values of the phase-shift for each point:

$$Dc = [\cos\Delta\varphi_{11}\ldots\cos\Delta\varphi_{1N}, \cos\Delta\varphi_{21}\ldots\cos\Delta\varphi_{2N}, \ldots, \cos\Delta\varphi_{N1}\ldots\cos\Delta\varphi_{NN}]^T \tag{3.17}$$

$$Ds = [\sin\Delta\varphi_{11}, \ldots, \sin\Delta\varphi_{1N}, \sin\Delta\varphi_{21}, \ldots, \sin\Delta\varphi_{2N}, \ldots, \sin\Delta\varphi_{N1}, \ldots, \sin\Delta\varphi_{NN}]^T \tag{3.18}$$

From Eq. 3.15 to 3.18, we acquire the next two equations (Eqs. 3.19 and 3.20) that will finally help us work out Dc and Ds, then the phase-shift of every pixel is solved and the corresponding depth map is recovered from it.

$$A = \begin{pmatrix} C_{11}^{1} & \cdots & C_{1N}^{1} \cdots C_{N1}^{1} & \cdots & C_{NN}^{1} \\ \vdots & \ddots & \vdots \quad\quad \vdots & \ddots & \vdots \\ C_{11}^{k} & \cdots & C_{1N}^{k} \cdots C_{N1}^{k} & \cdots & C_{NN}^{k} \\ C_{11}^{k+1} & \cdots & C_{1N}^{k+1} \cdots C_{N1}^{k+1} & \cdots & C_{NN}^{k+1} \\ \vdots & \ddots & \vdots \quad\quad \vdots & \ddots & \vdots \\ C_{11}^{P} & \cdots & C_{1N}^{P} \cdots C_{N1}^{P} & \cdots & C_{NN}^{P} \end{pmatrix} Dc \tag{3.19}$$

$$B = \begin{pmatrix} C_{11}^{1} & \cdots & C_{1N}^{1} \cdots C_{N1}^{1} & \cdots & C_{NN}^{1} \\ \vdots & \ddots & \vdots \quad\quad \vdots & \ddots & \vdots \\ C_{11}^{k} & \cdots & C_{1N}^{k} \cdots C_{N1}^{k} & \cdots & C_{NN}^{k} \\ C_{11}^{k+1} & \cdots & C_{1N}^{k+1} \cdots C_{N1}^{k+1} & \cdots & C_{NN}^{k+1} \\ \vdots & \ddots & \vdots \quad\quad \vdots & \ddots & \vdots \\ C_{11}^{P} & \cdots & C_{1N}^{P} \cdots C_{N1}^{P} & \cdots & C_{NN}^{P} \end{pmatrix} Ds \tag{3.20}$$

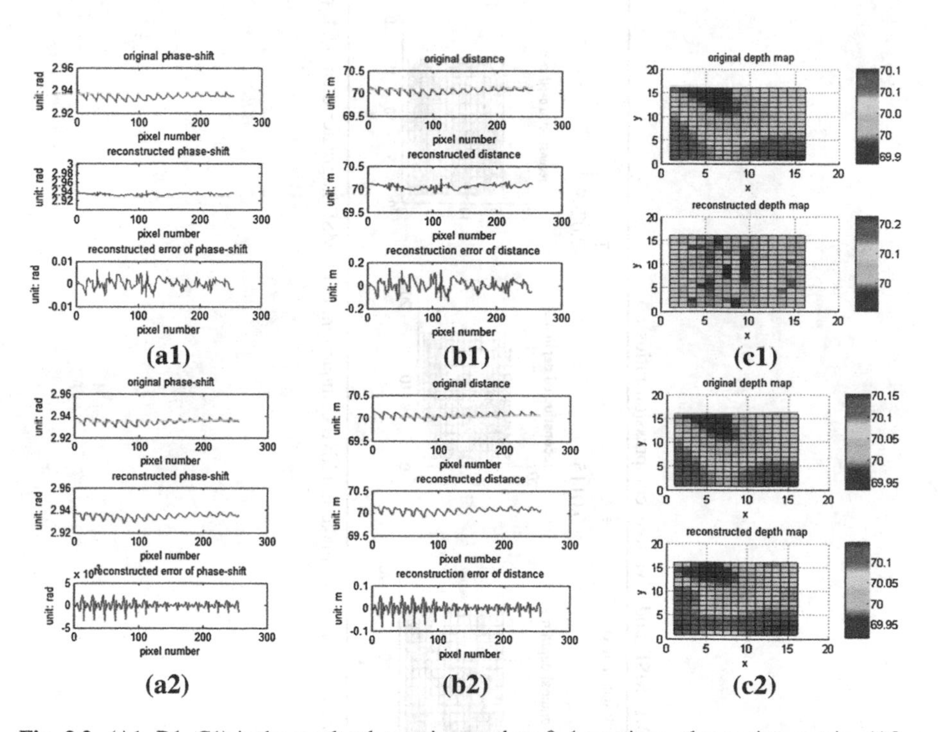

Fig. 3.2 (A1, B1, C1) is the result when using random 0–1 matrix as observation matrix, (A2, B2, C2) uses OHM matrix. **a** Both phase-shift, **b** distance are respectively depicted as well as reconstruction errors and **c** shows the depth maps of simulated terrain

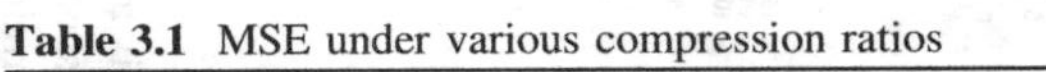

Table 3.1 MSE under various compression ratios

Compression ratio	$\frac{2}{1}$	$\frac{4}{1}$	$\frac{8}{1}$	$\frac{16}{1}$
MSE	0.0115	0.0209	0.0322	0.0414

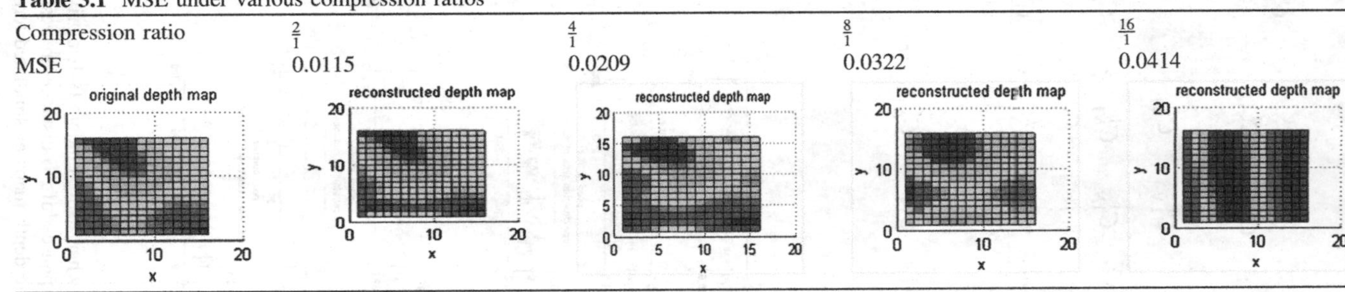

Depth maps are listed to make intuitive comparison, once MSE is greater than 0.04, reconstruction result can be incorrect

3.4 Simulation Results

We performed a series of simulations by MATLAB. The initial terrain data is produced by FBM (Fractal Brown Motion), and all simulations was performed in a distance of 70 m. Odd rows of Hadamard matrix (OHM) [5] and random 0–1 matrix is used as observation matrix Φ in Eq. 3.3 for they are easy to be represented by DMD. The sparsity basis is DWT (discrete wavelet transform) matrix and we choose BP to solve Eqs. 3.19 and 3.20. Moreover, recovery effects under various compression ratios are analyzed to find out a proper compression ratio.

Figure 3.2 shows an original depth map (denoted by color map) of 16×16 resolution and the reconstruction result of the two specific Φ, both were conducted with a 1/4 compression ratio

We can see in Fig. 3.2 (C2) that the reconstructed one has great similarity with the original one. Errors in (A2, B2) are much smaller than in (A1, B1), (B2) proves the ranging accuracy is less than 0.01 m which means a centimeter magnitude

The result proves that OHM works better than other matrix. Now keep OHM as Φ, change the number of observations (compression ratio), measure MSE of each experiment, we get the following results listed in Table 3.1.

MSE is slightly descending when compression ratio rises from 2/1 to 16/1. With a compression ratio lower than 16/1 the compressive laser range finding prototype could gain ranging accuracy in centimeter magnitude.

3.5 Conclusion

We propose a novel algorithm based on phase-shift method for 3D imaging using compressive laser range finding. In contrast to traditional scanning laser radar or nonscanning one with a sensor array, our system with a single photo-detector performs faster, and the ranging accuracy fulfills terrain measurement with a less particular demanding for components. Compared to CS laser radar that based on time-delay method, applying phase-shift method in our system is able to improve SNR to get a better observation result. Through a series of simulation experiments, the results turned out impressive for the reconstruction MSE is less than 0.05 and the ranging accuracy is in centimeter magnitude.

Acknowledgment The work is supported in part by the preliminary research fund under project 40405030103.

References

1. Aull BF, Loomis AH, Young DJ, Stern A, Felton BJ, Daniels PJ et al (2004) Three-dimensional imaging with arrays of Geiger-mode avalanche photodiodes. In: Proceedings of the SPIE—The international society for optical engineering semiconductor photodetectors
2. Candes EJ (2006) Compressive sampling. In: International congress of mathematicians, vol. III. European Mathematical Society, Zürich, pp 1433–1452
3. Colaco A, Kirmani A, Howland GA, Howell JC, Goyal VK (2012) Compressive depth map acquisition using a single photon-counting detector: parametric signal processing meets sparsity. In: IEEE conference on computer vision and pattern recognition (CVPR), 2012
4. Duarte MF, Davenport MA, Takhar D, Laska JN, Ting S, Kelly KF et al (2008) Single-pixel imaging via compressive sampling. Sig Process Mag IEEE 25(2):83–91
5. Wang F, Wei P, Ke J (2011) A specific measurement matrix in compressive imaging system. In: 2011 international conference on optical instruments and technology: optoelectronic imaging and processing technology. Beijing, China

Chapter 4
Gabor Orientation Histogram for Face Representation and Recognition

Jun Yi and Fei Su

Abstract Recently the Gabor-based features have been successfully used for face representation and recognition. In these methods, the face image is filtered with the multiscale multiorientation Gabor filter bank to generate multiple Gabor magnitude images (GMIs), and then the down-sampled GMIs or the LBP (local binary pattern) histograms of GMIs are stacked to form the feature. The stacking procedure makes the dimensions of these features very high, which causes extreme computing and storage load. In this paper, a novel Gabor-based feature termed Gabor orientation histogram (GOH) is proposed, which greatly reduces the feature dimension. Unlike stacking, GOH takes the structure underlying different GMIs into account by regarding the GMIs of different orientations at the same point as a whole, namely orientation vector, to represent the point. Moreover, GOH takes the structure of local region into account by calculating the orientation histogram based on the orientation vectors of points in the local region to describe the region, which is robust to local deformation and noises. The experimental results on the FERET and FRGC databases show that the proposed GOH reduces the feature extraction and recognition time significantly while retains the high recognition performance, which makes a progress toward the practical applications of Gabor-based features for face representation and recognition.

Keywords Gabor filter · Gabor feature · Face representation · Face recognition

---

J. Yi (✉) · F. Su
Beijing University of Posts and Telecommunications, Beijing, China
e-mail: yijun@bupt.edu.cn

F. Su
e-mail: sufei@bupt.edu.cn

A. A. Farag et al. (eds.), *Proceedings of the 3rd International Conference on Multimedia Technology (ICMT 2013)*, Lecture Notes in Electrical Engineering 278,
DOI: 10.1007/978-3-642-41407-7_4,

4.1 Introduction

One-dimensional (1D) Gabor function was proposed by Gabor in 1946 [4]. It achieves the theoretical lower limit for the uncertainty of time and frequency according to Gabor's uncertainty principle of 1D temporal signal. Daugman generalized Gabor's uncertainty principle from 1D temporal signal to two-dimensional (2D) spatial signal and proposed 2D Gabor function in 1985 [3]. Moreover, he showed that the 2D receptive-field profiles of simple cells in mammalian visual cortex were well described by the 2D Gabor function, which was further evaluated and verified by Jones and Palmer in 1987 [5]. Due to the nice mathematical property of Gabor function and its analogy to the biological mechanism, 2D Gabor filter has been widely used in image processing and analysis including face representation and recognition. Gabor-based features have achieved great success in the FERET evaluation [9] and excellent performance on the FERET database [10, 12]. However, Gabor-based features all treated the multiple outputs of different Gabor filters separately and just stacked the features of each output together. The stacking procedure multiplies the feature dimension, leading to extremely high feature dimension, which imposes high computation and storage load.

In this paper, a novel Gabor-based feature termed Gabor orientation histogram (GOH), which greatly reduces the feature dimension and retains the high performance, is proposed to overcome the problems caused by high feature dimension. Unlike stacking, the proposed GOH takes the structure underlying different Gabor magnitude images (GMIs) into account by regarding the GMIs of different orientations at the same point as a whole, namely orientation vector, to represent the point. With the orientation property of points, GOH calculates the orientation histogram in local regions, which takes the structure of local region into account and is robust to local deformation and noises.

GOH is motivated by the success of orientation histogram in SIFT descriptor [6] and HOG descriptor [2]. Moreover, there is a biological motivation that orientation plays an important role in visual perception and understanding, which has been proved by some discoveries such as the orientation selectivity [11] and cross-orientation inhibition [7] presented in the cat cortical visual cells.

The feature dimension of GOH has been greatly reduced relative to other Gabor-based features. The low dimension can reduce the computation and storage load significantly. Moreover, it avoids the small sample size problem and makes the applications of dimensionality reduction methods more flexible, further reducing the computation and storage load and enhances the face recognition performance. The experimental results on the FERET [9] and FRGC [8] databases show that the proposed GOH reduces the feature extraction and recognition time significantly while retains the high recognition performance.

The remaining part of the paper is organized as follows: Sect. 4.2 describes face representation with GOH. Face recognition with GOH is presented in Sect. 4.3, followed by the experimental results in Sect. 4.4. Finally, the conclusion and discussion are given in Sect. 4.5.

4.2 Face Representation with GOH

There are two main steps in our proposed face representation with GOH.

- Filter the face image with the Gabor filter bank and calculate the magnitudes of the complex responses, aiming to obtain the Gabor magnitude images.
- Establish orientation histogram based on the Gabor magnitude images.

The details are given in the following two subsections, followed by the section discussing the properties of GOH.

4.2.1 Gabor Magnitude Image

The Gabor filter bank we use follows the one adopted by Lades et al.(1993) and normally used for face recognition, which is defined as:

$$\Psi_{u,v}(x,y) = \frac{\| \mathbf{K}_{u,v} \|^2}{\sigma^2} \exp\left(-\frac{\| \mathbf{K}_{u,v} \|^2 \| \mathbf{z} \|^2}{2\sigma^2}\right) \left[\exp(i\mathbf{K}_{u,v} \cdot \mathbf{z}) - \exp\left(-\frac{\sigma^2}{2}\right)\right] \tag{4.1}$$

where $\| \mathbf{x} \|$ denotes the ℓ_2 norm of vector $\mathbf{x}$, $\cdot$ denotes the dot product, $\mathbf{z} = [x, y]$, $\mathbf{K}_{u,v} = [k_x, k_y] = [k_v \cos\phi_u, k_v \sin\phi_u]$, $k_v = k_{max}/f^v$, $v = 0,\ 1,\ \cdots,\ V-1$, $\phi_u = \pi u/U$, $u = 0,\ 1,\ \cdots,\ U-1$. v and u define the scale and orientation of the Gabor filter. V and U are the number of scale and orientation of the Gabor filter bank. The parameter σ determines the number of oscillations under the Gaussian envelope, k_{max} is the highest center frequency of the Gabor filter bank, and f is the spacing factor of the center frequency of the Gabor filters in the filter bank. We set these parameters as normally used for face recognition, i.e., $\sigma = 2\pi$, $k_{max} = \pi/2$, and $f = \sqrt{2}$.

The response of an image $I(x,y)$ for the Gabor filter bank is defined as:

$$G_{u,v}(x,y) = I(x,y) * \Psi_{u,v}(x,y) = \rho_{u,v}(x,y)e^{j\theta_{u,v}(x,y)} \tag{4.2}$$

where $*$ denotes the convolution operator, $\rho_{u,v}(x,y)$ and $\theta_{u,v}(x,y)$ are the magnitudes and phases of the response, respectively. Experimental results show that the phase is less discriminative than the magnitude in our method. Thus, as normally used for face recognition, only the magnitude is used in our method, which is referred as Gabor magnitude image (GMI).

4.2.2 Gabor Orientation Histogram

For each pixel in the face image, there are $U \times V$ magnitudes associated with U orientations and$V$ scales. Both the traditional jet and recent LGBP descriptors treat these scales and orientations separately and adopt the strategy of stacking. Here, we interpret the orientation in a new aspect. The orientation is viewed as an image attribute represented by an U dimensional vector termed as orientation vector for every scale. Each dimension of the orientation vector corresponds to an orientation of the Gabor filter bank, and the value of this dimension is the magnitude of that orientation under the scale. Thus, the orientation vector under each scale v is defined as:

$$\mathbf{o}_v(x,y) = \left[\rho_{0,v}(x,y),\ \rho_{1,v}(x,y),\ \cdots,\ \rho_{U-1,v}(x,y)\right] \tag{4.3}$$

Since every point has the orientation attribute, we can calculate orientation histograms for local regions. In our method, the image is segmented into K small blocks. For each block B_k, $k \in 1, 2, \cdots, K$, the orientation histogram is calculated as:

$$\begin{aligned} \mathbf{O}_v(B_k) &= \sum_{(x,y)\in B_k} \mathbf{o}_v(x,y) \\ \mathbf{h}_v(B_k) &= \mathbf{O}_v(B_k) / \parallel \mathbf{O}_v(B_k) \parallel_1 \end{aligned} \tag{4.4}$$

where $\parallel \mathbf{x} \parallel_1$ denotes the ℓ_1 norm of vector $\mathbf{x}$.

Note that $\mathbf{h}_v(B_k)$ just describes the orientation histogram of the block B_k under the scale v. We concatenate all the V scales to get a multiscale orientation histogram for the block, which is the GOH descriptor:

$$\mathbf{h}(B_k) = \left[\mathbf{h}_0(B_k),\ \mathbf{h}_1(B_k),\ \cdots,\ \mathbf{h}_{V-1}\ (B_k)\right] \tag{4.5}$$

Then the GOH descriptors of all the blocks are concatenated to form a GOH sequence (GOHS) as:

$$\mathbf{H} = \left[\mathbf{h}(B_1),\ \mathbf{h}(B_2),\ \cdots,\ \mathbf{h}(B_K)\right] \tag{4.6}$$

$\mathbf{H}$ is the final feature vector representing the image.

The procedure of face representation with GOH is illustrated in Fig. 4.1 and summarized in Algorithm 4.1.

Algorithm 4.1. Gabor Orientation Histogram

Input: The image $I(x,y)$

1. Filter $I(x,y)$ with the Gabor filter bank to get the GMIs $\rho_{u,v}(x,y)$ as in Eq. 4.2.
2. Organize the magnitudes of each point under the same scale and different orientations into orientation vector $\mathbf{o}_v(x,y)$ as in Eq. 4.3.
3. Segment the image into small blocks and calculate the orientation histograms of the blocks under each scale $\mathbf{h}_v(B_k)$ as in Eq. 4.4.
4. Concatenate the orientation histograms of each block under different scales into the multiscale orientation histograms $\mathbf{h}(B_k)$ as in Eq. 4.5.
5. Concatenate the multiscale orientation histograms of all the blocks into $\mathbf{H}$ as in Eq. 4.6.

Output: The GOH feature $\mathbf{H}$ of the image $I(x,y)$

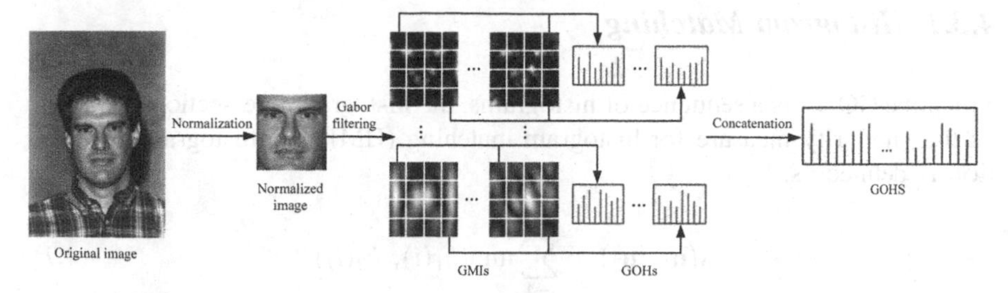

Fig. 4.1 The procedure of face representation with GOH

4.2.3 Properties of GOH

The procedure and algorithm of GOH are very simple, which makes it very efficient to compute. Moreover, GOH has some good properties as discussed below to guarantee its performance.

1. GOH inherits the multiscale property of the Gabor filter bank.
2. GOH uses the local histogram which is region-based and robust to local deformation and noises.
3. Some parameters including the number of scale and orientation of the Gabor filter bank and the block size, can be adjusted flexibly in GOH to change the feature dimension, which offer a tradeoff between the recognition performance and the computation and storage load. Moreover, the blocks can be flexibly arranged, such as overlapping them or organizing them in a hierarchy structure.
4. The histogram in GOH is different from other histograms. In GOH, each point contributes to all bins and the contributions are weighted according to the corresponding Gabor magnitudes. However, in LBP histogram, each point contributes to only one bin without weight, and in HOG, the contribution of each point is weighted according to its gradient magnitude as GOH, but each point contributes to only two neighbor bins.

4.3 Face Recognition with GOH

In this paper, we apply two methods for face recognition to verify the feature description ability and the potential discriminant ability of the proposed GOH feature.

4.3.1 Histogram Matching

Since the GOHS is a sequence of histograms, the histogram intersection is applied as the similarity measure for histogram matching (HM). The histogram intersection is defined as:

$$s(\mathbf{h}_1, \mathbf{h}_2) = \sum_{i=1}^{N} \min(h_1(i), h_2(i)) \tag{4.7}$$

where $\mathbf{h}_1$ and $\mathbf{h}_2$ are two histograms, $h_1(i)$ and $h_2(i)$ are ith bin of $\mathbf{h}_1$ and $\mathbf{h}_2$,respectively, and N is the number of histogram bins. This method utilizes no other information except the original GOH feature, which can test the feature description ability of the GOH.

4.3.2 NLDA and Cosine Distance

In this method, we first use NLDA [1] to transform the GOHS, and then apply the cosine distance (CD) on the transformed feature as the similarity measure:

$$d(\mathbf{v}_1, \mathbf{v}_2) = \frac{\mathbf{v}_1^{\mathrm{T}}\mathbf{v}_2}{\| \mathbf{v}_1 \| \| \mathbf{v}_2 \|} \tag{4.8}$$

where $\mathbf{v}_1$ and $\mathbf{v}_2$ are the two transformed features. NLDA can not only reduce the dimension of the feature, but also enhance the discriminating power of the feature. Therefore, it is used here to verify the potential of the GOH feature.

4.4 Experimental Results

To evaluate the performance of our proposed GOH, it is compared to LGBPH [12] which is a state-of-the-art Gabor-based feature. For a fair comparison, the same 5-scale 8-orientation Gabor filter bank is used, and all the face images are preprocessed and normalized to 80×88 size in our experiments as that in LGBPH methods [10, 12]. The block size is nonoverlap 8×8 for GOH and 8×4 for LGBPH. The number of histogram bins for LGBPH is set to 8. Therefore, the dimension of GOH is 4400 and the dimension of LGBPH is 70400.

Since we focus on the applicability of face recognition methods, the recognition performance, the storage requirements, and the computing time are all concerned, which are presented in the following subsections, respectively.

4.4.1 Recognition Performance

The recognition performance of GOH is evaluated by face identification on the FERET database [9] and face verification on the FRGC database [8].

GOH-based HM and NLDA were tested on the FERET database for face identification, compared to LGBPH-based ensemble of piecewise Fisher discriminant analysis (EPFDA) [10] which achieved the state-of-the-art performance on FERET to our knowledge. The best result of FERET97 [9] and the result of LGBPH-based HM in [12] were also given for reference. NLDA and EPFDA were trained with all the 1,621 samples in the training set. All methods were tested on four probe sets: Fb (1,195 samples with expression variation), Fc (194 samples with lighting variation), Dup.I (722 samples with 0–34 months of time interval), and Dup.II (234 samples with more than 18 months of time interval), with Fa (1,196 samples) as the templates. Table 4.1 summarizes the comparison with the feature dimension of each method, which is the reduced dimension in LDA methods.

Table 4.2 gives the comparision results of the GOH-based NLDA and LGBPH-based EPFDA on the FRGC database for face verification. NLDA and EPFDA were trained with 6,630 samples in the training set (for each of the 221 persons in the training set, randomly selecting 30 samples). For the genuine match, each sample of each person in the target set was matched with all other samples of the same person. For the impostor match, each sample of each person was compared with the randomly selected 5 samples of other persons. Then the Equal Error Rate (EER) was calculated with these matching scores. The imposter matches were randomly selected 100 times to get 100 EERs, the average of which was the final EER.

It is obvious that the feature dimension of the GOH-based methods is much lower than the LGBPH-based methods, thus GOH-based methods save a lot of storage for the face templates, which is a great advantage of GOH in the practical applications. It is very important in the scenario of face recognition on resource constrained implementation. In spite of the much lower feature dimension, the results show that the recognition performance of the GOH-based methods can rival

Table 4.1 The identification performance on FERET

Method	Dimension	Fb	Fc	Dup.I	Dup.II
FERET97 Best [9]	–	0.962	0.820	0.591	0.521
LGBPH-HM [12]	70400	0.94	0.97	0.68	0.53
GOH-HM	4400	0.92	0.96	0.66	0.63
LGBPH-EPFDA	11000	0.980	0.959	0.849	0.816
GOH-NLDA	428	0.985	0.964	0.823	0.820

Table 4.2 The verification performances on FRGC

Method	EER
LGBPH-EPFDA	0.043
GOH-NLDA	0.032

Table 4.3 The feature extraction and recognition (1:1000 matching) time (ms/image)

Method	Extraction time	Recognition time
LGBPH-EPFDA	110	502
GOH-NLDA	45	110

or even exceed the LGBPH-based methods. As we expected, NLDA indeed enhances the discriminating power and GOH-based NLDA achieves the best performance.

4.4.2 Feature Extraction and Recognition Time

To evaluate the computation cost, we tested the feature extraction time of GOH and LGBPH and the recognition time of GOH-based NLDA and LGBPH-based EPFDA, respectively, on a PC with 2.93 GHz Intel Core 2 Duo CPU and 2 GB RAM. The images were all of 80×88 size. The results are shown in Table 4.3. It is obvious that both the feature extraction time and recognition time of GOH are much less than LGBPH, which is a great advantage of GOH in the practical applications.

4.5 Conclusion and Discussion

The proposed GOH feature greatly reduces the feature dimension relative to other Gabor-based features while retains the high recognition performance. Thus, GOH efficiently solves the storage and computing problem suffered by other Gabor-based features, which makes a progress toward the practical applications of Gabor-based features for face representation and recognition.

In this paper, although GOH is used as a global descriptor, it can also be viewed as a local descriptor which describes the local region associated with the key point, in fact, and can be applied to represent and recognize objects other than face in general. The potential discriminating power of GOH will be further explored with more sophisticated recognition methods. The fusion of the scales of GOH needs to be addressed too.

References

1. Chen LF, Liao HY, Ko MT, Lin JC, Yu GJ (2000) A new LDA-based face recognition system which can solve the small sample size problem. Pattern Recogn 33(10):1713–1726
2. Dalal N, Triggs B (2005) Histograms of oriented gradients for human detection. In: Proceedings of IEEE conference on computer vision and pattern recognition, vol. 1. pp 886–893

3. Daugman J (1985) Uncertainty relation for resolution in space, spatial frequency, and orientation optimized by two-dimensional visual cortical filters. J Opt Soc Am A: Opt 2(7):1160–1169
4. Gabor D (1946) Theory of communication. part 1: the analysis of information. J Inst Electr Eng Part II I: Radio Commun Eng 93(26):429–441
5. Jones J, Palmer L (1987) An evaluation of the two-dimensional gabor filter model of simple receptive fields in cat striate cortex. J Neurophysiol 58(6):1233–1258
6. Lowe D (2004) Distinctive image features from scale-invariant keypoints. Int J Comput Vision 60(2):91–110
7. Morrone M, Burr D, Maffei L (1982) Functional implications of cross-orientation inhibition of cortical visual cells. i. neurophysiological evidence. In: Proceedings of the royal society B, vol. 216(1204). pp 335–354
8. Phillips P, Flynn P, Scruggs T, Bowyer K, et al (2005) Overview of the face recognition grand challenge. In: Proceedings of IEEE conference on computer vision and pattern recognition, vol. 1. pp 947–954
9. Phillips P, Hyeonjoon M, Rizvi S, Rauss P (2000) The FERET evaluation methodology for face-recognition algorithms. IEEE Trans Pattern Anal Mach Intell 22(10):1090–1104
10. Shan S, Zhang W, Su Y, Chen X, Gao W (2006) Ensemble of piecewise FDA based on spatial histograms of local (gabor) binary patterns for face recognition. In: Proceedings of 18th Int'l Conference on pattern recognition, vol. 4. pp 606–609
11. Watkins D, Berkley M (1974) The orientation selectivity of single neurons in cat striate cortex. Exp Brain Res 19(4):433–446
12. Zhang W, Shan S, Gao W, Chen X, Zhang H (2005) Local gabor binary pattern histogram sequence (LGBPHS): a novel non-statistical model for face representation and recognition. In: Proceedings on Int'l Conference on computer vision, vol. 1. pp 786–791

Chapter 5
A Simplify Linear Multi-User Detection

Bingchao Liu, Jingjing Liang, Li Fang and Daoben Li

Abstract The high complexity of the maximum likelihood (ML) multi-user detection (MUD) restricts its application. The linear detector is a suboptimal algorithm with lower complexity, which is also the basis and criterion of many other algorithms, such as interference cancellation algorithms and adaptive receiver. In order to avoid the stochastic matrix inversion, which is the critical part of the conventional detector, a novel simplify algorithm, which has the same performance compare with the conventional detector but lower complexity, is proposed in this paper. And lower complexity linear equalizers can be derived by using the similar method.

Keywords MUD · Linear equalizers · Lower complexity

5.1 Introduction

With the development of communication technology, better communication quality and higher bit rate data service become increasingly important to a communication system. Multiplex access technology has became a key technology in

B. Liu (✉) · J. Liang · L. Fang · D. Li
School of Information and Communication Engineering, Beijing University of Posts and Telecommunications, Beijing 100876, China
e-mail: liubingchao2272@sina.com

J. Liang
e-mail: ljj5456@163.com

L. Fang
e-mail: mrs.fangli@gmail.com

D. Li
e-mail: lidaoben@vip.sohu.net

A. A. Farag et al. (eds.), *Proceedings of the 3rd International Conference on Multimedia Technology (ICMT 2013)*, Lecture Notes in Electrical Engineering 278,
DOI: 10.1007/978-3-642-41407-7_5,

the modern communication system. And it requires all the signals from different users are still orthogonal at receiver side, however, the complex time-varying multipath fading characteristics of mobile communication channel not only induces a very large power penalty on the performance of modulation over wireless channels, but also reduces the system spectrum efficiency.

Take a code division multiple access (CDMA) system as an example, the time diffusion of the wireless channel causes the inter-symbol interference (ISI) for single user system, which can be eliminated by using maximum likelihood sequence detector (MLSD), and the nonorthogonality in the signature waveforms caused the multiple access interference (MAI) among multiplexed signals, which should be overcame by using the multi-user detector.

The idea of multi-user detection (MUD) is first proposed by Schneider in 1979 [1]. R. Kohno introduced a kind of MUD using side information of other users' signals in 1983 [2]. And the optimal MUD was introduced by Verd in 1986 [3].

The optimum solution of MUD is an exponentially complex task, whose complexity is $2^{Q(L-1)}$(Q-modulation order, L-number of users). Moreover, the problem has been shown to be NP-complete [4].

Therefore, there is a need to look for sub-optimum approaches to multi-user detection. There are three kinds of sub-optimum algorithms. The first one is linear multi-user detectors, such as decorrelating detector and the minimum mean square error (MMSE) detector [5, 6]. They do not have the optimal performance but the lowest complexity.

The second category is the nun-linear detector, including successive interference cancellation (SIC) [7] and parallel interference cancellation (PIC) [7] receiver, which can improve the performance of linear detector, but still have the problem of error propagation.

The third category is adaptive multi-user detectors. Those algorithms are adaptive in nature and can adapt to time-varying channel conditions through training mechanisms. Their adaptive nature makes them more suitable for deployment in practical systems. Examples of this class of algorithms include recursive least square (RLS), least mean square (LMS), and gradient descent based MMSE optimisation.

Linear detection is the basis and criterion of lots of useful algorithms, such as sphere decoding, MMSE-SIC, and MMSE-LMS receivers. The key point of the linear detector is the matrix inversion. Take the zero-forcing (ZF) detector as an example, the inversion of the random matrix is required, so its computation complexity will be higher with the increase of the symbol block length. In order to avoid the stochastic matrix inversion, a simplify linear multi-user detector is proposed.

This remainder of the paper is organized as follows. The system model is described in Sect. 5.2, followed by the conventional detector in Sect. 5.3 and the novel algorithm is derived in Sect. 5.4. Simulation results are analyzed in Sect. 5.5. Finally, discussion and conclusions are given in Sect. 5.6.

5.2 System Model

Consider a communication system with L users. The mathematical model of transmitted signal of the lth user in equivalent lowpass waveform can be expressed as:

$$x_l(t) = \sum_n b_l(n) g_l(t - nT) \tag{5.1}$$

where $g_l(t)$ is the pulse shaping of user-l satisfying $\int_0^T |g_l(t)|^2 dt = 1,\ l = 0,\ 1, \cdots, L-1$. $b_l(n)$ is user-l's transmitted sequence, T is the symbol duration.

Assume the lth channel response is $h_l(t, \xi), \xi \in [0, \Delta]$, satisfying $\int_0^\Delta |h_l(t, \xi)|^2 d\xi = 1$, Δ is the maximum delay spreading of wireless channel. The channel is assumed to be approximately stationary over a symbol period. The received signal can be expressed as:

$$v(t) = \sum_{l=0}^{L-1} \sum_n b_l(n) \int_0^\Delta h_l(t, \xi) g_l(t - nT - \xi) d\xi + n(t) \tag{5.2}$$

where $n(t)$ is white Gaussian noise with single side power spectrum density N_0.

Because the two-side bandwidth of $g_l(t)$ is BHz, according to the sampling theory, Eq. (5.2) can be described by the points whose sampling interval is $1/B$.

$$v(t) = \sum_{l=0}^{L-1} \sum_n b_l(n) \sum_{i=0}^{K-1} g_i^l(t - nT) h_i^l + n(t) \tag{5.3}$$

where $K \triangleq \lfloor B\Delta + 1 \rfloor$, $\lfloor \Delta \rfloor$ stands the largest integer not less than Δ.

$$g_i^l(t) \triangleq g_l\left(t - \frac{i}{B}\right) \tag{5.4}$$

$$h_i^l \triangleq \int_{\frac{i}{B}}^{\frac{i+1}{B}} h_l(t,\ \xi) d\xi \tag{5.5}$$

5.3 The Conventional Detector

For equal apriori probabilities of transmitted symbol, the optimum multi-user detector under Gaussian interference is the minimum Euclidean distance receiver, which can be expressed as:

$$\min_{\{\hat{b}_l(n)\}} \int_{nT}^{(n+1)T} \left| v(t) - \sum_{l=0}^{L-1} \sum_{i=0}^{K-1} b_l(n) g_i^l(t - nT) h_i^l \right|^2 dt \tag{5.6}$$

That is, choosing $\{\hat{b}_l(n)\}$ such that $\sum_{l=0}^{L-1}\sum_{i=0}^{K-1} b_l(n)g_i^l(t-nT)h_i^l$ is closest to the received signal in the mean square sense.

To simplify, it can be rewrite in matrix form. Let

$$\mathbf{h}_l \triangleq \left[h_0^l, h_1^l, \cdots, h_{K-1}^l\right]^T \tag{5.7}$$

$$\mathbf{g}_l(t) \triangleq \left[g_0^l(t),\ g_1^l(t), \cdots, g_{K-1}^l(t)\right]^T \tag{5.8}$$

$$\mathbf{G}(t) \triangleq \begin{bmatrix} \mathbf{h}_0^T\mathbf{g}_0(t) \\ \mathbf{h}_1^T\mathbf{g}_1(t) \\ \vdots \\ \mathbf{h}_{L-1}^T\mathbf{g}_{L-1}(t) \end{bmatrix} \tag{5.9}$$

$$\mathbf{b}(n) \triangleq [b_0(n),\ b_1(n), \cdots, b_{L-1}(n)]^T \tag{5.10}$$

where Δ^T stands the transpose of Δ

Then the maximum likelihood detector can be expressed in matrix form as:

$$\min_{\hat{\mathbf{b}}(n)} \int_{nT}^{(n+1)T} \left|v(t) - \mathbf{b}^T(n)\mathbf{G}(t-nT)\right|^2 dt \tag{5.11}$$

Define

$$\mathbf{y}_l \triangleq \left[y_0^l,\ y_1^l, \cdots, y_{K-1}^l\right]^T \tag{5.12}$$

where

$$y_i^l \triangleq \int_{nT}^{(n+1)T} v(t)g_i^{l*}(t-nT)dt \tag{5.13}$$

and

$$\mathbf{Y} \triangleq \begin{bmatrix} \mathbf{h}_0^H\mathbf{y}_0 \\ \mathbf{h}_1^H\mathbf{y}_1 \\ \vdots \\ \mathbf{h}_{L-1}^H\mathbf{y}_{L-1} \end{bmatrix} \tag{5.14}$$

where ΔH stands the conjugate transpose of Δ.

$\mathbf{Y}$ is a complex Gaussian vector, which is a sufficient statistic for $\mathbf{b}$.

Define

$$r_{ii'}^{(l,l')} \triangleq \frac{1}{2}\int_0^T g_i^l(t)g_{i'}^{l'*}(t)dt \tag{5.15}$$

$$\mathbf{r}_{ll'} \triangleq \left[r_{ii'}^{(l,l')} \right]_{K \times K} \tag{5.16}$$

$$\mathbf{R} \triangleq \left[\mathbf{h}_l^H \mathbf{r}_{ll'} \mathbf{h}_{l'} \right]_{L \times L} \tag{5.17}$$

where Δ^*stands the conjugation of Δ. Then Eq. (5.17) can be further simplified as (For simplicity, the variable n will be omitted)

$$\hat{\mathbf{b}} = \max_{\hat{\mathbf{b}}} \left\{ \mathbf{b}^H \mathbf{Y} + \mathbf{b}^T \mathbf{Y}^* - \mathbf{b}^H \mathbf{R} \mathbf{b} \right\} \tag{5.18}$$

In order to find the optimal solution of (18), **R** and **Y** should be calculated firstly. **R** is a stochastic matrix. $\mathbf{Y} = [y_0, y_1, \cdots, y_{L-1}]^T$, and the lth element of **Y** is

$$y_l = \sum_{i=0}^{K-1} \int_{nT}^{(n+1)T} v(t) h_i^{l*} g_i^{l*}(t - nT) dt \tag{5.19}$$

which is just the summation of the output of user-l's matched filters.

Let L be the number of simultaneous users, M the number of modulation level, the total number of $\{\mathbf{b}\}$ is M^L. If M or L is too larger, the complexity of exhaustive search method is too high to implement.

It can be seen that (5.18) is the quadratic form of b. Calculating the first order gradient of the right side of Eq. (5.18):

$$\nabla_{\mathbf{b}} \left(\mathbf{b}^H \mathbf{Y} + \mathbf{b}^T \mathbf{Y}^* - \mathbf{b}^H \mathbf{R} \mathbf{b} \right) = \mathbf{Y} - \mathbf{R} \mathbf{b} \tag{5.20}$$

then calculating the second order gradient of (5.20):

$$\nabla_{\mathbf{b}} \left[\nabla_{\mathbf{b}^*} \left(\mathbf{b}^H \mathbf{Y} + \mathbf{b}^T \mathbf{Y}^* - \mathbf{b}^H \mathbf{R} \mathbf{b} \right) \right] = -\mathbf{R} \tag{5.21}$$

Because **R** is a positive definite matrix, Eq. (5.18) has the unique maximum solution, which is

$$\hat{\mathbf{b}} = \mathbf{R}^{-1} \mathbf{Y} \tag{5.22}$$

This is just the direct solution of MUD, obviously, it is also the zero-forcing solution of multi-user detector. While (5.22) needs the inversion operation of stochastic matrix in the time-varying channel, if L or K is too larger, the complexity of the stochastic matrix inversion operation is also too high to implement.

5.4 The Novel Algorithm

To overcome the problem described in the previous section, a novel algorithm is proposed to avoided the matrix inversion.

Define

$$\mathbf{G}'(t) \triangleq \left[g_0^0(t), \cdots, g_{K-1}^0(t), \cdots, g_0^{L-1}(t), \cdots, g_{K-1}^{L-1}(t)\right]^T \tag{5.23}$$

$$\mathbf{H} \triangleq \mathrm{diag}\left\{h_0^0, \cdots, h_{K-1}^0, \cdots, h_0^{L-1}, \cdots, h_{K-1}^{L-1}\right\} \tag{5.24}$$

$$\mathbf{b}'(t) \triangleq \left[\underbrace{b_0(n), \cdots, b_0(n)}_{\text{repet } K \text{ times}}, \cdots, \underbrace{b_{L-1}(n), \cdots, b_{L-1}(n)}_{\text{repet } K \text{ times}}\right]^T \tag{5.25}$$

In this case, (5.2) could be rewrite as:

$$\min_{\hat{\mathbf{b}}'(n)} \int_{nT}^{(n+1)T} \left|v(t) - \mathbf{b}'^T(n)\mathbf{H}\mathbf{G}'(t - nT)\right|^2 dt \tag{5.26}$$

which is equivalent to

$$\begin{aligned} \max_{\hat{\mathbf{b}}'} (n) \Bigg\{ & \mathbf{b}'^H(n)\mathbf{H}^* \int_{nT}^{(n+1)T} v(t)\mathbf{G}'^*(t - nT)dt \\ & + \mathbf{b}'^T(n)\mathbf{H} \int_{nT}^{(n+1)T} v^*(t)\mathbf{G}'(t - nT)dt \\ & - \mathbf{b}'^H(n)\mathbf{H}^* \int_{nT}^{(n+1)T} \mathbf{G}'^*(t - nT)\mathbf{G}'^T(t - nT)dt\mathbf{H}\mathbf{b}'(n) \Bigg\} \end{aligned} \tag{5.27}$$

If we define

$$\mathbf{Y}'(n) \triangleq \left[y_0^0(n), \cdots, y_{K-1}^0(n), \cdots, y_0^{L-1}(n), \cdots, y_{K-1}^{L-1}(n)\right]^T \tag{5.28}$$

where

$$y_i^l(n) \triangleq \int_{nT}^{(n+1)T} v(t) g_i^{l*}(t - nT)dt \tag{5.29}$$

and

$$\mathbf{R}' \triangleq [\mathbf{r}_{ll'}]_{L \times L} \tag{5.30}$$

which is a block matrix of size $L \times L$, where $\mathbf{r}_{ll'}$ has been defined as Eq. (5.16). $\mathbf{r}_{ll'}$ is a square matrix of size $K \times K$, and $\mathbf{R}'$is the correlation matrix of different user's signals, which is a determinate matrix but a stochastic matrix. Equation (5.18) can be modified for (For simplicity, the variable n will be omitted.)

$$\max_{\hat{\mathbf{b}}'} \left\{\mathbf{b}'^H\mathbf{H}^*\mathbf{Y}' + \mathbf{b}'^T\mathbf{H}\mathbf{Y}'^* - \mathbf{b}'^H\mathbf{H}^*\mathbf{R}'\mathbf{H}\mathbf{b}'\right\} \tag{5.31}$$

It is easy to prove that the solution of Eq. (31) is equal to the solution of Eq. (5.18), which is also the maximum likelihood exhaustive solutions.

Equation (5.31) is also the quadratic form of $\mathbf{b}'$, and $\mathbf{H}^*\mathbf{R}'\mathbf{H}$ is a positive defined form, so it has the unique maximum solution.

Calculating the first order gradient of the right side of Eq. (5.31):

$$\nabla_{\mathbf{b}'}\left(\mathbf{b}'^{H}\mathbf{H}^*\mathbf{Y}' + \mathbf{b}'^{T}\mathbf{H}\mathbf{Y}'^{*} - \mathbf{b}'^{H}\mathbf{H}^*\mathbf{R}'\mathbf{H}\mathbf{b}'\right) = \mathbf{H}^*\mathbf{Y}' - \mathbf{H}^*\mathbf{R}'\mathbf{H}\mathbf{b}' \tag{5.32}$$

Let the right side of (5.32) equal to 0, it can be get

$$\hat{\mathbf{b}'} = \mathbf{H}^{-1}\mathbf{R}'^{-1}\mathbf{Y}' \tag{5.33}$$

Formula (5.33) is the proposed maximum likelihood direct solution of (5.31). $\mathbf{b}'$ and $\hat{\mathbf{b}'}$ are complex vectors of size *KL*, which are not the required complex vector of size *L*. In order to get the required $\hat{\mathbf{b}}$. One of the reasonable method is to take the algebraic average or the weighted average of the *K* duplications of the *l*th symbol.

Matrix inversion is needed for the two Algorithm, however, $\mathbf{R}'$ is a stochastic matrix in the previous algorithm but a determinate matrix for the latter one. In the proposed algorithm, $\mathbf{R}'^{-1}$ is a known matrix, which can be calculated in advance, and $\mathbf{H}$ is a diagonal matrix, whose inversion is very simple to be calculated.

ZF detector is discuss in this paper, however, MMSE detector and other algorithms can be got by using the similar method. And ZF and MMSE equalizers can be also derived for eliminating ISI.

5.5 Simulation Results

In this section, simulation results are provided to illustrate the performance of the proposed algorithm. A eight-user CDMA system is considered, and the perfect channel estimation for MUD of base station (BS) and mobile station (MS). The spreading codes for every MS in the CDMA system are using m code of length 64. Binary phase shift keying (BPSK) signaling is assumed.

The ITU-VA channel model [9] is used in the simulations. The system bandwidth is 1.25 MHz, and the chip rate is 1.2288 Mcps. The movement speed is assumed 30 km/h.

Figure 5.1 presents the bit error rate (BER) comparison among different receivers of up-link and down-link, respectively.

It can be seen that all the performance of different receivers is almost the same, except the direct algebraic average solution.

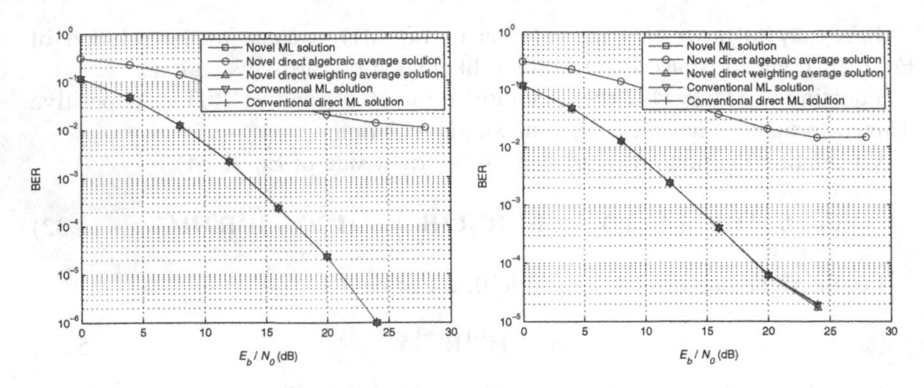

Fig. 5.1 BER performance of down-link and up-link using different receivers

5.6 Conclusions

In order to avoid the stochastic matrix inversion of the direct maximum likelihood MUD, a novel direct ML-MUD is proposed, which avoid the stochastic matrix inversion without performance degradation. And simplify MMSE-MUD, MMSE or ZF equalizer can be easily derived by using the similar method.

References

1. Schneider KS (1979) Optimum detection of code division multiplexed signals. IEEE Trans Aeros Electron Syst AES 15(1):181–185
2. Imai KRH, Hatori M (1983) Cancellation Technique of Co-channel interference in Asynchronous Spread Spectrum Multiple Access system. ICICE Trans Comm 65-A:416–423
3. Verd S (1986) Minimum Probability of error for asynchronous gaussian multiple-access channels. IEEE Trans Inform Theory IT-32:85C96
4. Verd S (1989) Computational complexity of optimal multiuser detection. Algorithmica 4:303–312
5. Lupas R, Verd S (1989) Linear multiuser detectors for synchronous code-division multiple-acess channels. IEEE Trans Inform Theory 35:123–136
6. Lupas R, Verd S (1990) Near-far resistance of multiuser detectors in synchronous channels. IEEE Trans Commun 38:496–508
7. Varanasi MK, Aazhang B (1990) Multistage detection in asynchronous code-division multiple-acess communications. IEEE Trans Commun 38:509–519
8. Divsalar D, Simon M, Raphaeli D (1995) Improved parallel interference cancellation for CDMA. IEEE Person Commun 2(2):46–58
9. Recommendation ITU-RM.1225 (2000) Guidelines for evaluation of radio transmission on technologies for IMT C

Chapter 6
Automatic Mosaic Method of UAV Water-Area Images Based on POS Data

Yaping Wang, Yijin Chen and Donghai Xie

Abstract This paper mainly studies the automatic mosaic method of Unmanned Aerial Vehicle (UAV) large water-area images. During the mosaic process, the monitoring water area was relatively larger and the image gradation changed small, both of which caused little effective feature points and can be detected. In order to solve this problem, Scale Invariant Feature Transform (SIFT) and Harris algorithms were synergistic used to extract the feature points, when there was not only water area but also one small land in images. Meanwhile, POS directional data was used to help the geometric correction of images in order to achieve image mosaic when there was only water area. Finally, this paper takes oil spills monitoring of offshore surface as experimental object to describe the UAV practical application value in terms of water resources monitoring and emergency relief.

Keywords UAV water-area images · SIFT · Harris · POS · Automatic mosaic

6.1 Introduction

Water quality monitoring is an important basis of water quality assessment and water pollution control. The real-time monitoring information of water quality disaster can help the local government and relevant departments to make correct decision quickly and effectively, which have great significance in disaster prevention and mitigation. At present, the conventional water quality monitoring

---

Y. Wang (✉) · Y. Chen
College of Geoscience and Surveying Engineering, China University of Mining and Technology, Beijing, China
e-mail: wangyp326@163.com

D. Xie
College of Resource Environment and Tourism, Capital Normal University, Beijing, China

A. A. Farag et al. (eds.), *Proceedings of the 3rd International Conference on Multimedia Technology (ICMT 2013)*, Lecture Notes in Electrical Engineering 278,
DOI: 10.1007/978-3-642-41407-7_6,

method in our country is that monitoring some fixed point or fixed sections, sampling, and analyzing over many years. Therefore, with the restriction of human resources, material and climatic conditions, and so on, the traditional detection methods are usually costly. Moreover, these methods are specifically for small areas water quality parameter, not reflect the overall large-scale distribution of water quality parameters [1]. Remote sensing technology has many advantages, such as a wide range monitoring quickly, low cost, and long-term dynamic monitoring. Therefore, the emergence and development of remote sensing technology has opened up new avenues for large range of water quality monitoring and research [2, 3]. Nowadays, China has gradually formed a complete set of three-dimensional water bodies (including terrestrial water and oceans) disaster surveillance monitoring system. The system has a full range of monitoring objects and diversification means, including satellite and aerial remote sensing, ship monitoring, buoys monitoring, shore stations monitoring, and so on [4, 5].

In these monitoring methods, Unmanned Aerial Vehicle (UAV) remote sensing monitoring has more and more widespread concern. Compared with traditional remote sensing images, high-resolution images obtained by UAV are superior in many aspects, such as low cost, rapidness, convenience, and can provide disaster investigation and emergency command with real-time information [6, 7]. However, when the UAV is used to monitor terrestrial-surface water or marine water for the image gradation changed small, the traditional feature detection methods cannot detect enough feature points to achieve image mosaic automatically.

We researched the features of UAV images and proposed an optimization automatic mosaic method of large water area. The rest of the paper is organized as follows. We briefly introduce some related algorithm principles and the key technology in Sect. 6.2, including the SIFT, Harris, RANSAC, Affine Transformation, and POS. In Sect. 6.3, some details of our proposed automatic mosaic method are given, followed by one application case about oil spills monitoring of offshore surface. Finally, we present our conclusions in Sect. 6.4.

6.2 Related Algorithm Principles and Key Technology

6.2.1 SIFT and Harris Feature Detection Algorithm

There are many feature point extraction methods in the field of image registration, such as: Moravee, Harris, SUSAN, Scale Invariant Feature Transform (SIFT), and some improved algorithm. In our research, when there was not only water area but also one small land in UAV images, SIFT algorithm was used to extract the feature points, together with Harris algorithm.

The core idea of SIFT is: use Difference-of-Gaussian (DOG) function to construct scale space and all the feature points detected are the extreme points of DOG function. Therefore, the result of SIFT algorithm may not always be clear terrain

Fig. 6.1 Feature detection with SIFT and Harris together

points on the image [8]. Harris algorithm is simple and only uses the first-order differential gray. And this algorithm also has a high stability and robustness [9], so that it can be used as a supplement in SIFT feature detection.

In Fig. 6.1, the mulberry arrows represented the main gradient direction of SIFT feature vectors and the red points represented the Harris corner points. According to statistics, 35 pairs of feature points were extracted by SIFT while 19 pairs of corner points were added by Harris algorithm. This method greatly increased the namesake points used for matching. However, these feature points still contained some false match points and the repeated points.

6.2.2 Eliminate the Error Matching Points by RANSAC

During the matching process, Random Sample Consensus (RANSAC) algorithm was used to delete the error points. It can effectively remove errors points about 50 % and the main steps are as follows:

- Calculate the distance l between each pair of namesake feature points,

$$l = \sqrt{dx^2 + dy^2}, \quad dx = x - x', \quad dy = y - y' \tag{6.1}$$

wherein, (x, y) and (x', y') were one pair of the namesake feature points.

- Threshold value was $3\sigma_O$. In one pair of namesake points, if $l > 3\sigma_O$, delete this pair.

$$\sigma_0 = \sqrt{l^T pl/n} \tag{6.2}$$

wherein, σ_O was the Mean Error, n was the number of the feature point pairs and P was the Unit Matrix [10].

6.2.3 Image Mosaic by Affine Transformation

Assumed the reference image and registered image are, respectively, $f(x, y)$ and $f'(x, y)$. The pixel coordinates of the feature points are (x, y) and (x', y'). The geometric relationship between them is:

$$\begin{bmatrix} x' \\ y' \end{bmatrix} = \begin{bmatrix} a_1 & a_2 \\ a_3 & a_4 \end{bmatrix} \begin{bmatrix} x \\ y \end{bmatrix} + \begin{bmatrix} b_1 \\ b_2 \end{bmatrix} \tag{6.3}$$

Formula (6.3) shows Affine Transformation Model. Wherein, a_1, a_2, a_3, a_4, b_1, b_2 are unknown parameters. If these six unknown parameters are calculated out, at least three pairs of feature points are needed. In fact, it often expands the number of pairs to reduce the influence of various factors and improve the accuracy of calculated parameters.

6.2.4 Image Mosaic with POS Data

Position and Orientation System (POS) is produced in the 1990s and it is composed of Inertial Measurement Unit (IMU) and Global Positioning System (GPS). UAV is equipped with POS systems to record UAV flight position and attitude information. However, photographing by UAV would be affected by many factors, such as: wind weather conditions, airplane load limiting, and other factors. These factors lead to the UAV cannot be equipped with high-precision GPS and raw POS data, often contains a lot of gross error. The direct utilization of raw POS data cannot meet the actual needs of geographical positioning. When it is used to make images stitching without any correction, it will arise a larger offset and dislocation in the final image. However, to the automatic stitching of water-area images, using some methods such as using image gray value, manually stitching, and so on, good results cannot be acquired, and the stitching is very slow. If only research the global area from the macro perspective, mosaicking based on the POS has a great advantage. This method is faster, so you may get the global information at the first time. For incident investigation and response to emergency, it is absolutely significant.

The overlap rate of UAV captured images is higher. Therefore, if making an appropriate "thinning" treatment to these images (delete some images on one principle), it wouldn't affect the splicing effect; on the contrary, the splicing accuracy can be improved to some extent [11]. Some images in the position of take-off, landing, and going around a turn, which caused a greater distortion, are needed to be removed. As the Table 6.1 shows that the data of NO.121-124 have some terrible error: three angle values (ω, φ, κ) are in larger change. In fact, the unmanned aerial vehicle was on the position of turning at that time. So these data should be removed. In addition, the overlap rate (55 % $\leq$ heading overlap rate < 90 %, 20 % $\leq$ side overlap rate < 50 %), swing angle (left, right $\leq$ 12°),

Table 6.1 UAV POS directional data

Num	Week	Time	Longitude	Latitude	Elevation	Omega	Phi	Kappa
114	2	15:35:01	121.93368	39.03889	381.867	−0.9	1.5	43.9
115	2	15:35:02	121.93423	39.03934	381.880	−0.7	1.9	43.8
116	2	15:35:04	121.93478	39.03979	382.233	−0.5	1.8	43.7
117	2	15:35:05	121.93533	39.04024	382.249	−0.9	2.8	43.7
118	2	15:35:07	121.93588	39.04069	382.221	−0.6	2.4	43.9
119	2	15:35:08	121.93643	39.04113	382.115	0	2	43.9
120	2	15:35:10	121.93698	39.04158	382.136	−0.4	1.7	43.9
121	2	15:35:11	121.93753	39.04202	382.298	−1.9	38	44.6
122	2	15:35:13	121.93811	39.0424	378.851	0.6	40.1	56.0
123	2	15:35:14	121.93872	39.04264	370.463	5.8	40.3	70.7
124	2	15:35:16	121.9394	39.04272	360.849	12.8	40.5	92.0

bending ($\leq 5°$), and the height difference between adjacent photos are also need to be checked and the abnormal data records and are removed.

The main idea of this method using POS to mosaic images is as follows: First, some conditions are defined. Each image is projected on the Earth Ellipsoid (assuming it is a plane). The latitude and longitude of photography center is known by the POS data. The dimensions (width and length) of an image are known and the size and resolution of each image are same in one strip. Then, with the above information, the geographic coordinates of each image can be calculated. The coordinates of photography center point and the four corner points are especially important. Finally, according to the coordinates of corner points of each image, all images are arranged in the sequence. And the whole strip is mosaicked by certain overlay principles. This process involves the transformation from spatial coordinate system to ground coordinate system, and then converse. In other words, each image data need to be converted between the two coordinate systems two times.

The formulas used were as follows:

The first step, transform the plane coordinates to the image space coordinates by formula (6.4)

$$\begin{aligned} x &= ix - wd/2 \\ y &= ht/2 - iy \\ z &= -f \end{aligned} \tag{6.4}$$

ix, iy are the image column number, (x, y, z) is the coordinate in the image space coordinate system $S-xyz$, f is the focal length, wd, ht are the length and width of the image.

According collinear equation,

$$\begin{aligned} X - X_S &= (Z - Z_S)\frac{a_1x + a_2y - a_3f}{c_1x + c_2y - c_3f} \\ Y - Y_S &= (Z - Z_S)\frac{b_1x + b_2y - b_3f}{c_1x + c_2y - c_3f} \end{aligned} \tag{6.5}$$

wherein, a_i, b_i, c_i is calculated based on $(\omega, \varphi, \kappa)$ (Xs, Ys, Zs) is the coordinate of Photography Center Point and (X, Y, Z) is the coordinate of one point on image in the ground coordinate system. (Xs, Ys, Zs) is calculated by the formula (6.6) [12].

$$\begin{bmatrix} X_S \\ Y_S \\ Z_S \end{bmatrix} = \begin{bmatrix} (N+H)\cos B\cos L \\ (N+H)\cos B\sin L \\ [N(1-e^2)+H]\sin B \end{bmatrix} \tag{6.6}$$

(B, L, H) is the value of Latitude, Longitude, and Elevation. N is the prime vertical radius of curvature.

Figure 6.2 shows the mosaic picture for water area based the POS. On the left, there are two adjacent UAV images. The mosaic result is on the right. The maritime surveillance vessel is in red-circled part and the bright gray area is spilled oil.

6.3 Experimental Details and Application Case

6.3.1 Experiment

In this experiment, the registration of two images (one pair) was taken for example to explain the mosaic of one air strip. First, only the images with high flight quality were selected as data source. In other words, based on POS data, according to some principles in Sect. 6.2.4, other images were selected after the appropriate "thinning" process.

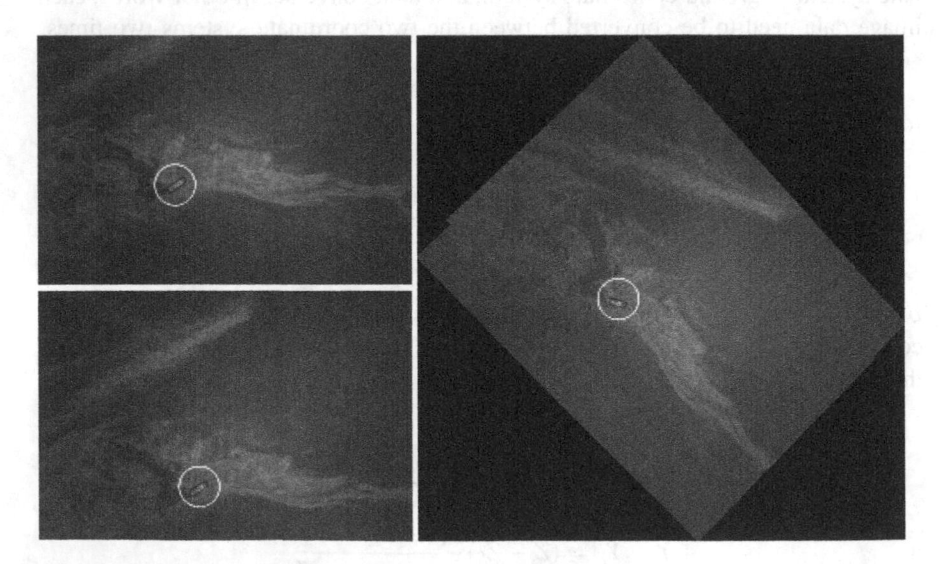

Fig. 6.2 Water-area image mosaic with POS

Generally speaking, there are four steps in the traditional mosaic methods, including detecting the feature points, eliminating the error matching points, calculating the transformation model, and resampling the mosaic images [6]. However, for the mosaic of water-area images, these methods aren't entirely applicable. Two special cases are needed to be considered. One case is that in the images there is not only water area but also one small land. The other is that the area is only water area. The detailed process is as follows and showed in Fig. 6.3.

First CASE: There is not only water area but also one small land in UAV images. We made some improvements based on the traditional feature detection algorithm.

1. Detecting the feature points:
   According to the basic algorithms in Sect. 6.2.1, SIFT and Harris algorithm were synergistic used to extract the feature points. We found this method was effective to deal with the problem that the monitoring water area was relatively larger and the image gradation changed small.

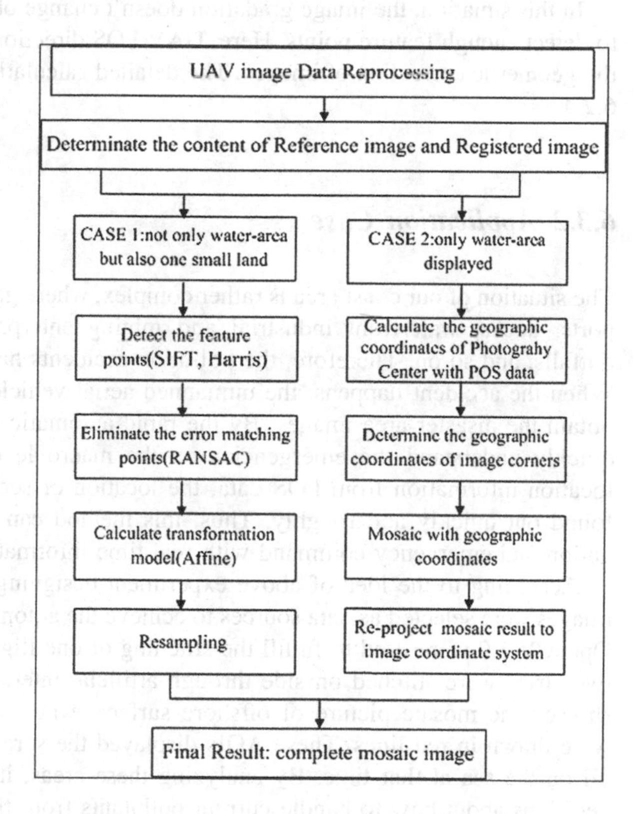

Fig. 6.3 The mosaic process with two images

2. Eliminating the error matching points:
   Here, RANSAC algorithm in Sect. 6.2.2 was used, which was absolutely effective to eliminate many "outliers" in practice.
3. Calculating the transformation model:
   Affine transformation was used to achieve two images mosaic. Actually, this transformation model is relatively simple and easy; its transitivity is good for more images mosaic. Therefore, in this experiment the overall mosaic of air strip was used with this transformation. From left image to right image, it turned to be selected as the reference image and the image to be registered. And then the registration of whole strip was finished.
4. Resampling the mosaic image:
   Resampling is the process of transforming a discrete image which is defined at one set of coordinate locations to a new set of coordinate points. Resampling can be divided conceptually into two processes: interpolation of the discrete image to a continuous image and then sampling the interpolated image. This step is related to the transformation model closely.

Second CASE: The area displayed in the UAV images is only water.

In this situation, the image gradation doesn't change obviously. So it is difficult to detect enough feature points. Here, UAV POS directional data was used to help the geometric correction of images. The detailed calculation process is in the Sect. 6.2.4.

6.3.2 Application Case

The situation of our coast area is rather complex, where gathered a large number of ports, docks, waterfront industrial and mining enterprises, aquaculture, rivers outfalls, and so on. Therefore, the pollution accidents happened more frequently. When the accident happens, the unmanned aerial vehicle can be used to quickly obtain the disaster area images. By the rapid automatic stitching images, we can quickly understand the emergencies on the macro level. Meanwhile, with the location information from POS data, the location of serious disaster area can be found out quickly and roughly. Thus, this method can provide disaster investigation and emergency command with real-time information.

According to the idea of above experiment designing, two strip data of UAV images were selected as data sources to achieve the automatic splicing. VC 6.0 and OpenCV 1.0 were used to fulfill the stitching of one flight strip images. Then the two strips were stitched on side through artificial interacting method. Figure 6.4 showed the mosaic picture of offshore surface, where Areas of Interest (AOIs) were drawn in red lines. These AOIs displayed the spread situation about spilled oil on the sea at that time. By analyzing these areas, it helped to quickly make decisions about how to handle current pollutants from the macro level.

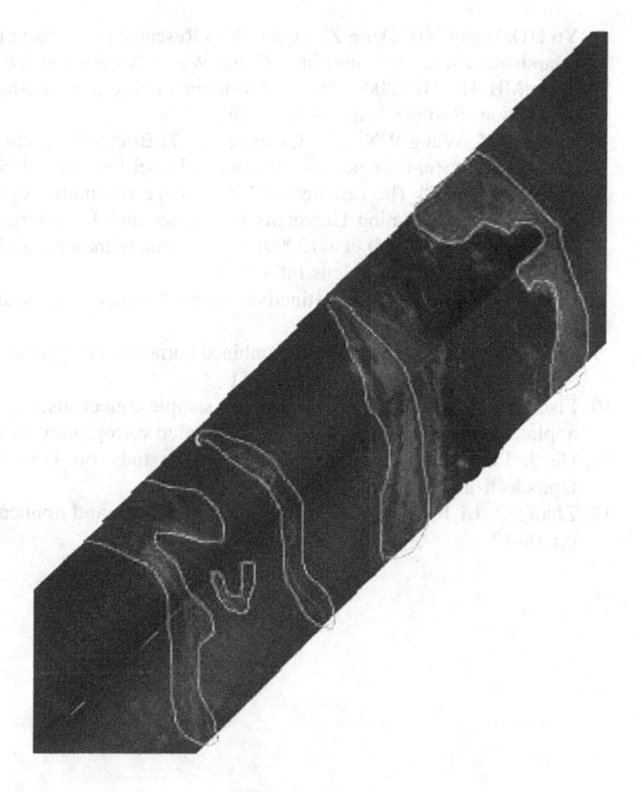

Fig. 6.4 Monitoring the oil spilling with the UAV mosaic image

6.4 Conclusions

In this paper, based on UAV POS positioning data, two traditional feature detection algorithms of SIFT and Harris were used to achieve automatic mosaic of a large-scale water-area images. The obtained mosaic image can help researcher's rapid learning the AOI situation from the macro level and make decisions correctly and quickly. The application case showed that UAV remote sensing technology has great practical value in the field of water quality monitoring.

Acknowledgments This research was supported by the Fundamental Research Funds for the Central Universities under the grant number No. 2010YD06.

References

1. Hang YZ, NIE YP, LIN QZ et al (2000) Surface water quality monitoring using remote sensing. Remote Sens Technol Appl 15(4):214–219
2. Zhang H, Zheng GM, LI ZW et al (2005) Multi-temporal remote sensing information model for pollution monitoring of inland water. Environ Monit China 21(5):63–68

3. Yu HD, Wang YH, Deng ZL et al (2008) Research progress in remote sensing technology for inland water quality monitoring. China Water Wastewater 24(22):12–16
4. Yang MH, HU HP (2000) On the development of remote sensing and agricultural information acquisition. Remote Sens Inf 4:44–46
5. Zhang WZ, Wang WX, Wei LT et al (2007) Brief introduction of remote sensing dynamic monitoring system for oceanic disaster in Hebei Province. J Nat Disasters 16(3):76–80
6. Wang YP (2010) The research of UAV image automatic registration and mosaic methods. Master thesis, Kunming University of Science and Technology
7. Jin W, Ge HL, DU HQ et al (2009) A review on unmanned aerial vehicle remote sensing and its application. Remote Sens Inf 1:88–92
8. David G, Lowe (2004) Distinctive image features from scale-invariant key points. Int J Comput Vis 60(2):91–110
9. Harris C, Stephens M (1988) A combined corner and edge detector. In: Proceedings of the 4th Alvey vision conference, pp 147–151
10. Fishier MA, Boles RC (1981) Random sample concensus: a paradigm for model fitting with applications to image analysis and automated cartography. Commun ACM 24(6):381–395
11. He J, Li YS, Lu H (2011) Experimental study on UAV image stitching error. Laser Optoelectron Prog 48(12):64–68
12. Zhang Q, Li JQ (2007) GPS surveying principles and applications. Science Press, Beijing, pp 16–17

Chapter 7
Video Copy-Move Forgery Detection and Localization Based on Structural Similarity

Fugui Li and Tianqiang Huang

Abstract Copy-move forgery is one of the most common types of video forgeries. To detect such forgery, a new algorithm based on structural similarity is proposed. In this algorithm, we extend structural similarity to measure the similarity between two frames of a video. Since the value of similarity between duplicated frames is higher than that between the normal inter-frames, a temporal similarity measurement strategy between short sub-sequences is put forward to detect copy-move forgery. In addition, we can obtain an accurate forgery localization. Extensive experimental results evaluated on 15 videos captured by the digital camera and mobile camera in stationary and moving mode show that the precision of this algorithm can reach 99.7 % which is higher than a previous relevant study.

Keywords Video forgery · Copy-move detection · Copy-move localization · Structural similarity

7.1 Introduction

With the wide use of a variety of digital multimedia devices as well as the development of powerful video editing tools (such as Adobe Premiere Pro and Adobe After Effects, etc.), it is becoming easy for common users to edit and process videos without leaving any visual clues. When a large number of edited and forged videos appear on the video sharing sites, the news, scientific discovery

F. Li (✉) · T. Huang
School of Mathematics and Computer Science, Fujian Normal University,
Fuzhou 350007, China
e-mail: leaf304@163.com

T. Huang
e-mail: fjhtq@fjnu.edu.cn

A. A. Farag et al. (eds.), *Proceedings of the 3rd International Conference on Multimedia Technology (ICMT 2013)*, Lecture Notes in Electrical Engineering 278,
DOI: 10.1007/978-3-642-41407-7_7,

and court exhibits, there is no doubt that they will have a significant adverse effects on the stability of society and the state. Therefore, digital video forensics has become a very important research issue [1].

Video forensics can be classified into two different categories: active forensics and passive forensics. For active forensics, some pre-embedded specific information which could not be perceived in the video is needed, such as digital watermark and digital signature. In this case, one can determine whether the video is tampered or not by detecting the integrity of the information. While there is no requirement on specific information for passive forensics just by analyzing some inherent properties of videos. Recently, more attention was drawn to passive forensics. For an MPEG video, it is usually resaved in MPEG format after tampering operations. In the literature, there are already different kinds of methods for detecting video forgeries in MPEG format. In [2, 3], the authors proposed methods to detect video forgeries based on double compression and double quantization. The authors of [4] proposed a feature curve to reveal the compression history of an MPEG video file with a given GOP structure, and used the temporal patterns of block artifacts as evidence to detect tampering, Su et al. [5] utilized the motion-compensated edge artifacts (MCEA) for detecting of video forgery with the type of frame-deletion. Meanwhile, Dong et al. [6] exploited the MCEA difference between adjacent P frames, and judged whether there are any spikes in the Fourier transform domain after double MPEG compression to detect video forgery. A scheme of tampering detection using statistics of motion vectors produced by inter-frame prediction was proposed in [7]. Huang et al. [8] employed the contents continuity between frames and bidirectional motion vectors for the frame deletion and insertion tampering. These detection methods are based on analyzing coding theory of the MPEG format video. However, for the unity of the video format, these detection methods are very limited. Recently, antiforensic techniques have also been reported in [9, 10] against some of the existing forensic techniques.

In addition to the type of frame insertion and deletion in video tampering, copy-move tampering is also a common type of video tampering, containing two types: spatial tampering and temporal tampering. In spatial tampering, a region may be pasted to a different location on the same frame or other frames. In this way, the tampering aims to replace or hide the undesired object will be achieved. While in temporal tampering, multiframe is replaced by the copy of previous ones, having the scenes replaced without affecting the continuity of the video, or pasted to a different location having the scenes occurred ahead or delay. Now, different approaches are developed to detect video copy-move forgery, and all of them are based on the same concept that a copy-move forgery brings a correlation between the original frames and the duplicated ones. In allusion to the existing detecting approaches, high calculating complexity and high false alarm rate still exist. Wang et al. [11] used the similarity in the temporal and spatial correlation matrices, embodying the correlations of short sub-sequences, as evidence of detect duplicated frames in a full-length video. In [12], the authors divided the video frames into different areas, by calculating pattern noise and correlation in the temporal adjacent and spatial overlapping blocks, proposed a method to detecting tampered video with regional copy-move. Meanwhile, Kobayashi et al. [13] proposed an approach to detect suspicious regions in video captured with a static scene by using noise characteristics, but for the regions from the video itself, the algorithm would be constrained, in addition, the noise characteristics may not be estimated correctly under low compression rates. In [14], the authors utilized the Histogram of Oriented Gradients (HOG) feature matching and video compression properties for the detecting of temporal copy-move tampering in videos, but the high dimensional features of HOG lead to a higher complexity of the algorithm. Lin et al. [15] presented a coarse-to-fine approach for detecting frame duplication forgery in the temporal, but many candidates are selected for the videos, which makes the computation time significantly longer in the fine search.

In this paper, a video copy-move forgery detection algorithm based on structural similarity is proposed. In this algorithm, a full-length video sequence is divided into short overlapping sub-sequences, and then the structural similarity is extended to measure the similarity between two frames of a video. Finally, similarities between the sub-sequences in the temporal domain are measured to find out pairs of sub-sequence where replication relationship exists. Moreover, those pairs of sub-sequence are combined into a complete duplicated sequence and the location of the duplicates is located. Such an algorithm allows us to see whether a copy-move attack has occurred or not and furthermore obtain an accurate forgery localization. Extensive experimental results evaluated on 15 videos captured by the digital camera and mobile camera in stationary and moving mode show that the precision of this algorithm can reach 99.7 % which is higher than a previous relevant study.

The rest of the paper is organized as follows. Section 7.2 gives a brief introduction to the structural similarity. Section 7.3 shows the details of the proposed detection method. Experimental results and analysis are presented in Sect. 7.4. The conclusions are finally drawn in Sect. 7.5.

7.2 Structural Similarity

Considering the perceptual features of the human visual system, Wang et al. [16] introduced a structural similarity (SSIM)-based quality metric. The SSIM metric measures the similarity with three statistical components, which are luminance comparison, contrast comparison, and structural comparison. Let Y be the distorted image of X, for any two pixels $x \in X$ and $y \in Y$, the SSIM metric is as follows,

$$l(x,y) = \frac{2\mu_x\mu_y + C_1}{\mu_x^2 + \mu_y^2 + C_1} \tag{7.1}$$

$$c(x,y) = \frac{2\sigma_x\sigma_y + C_2}{\sigma_x^2 + \sigma_y^2 + C_2} \tag{7.2}$$

$$s(x,y) = \frac{\sigma_{xy} + C_3}{\sigma_x \sigma_y + C_3} \tag{7.3}$$

Combining the three comparison functions of Eqs. (7.1)–(7.3) produces a general form of the SSIM index:

$$\mathrm{SSIM}(x,y) = [l(x,y)]^{\alpha} [c(x,y)]^{\beta} [s(x,y)]^{\gamma} \tag{7.4}$$

Here, parameters α, β and γ adjust the relative importance of three components. Usually, $\alpha = \beta = \gamma = 1$, $C_3 = C_2/2$, producing a specific form of the SSIM index:

$$\mathrm{SSIM}(x,y) = \frac{(2\mu_x \mu_y + C_1)(2\sigma_{xy} + C_2)}{(\mu_x^2 + \mu_y^2 + C_1)(\sigma_x^2 + \sigma_y^2 + C_2)} \tag{7.5}$$

where μ_x and μ_y are the means of the local windows, which are with a size of 11×11, centered at x and y, respectively, σ_x and σ_y are the standard variance, σ_{xy} is the covariance of the two windows, C_1, C_2, C_3 are small constants to make sure the denominator not being zero.

Then the mean SSIM (MSSIM) is used to evaluate the overall image quality,

$$\mathrm{MSSIM}(x,y) = \frac{1}{M} \sum_{i=1}^{M} \mathrm{SSIM}(x_i, y_i) \tag{7.6}$$

where, M is the number of local windows in the image, the higher the MSSIM is, the better quality of the distorted image will be.

7.3 Proposed Method

In temporal tampering, multiframe is replaced by a copy of previous ones, having the scenes replaced without affecting the continuity of the video, or pasted to a different location having the scenes occurred ahead or delay. In this section, the proposed method for detecting copy-move forgery and locating the duplicated frames is presented in detail. Therefore, our method includes three parts: (1) inter-frame similarity measurement; (2) forgery detection; and (3) forgery localization.

7.3.1 Inter-Frame Similarity Measurement

As described in Sect. 7.2, structural similarity can be used to evaluate image quality, the higher the MSSIM is, the better the quality of the distorted image will be, that is, the image X is more similar to image Y. For a video sequence, it is just a successive images in the temporal domain. Thus, we extend structural similarity to measure the similarity between two images. If a video has been tampered by copy-move in the temporal domain, duplicated frames will exist in it which makes the value of similarity between duplicated frames higher than that between the normal inter-frames. Here, we definite a threshold to judge whether a video has duplicated frames. When the MSSIM is higher than threshold $\tau = 0.994$, we consider the two images have the relationship of replication, where τ is experiment threshold.

Figure 7.1 shows the procedure of similarity measure between two images. For any of the two images I_1 and I_2 in the video sequence, first, we convert images from color to grayscale for reducing computation time, and the luminance information of gray-scale images are extracted. Then we remove the luminance information of the images to calculate the contrast information. Finally, we divide by the contrast information to calculate structural information. As described in Sect. 7.2, the MSSIM will be obtained to measure the similarity between two images.

For all video sequence with different content and captured by different equipment, it is difficult or impossible to obtain an ideal threshold τ. Therefore, through measuring the similarities between adjacent frames of the normal videos, we can estimate the maximum possible value of similarity in the video sequence.

Generally, video frame rates vary with different capturing equipment. For instance, videos taken by mobile cameras are usually at 15 or 20 fps, and those taken by digital cameras are at 24 or 30 fps. The higher the video frame rate, the more the frames will be per second, thus it will make the higher similarities between adjacent frames. Figure 7.2 shows the similarities between adjacent frames of the normal videos. In Fig. 7.2a, videos are taken by mobile camera at 15 fps and digital camera at 30 fps in the same scene with stationary cameras. For the difference of frame rate, we can see that the similarities between adjacent frames in digital camera are relatively flat and the values are nearly to 1, and the maximum value is 0.9931. It means that if the frame rate is higher, the similarities will also be higher. Similarly, In Fig. 7.2b, videos are taken by digital camera at

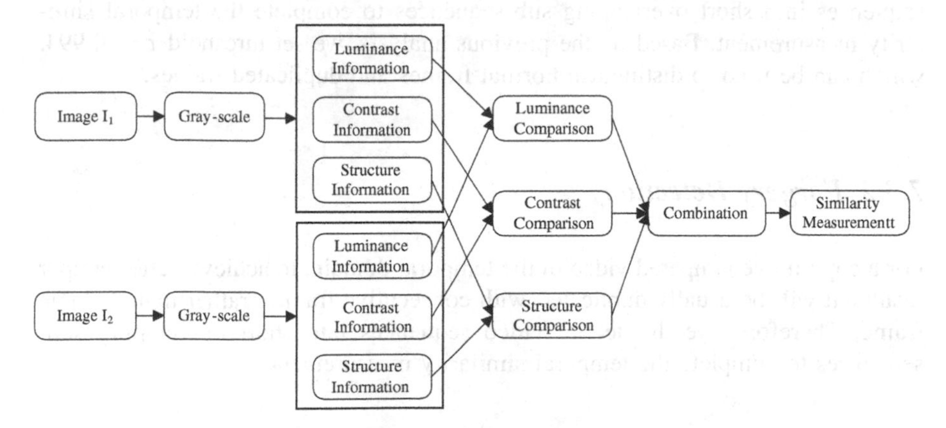

Fig. 7.1 Similarity measurement between two images

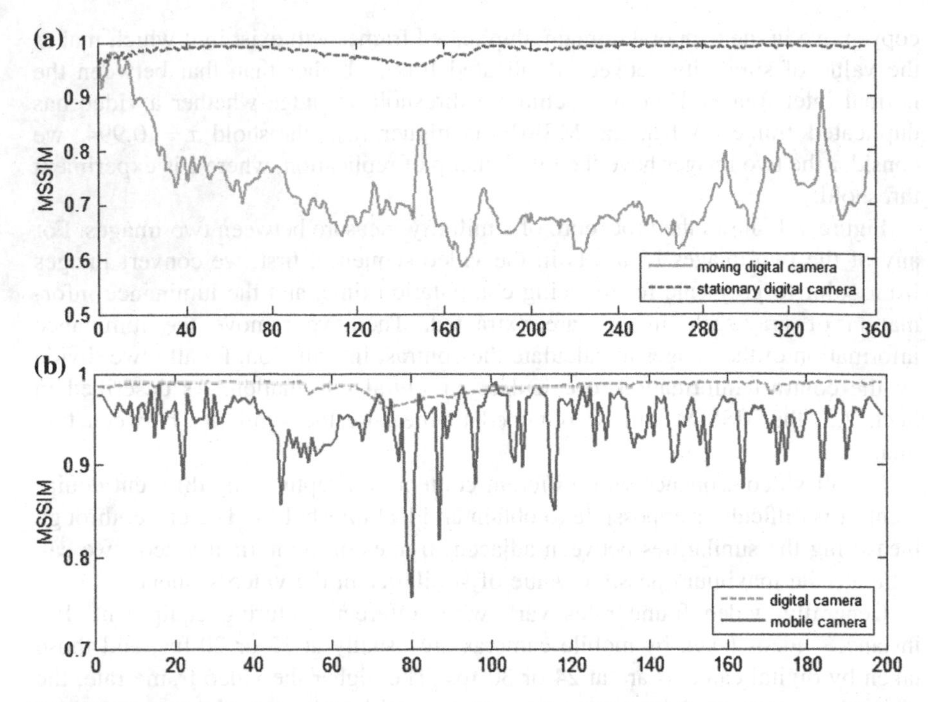

Fig. 7.2 Similarities between adjacent frames of normal videos. **a** similarities between adjacent frames in videos taken by different cameras, **b** similarities between adjacent frames in videos taken by digital camera

30 fps in the same scene with stationary camera and moving camera. For the reason that the content are changing slowly in video taken by stationary camera, the similarities between adjacent frames are higher and the maximum value is 0.9934.

For a copy-move tampered video, it will make similarities between duplicated frames higher than that between normal frames. Meanwhile, we divide the video sequences into short overlapping sub-sequences to complete the temporal similarity measurement. Based on the previous analysis, we set threshold $\tau = 0.994$, which can be used to distinguish normal frames and duplicated frames.

7.3.2 Forgery Detection

For a copy-move tampered video in the temporal domain, to achieve better tamper results, it will be usually duplicated with consecutive frames, rather than a single frame. Therefore, we divide the video sequences into short overlapping sub-sequences to complete the temporal similarity measurement.

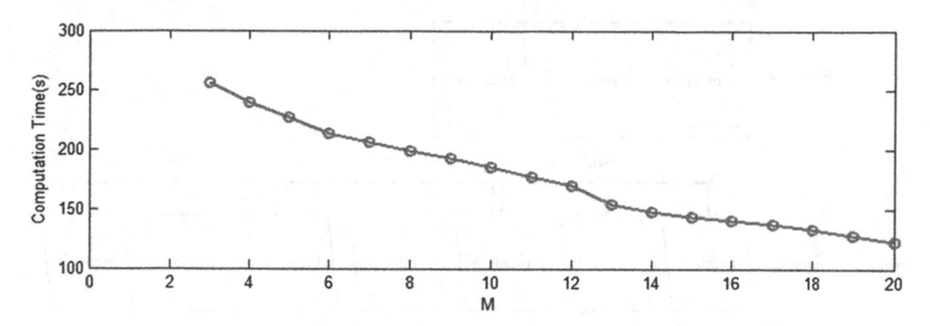

Fig. 7.3 The computation time of a test video with different length of sub-sequence

7.3.2.1 Sub-sequence Partition

A video sequence consists of many continuous images, which can be expressed as $F = I(x, y, t), x \in [0, W - 1], y \in [0, H - 1], t \in [0, N - 1]$, where W and H represent the size of a frame, N is the length of video sequence. First, converting each frame from color to grayscale. Then, dividing a full-length grayscale video sequence into short overlapping sub-sequences, we suppose M is the length of sub-sequence.

For the unknown of whether the video has be tampered, as well as the unknown length of the duplicated sequence. Therefore, when the M is larger than the length of duplicated sequence, it will increase false detection rate of the algorithm, while the M is low, it will increase the times of similarity measurement in each sub-sequence and lead to high time complexity of the algorithm. Figure 7.3 illustrations the computation time of a test video with different length of sub-sequence. With the increasing of M, the computation time shows a decreasing trend.

In [11], the authors selected 30 frames as the length of sub-sequence, but for the number of duplicated frames less than 30 frames, the method will be failed. Considering frame rate of the current digital products and that few numbers of tampered frames have few effects on understanding the content of the video, we set $M = 15$ as the length of sub-sequence.

7.3.2.2 Temporal Similarity Measurement Strategy

In this subsection, a temporal similarity measurement strategy between short sub-sequences is put forward. We set $M = 15$ as the length of sub-sequence, by sliding a frame to get a new sub-sequence, therefore, there is $N-1$ sub-sequence in total. Supposing the first sub-sequence is Seq^1, so the last sequence is Seq^{N-1}, where the superscript number of the sub-sequence equals to the first frame's number in each sub-sequence. An example of detailed temporal similarity measurement strategy is described in Fig. 7.4.

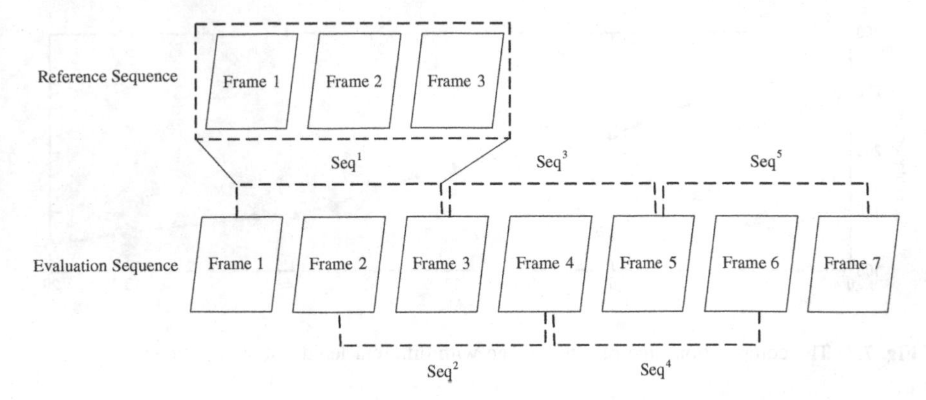

Fig. 7.4 An example of temporal similarity measurement between sub-sequences

Due to the length of sub-sequence is M, we need to measure similarity between each reference sequence and evaluation sequence for M times. If the two sub-sequences exist replication relationship, it means that the value of M times similarity measurement is higher than threshold τ. When the value of the pervious measurement is higher than threshold τ, we only conduct the next measurement, thus it will decrease the times of measurement and improve the efficiency of the algorithm. Conversely, it will keep jumping to next measurement of another two sub-sequences until the measurements of the entire video sequence are completed.

The steps are as follows. First define a value K and initialize it with zero, which is used to record the number of inter-frame similarity value higher than the predefined threshold τ between two sub-sequences. Then, if getting a match, the value will be updated as $K = K+1$. When K equals to M, we consider the current evaluation sequence as a replica of the reference sequence. Finally, we record the superscript number of the two sub-sequences in matrix A. If the matrix A is an empty matrix, we consider that it is a normal video.

7.3.2.3 Merging Duplicated Sub-Sequences

When a duplicated sequence is divided into sub-sequences for detecting copy-move forgery, we need to merge several duplicated sequences to form a complete duplicated sequence. Moreover, we need to remove the false duplicated sequences, therefore, a simple and effective merging strategy is designed.

For a copy-move tampered video, the differences are equal between the superscript of the reference sequence and evaluation sequence. However, due to the influence of adjacent sub-sequences, a reference sequence may be matched with two or more evaluation sequences. Therefore, we select the maximum frequency of distance as copy-move tamper distance to merge these sub-sequences to form a new sequence (see Fig. 7.6).

7.3.3 Forgery Localization

Through the above steps, we can get the merged two copy-move sequences, but we cannot distinguish which sequence is the original sequence, and which sequence is the duplicated sequence. For an original sequence, based on the continuity of the content in a video, both the first and last frame of the sequence are highly similar to the adjacent frames, but the duplicated sequence destroyed the continuity of video, so the value of similarity between them will be relatively low. We suppose the pair of forgery frames are $i-j$ and $i+m-j+m$, by calculating the similarity between the first and last frame of two sequences with the adjacent frame respectively, to distinguish the position of the original sequence and the duplicated sequence. The calculation formulas are as follows:

$$\mathrm{MSSIM}_i = \mathrm{SSIM}(i, i-1) \tag{7.7}$$

$$\mathrm{MSSIM}_j = \mathrm{SSIM}(j, j+1) \tag{7.8}$$

$$\mathrm{MSSIM}_{i+m} = \mathrm{SSIM}(i+m, i+m-1) \tag{7.9}$$

$$\mathrm{MSSIM}_{j+m} = \mathrm{SSIM}(j+m, j+m+1) \tag{7.10}$$

If $\mathrm{MSSIM}_i + \mathrm{MSSIM}_j > \mathrm{MSSIM}_{i+m} + \mathrm{MSSIM}_{j+m}$, we think that the original sequence is $i-j$ and the duplicated sequence is $i+m-j+m$, otherwise, the original sequence is $i+m-j+m$ and the duplicated sequence is $i-j$, thus we have located the location of duplicated sequence.

7.4 Experimental Results and Analysis

In our experiment, we selected 15 test videos captured by the digital camera and mobile camera in stationary and moving mode. Each frame is 640 × 480 pixels in size, and the frame rate is 30 and 15 fps. To evaluate the performance of the proposed algorithm, we created duplicated frames with different lengths ranging from 35 to 200, and used MPEG-VCR and Adobe Premiere Pro CS4 to tamper the videos. Table 7.1 shows the details of 15 test videos. The computer used in experiments is configured as 3.06 GHz Intel processor and the operating environment is MATLAB R2010b.

A simple example of video copy-move tamper is shown in Fig. 7.5, the video sequences in the first row are captured normally, but the video sequences in the second row are tampered with the type of copy-move forgery. As the example shows, the location of video frame 4 and 5 are replaced by frame 1 and 2, which make the car disappeared without leaving any visual clues.

Table 7.2 shows the detection results of the proposed method for the 15 test videos in the experiments. Obviously, the proposed method is able to detect temporal copy-move tampering correctly and locate the location of the duplicates

Table 7.1 Test videos

Test videos	Equipment	Length	Resolution	Tamper location
Video 1	Digital camera	679	640×480	No
Video 2		374		66–150 are copied to 251–335
Video 3		535		101–200 are copied to 301–400
Video 4		247		57–111 are copied to 170–224
Video 5		222		31–80 are copied to 111–160
Video 6		791		121–320 are copied to 521–720
Video 7		500		136–235 are copied to 357–456
Video 8		320		86–125 are copied to 233–272
Video 9		504		101–194 are copied to 348–441
Video 10	Mobile camera	169		31–70 are copied to 121–160
Video 11		291		No
Video 12		318		71–140 are copied to 201–270
Video 13		362		47–100 are copied to 269–322
Video 14		348		51–150 are copied to 231–330
Video 15		264		21–55 are copied to 96–130

Fig. 7.5 A simple example of a forged video sequence

exactly, and the results of videos 10–15 have exemplified it which are taken by mobile camera with the frame rate 15 fps. For video 4 and video 7, there are relatively a few miss-detected replicas. The main reason is that the first and last frame of duplicated sequences will be affected by the adjacent frames. For video 7 and video 9, the scenes are changing slowly which are taken by the stationary digital camera, therefore, the number of duplicated pairs is higher than in ideal state. Therefore, it is necessary to remove the miss-detected duplicated pairs, as described in Sect. 7.3.2.3. Compared with the method in [15], the computation time is shorter which will be more acceptable. With the increasing length of the video sequence, the computation time will be longer. For the reason that the longer the video sequence is, the more the measurement times will be needed, this is exemplified by the results for video 1, video 3, and video 6 in Table 7.2.

Figure 7.6a shows the number of duplicated pairs between reference sequence and evaluation sequence in the experiments. Generally, we should get a continuous values of duplicated pairs between reference sequence and evaluation sequence,

Table 7.2 Detection results

Test videos	Results	No. of duplicated pairs	Computation time (s/frame)
Video 1	Normal video		11.303
Video 2	Originals: 66–150 Duplicates: 251–335	87	5.921
Video 3	Originals: 101–200 Duplicates: 301–400	86	8.824
Video 4	Originals: 57–112 Duplicates: 170–226	68	3.682
Video 5	Originals: 31–80 Duplicates: 111–160	66	3.209
Video 6	Originals: 121–320 Duplicates: 521–720	186	13.464
Video 7	Originals: 136–236 Duplicates: 357–457	144	8.141
Video 8	Originals: 86–125 Duplicates: 233–272	44	4.876
Video 9	Originals: 101–194 Duplicates: 348–441	134	8.289
Video 10	Originals: 31–70 Duplicates: 121–160	26	2.318
Video 11	Normal video		4.384
Video 12	Originals: 71–140 Duplicates: 201–270	56	5.455
Video 13	Originals: 47–100 Duplicates: 269–322	40	5.676
Video 14	Originals: 51–150 Duplicates: 231–330	86	5.487
Video 15	Originals: 21–55 Duplicates: 96–130	21	3.948

which can be fitted by a straight line, just like the video 3. But the content of the videos which captured by stationary camera will be changed slowly, and even have the situation of some frames with the same content, thus making some original frames will have a higher value of similarities in the video. For video 9, it appears a reference sequence may be matched with two or more evaluation sequences. According to the merging strategy described in Sect. 7.3.2.3, we can use it to remove the false pairs, as the Fig. 7.6b shows.

To evaluate the proposed method, we utilize the precision and recall which can be expressed as:

$$\text{Precision} = N_c/(N_c + N_f) \tag{7.11}$$

$$\text{Recall} = N_c/(N_c + N_m) \tag{7.12}$$

where N_c is the numbers of correct detections, N_f is the numbers of false alarms, N_m is the numbers of missed detections.

We give a comparison of our results against the results reported in [15] for frame duplication. As shown in Table 7.3, the precision of our algorithm can reach 99.7 % which is higher than the method in [15], the main reason is that we extend the structural similarity as a feature and measurement tool in the temporal similarity measurement strategy. From the procedure of structural similarity, the inter-frame similarity measurement actually contains the spatial similarity measurement between two frames. In [15], although the duplicated frames which are tampered in the videos can be found by the coarse-to-fine search strategy, due to the low accuracy of the search, more frames will be miss-detected as duplicated frames. Meanwhile, this strategy will also make the computation time more complex. In

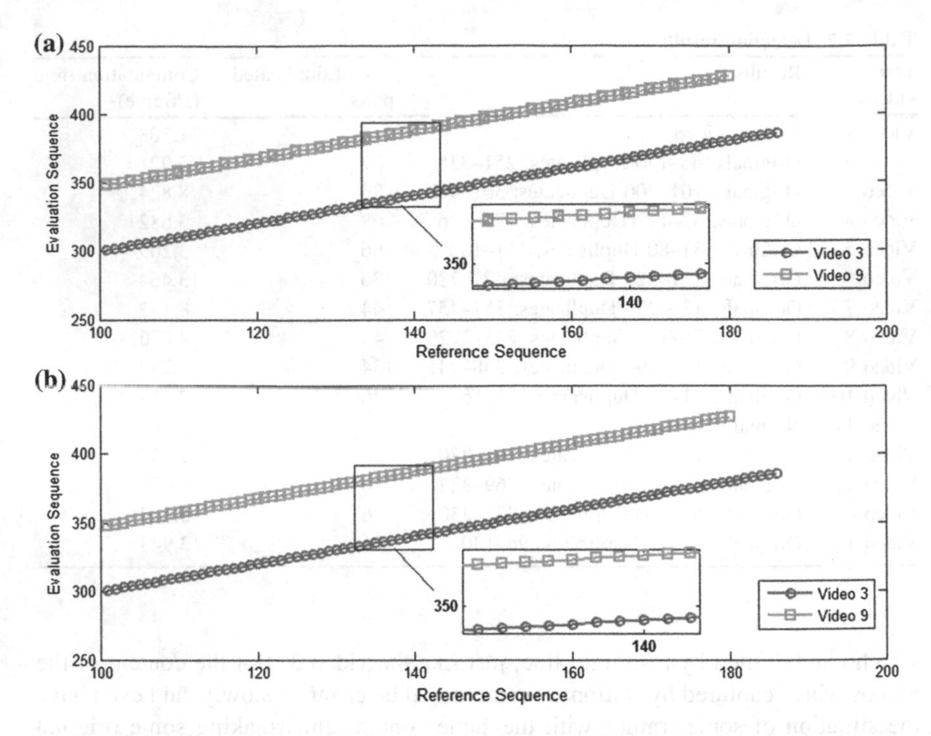

Fig. 7.6 Illustrations of pairs of duplicates. **a** Original pairs of duplicates, **b** pairs of duplicates after merging strategy

Table 7.3 Comparison with Lin et al.'s [15] method

Method	Precision	Recall	Location of duplicates
Ref. [15]	0.849	1	No
Proposed	0.997	1	Yes

addition, for different kinds of test videos, the proposed method can not only detect the duplicates but also locate the location precisely. Therefore, we can exclude the interference of the duplicated frames and understand the content of the videos correctly. This means that the proposed method shows a better performance than the method in [15].

7.5 Conclusions

In this paper, we have proposed a new algorithm for detecting copy-move tampered digital videos based on structural similarity. In this algorithm, a full-length video sequence is divided into some short overlapping sub-sequences, and then the structural similarity is extended to measure the similarity between two frames of a video. Finally, similarities between the sub-sequences in the temporal domain are measured to find out pairs of sub-sequence where replication relationship exists. Moreover, those pairs of sub-sequence are combined into a complete duplicated sequence and the location of the replica is located.

Extensive experimental results evaluated on 15 videos captured by the digital camera and mobile camera in stationary and moving mode show that the precision of this algorithm can reach 99.7 % which is higher than a previous relevant study. The algorithm is able to detect and locate the location of the duplicates correctly. But one limitation is that it have a few duplicates miss-detected for the videos have a long time still scenes. Future work will be mainly dedicated to investigating how to reduce the computation time. In particular, integration with other forensics techniques applied onto video copy-move forgery is envisaged.

Acknowledgments This work was supported by the National Natural Science Foundation of China (Grant No. 61070062), Industry-university Cooperation Major Projects in Fujian Province (Grant No. 2012H6006), Program for New Century Excellent Talents in University in Fujian Province(Grant No. JAI1038).

References

1. Milani S, Fontani M, Bestagini P et al (2012) An overview on video forensics. APSIPA Trans Signal Inf Process 1:e2. doi:10.1017/ATSIP.2012.2
2. Wang W, Farid H (2006) Exposing digital forgeries in video by detecting double MPEG compression. In: Proceedings of the 8th workshop on multimedia and security. doi: 10.1145/1161366.1161375
3. Wang W, Farid H (2009) Exposing digital forgeries in video by detecting double quantization. In: Proceedings of the 11th ACM workshop on multimedia and security. doi: 10.1145/1597817.1597826
4. Luo W, Wu M, Huang J (2008) MPEG recompression detection based on block artifacts. In: Proceedings of the SPIE on security, forensics, steganography and watermarking of multimedia imaging. doi:10.1117/12.767112
5. Su Y, Zhang J, Liu J (2009) Exposing digital video forgery by detecting motion-compensated edge artifact. In: Proceedings of international conference on computational intelligence and software engineering. doi: 10.1109/CISE.2009.5366884
6. Dong Q, Yang G, Zhu N (2012) A MCEA based passive forensics scheme for detecting frame-based video tampering. Digit Invest 9(2):151–159
7. Qin Y, Sun G, Zhang X (2009) Exposing digital forgeries in video via motion vectors. J Comput Res Dev. 46(Suppl.):227–233 (in Chinese)
8. Huang T, Chen Z (2011) Digital video forgeries detection based on bidirectional motion vectors. J Shandong Univ (Engineering Science) 41(4):13–19 (in Chinese)
9. Stamm MC, Liu KJR (2011). Anti-forensics for frame deletion/addition in MPEG video. In: Proceedings of 2011 IEEE international conference on acoustics, speech and signal processing (ICASSP). doi: 10.1109/ICASSP.2017.5946872
10. Stamm MC, Lin WS, Liu KJR (2012) Temporal forensics and anti-Forensics for motion compensated video. IEEE Trans Inf Forensics Secur 7(4):1315–1329
11. Weihong W, Hany F (2007) Exposing digital forgeries in video by detecting duplication. In: Proceedings of the 9th workshop on multimedia and security. doi: 10.1145/1288869.1288876

12. Chih-Chung H, Tzu-Yi H, Lin C-W, Chiou-Ting H (2008) Video forgery detection using correlation of noise residue. In: Proceedings of 2008 IEEE 10th workshop on multimedia signal processing. doi: 10.1109/MMSP.2008.4665069
13. Kobayashi M, Okabe T, Sato Y (2010) Detecting forgery from static-scene video based on inconsistency in noise level functions. IEEE Trans Inf Forensics Secur 5(4):883–892
14. Subramanyam AV, Emmanuel S (2012) Video forgery detection using HOG features and compression properties. In: Proceedings of 2012 IEEE 14th international workshop on multimedia signal processing (MMSP). doi: 10.1109/MMSP.2012.6343421
15. Lin G-S, Chang J-F (2012) Detection of frame duplication forgery in videos based on spatial and temporal analysis. Int J Pattern Recognit Artif Intell 26(7):1–18
16. Wang Z, Bovik AC, Sheikh HR, Simoncelli EP (2004) Image quality assessment: from error visibility to structural similarity. IEEE Trans Image Process 13(4):600–612

Chapter 8
Underdetermined Blind Recovery of Communication Signals Based on Minimum Euclidean Distance in Time-Frequency Domain

Zhaoyang Peng and Wenli Jiang

Abstract To recover the source signals in underdetermined case is a challenging problem, especially when the source signals are non-disjoint in Time-Frequency (TF) domain. The conventional algorithms such as subspace-based complete the blind recovery utilizing the sparsity of the original signals with the assumption that the number of active sources at any TF point is strictly less than that of sensors. Moreover, the processed signals are speech or image signals usually because these signals are sparse in TF domain. But for communication signals, the overlapping amount become more serious in TF domain, so that the performances of conventional algorithms are weaken. Considering the continuity of communication signals in TF domain, this paper proposes a new method to recover communication signals which relaxes the sparsity condition of sources in TF domain. The method allows that the number of active sources at any TF point simultaneously equals to the number of sensors. We can identify the active sources and estimate their corresponding TF values at any TF point by calculating the Euclidean distances between the detected point and all single source points. The computer simulations show that the proposed estimation algorithm is more efficient than the previous algorithms.

Keywords Underdetermined blind separation · Time-frequency domain · Single source point · Euclidean distance

Z. Peng (✉) · W. Jiang
College of Electronic Science and Engineering, National University of Defense Technology, Changsha 410073 Hunan, People's Republic of China
e-mail: pengzhaoyang1983@163.com

A. A. Farag et al. (eds.), *Proceedings of the 3rd International Conference on Multimedia Technology (ICMT 2013)*, Lecture Notes in Electrical Engineering 278,
DOI: 10.1007/978-3-642-41407-7_8, © Springer-Verlag Berlin Heidelberg 2014

8.1 Introduction

The objective of blind source separation (BSS) is to estimate the original signals from their mixtures without any prior information about the sources or the mixing process. When the number of sources is more than that of sensors, the problem is called underdetermined BSS (UBSS). In practical field, because the number of the sources is unknown and the number of the sensors is finite, the problem of UBSS become more and more prevalent. It is important to research the UBSS problem.

Today, sparse component analysis (SCA) is the main method to solve the problem of UBSS [1–4], which usually consists of two steps: estimate the mixing matrix first and then reconstruct the sources [4]. The mixing matrix usually is estimated in advance via clustering algorithms such as k-means method [1], fuzzy c-means method [2], K-SVD method [3]. After the mixing matrix has been estimated, some algorithms such as geometry shortest path algorithm [6], minimizing l_1 norm algorithm [7], maximum a posteriori (MAP) method [8], expectation-maximization (EM) algorithm and Bayesian method can be used to achieve the underdetermined blind separation [9, 10] for the signals which are sparse in the time domain. TF distribution can be obtained using different methods, which can be generally divided into two categories [4]: linear TF distribution [5, 11] and quadratic TF distribution [5, 12–14]. A. Aïssa-El-Bey proposes double methods based on subspace to estimate the underdetermined mixed signals in [5], which allow sources are non-disjoint in TF domain as long as the number of active sources at any point is less than the number of sensors. In [11], the linear TF distribution of mixing signals were calculated by wavelet transform. D. Peng proposes a new underdetermined blind separation method based on quadratic TF distribution in [12]. In [13, 14], the method based on quadratic TF distribution is developed to separate more signals in given sensor as long as $N \leq 2M - 1$, where N and M are the number of signals and sensors respectively. However, all methods based on quadratic TF distribution assume the mixing matrix must satisfy some special constraints condition. Moreover, the quadratic TF distribution may introduce the so-called 'cross-terms', so they assume that there is almost no superimposition between auto-source point and cross-source point in the time-frequency plane but it is difficult to satisfy for communication signals.

Another important classification of UBSS method is based on underdetermined independent component analysis (UICA), which utilizes the uncorrelated property of the sources [15–17]. They identify the active sources and then estimate their corresponding TF values in any TF neighborhood by measuring the diagonalization degree of covariance matrix. These methods don't require the sources to be sparse in TF domain, but the sources must be independent mutually. The performance of these methods is influenced by the technique of TF plane partition. This is another severe drawback of UICA.

In this paper, we research the problem of UBSS using linear TF distribution and considering the mixing matrix has been estimated or known. We propose a source recovery algorithm based on minimum Euclidean distance in TF domain

(EDTF-UBSS), which relaxes the condition on sparsity of sources in TF domain, allowing the sources are non-disjoint in TF domain as long as the number of active sources in any TF neighborhood does not exceed that of sensors. First, the proposed algorithm finds the single sources points for each source. Then we identify the active sources at every support TF points by calculating the Euclidean distances between this point and all single source points and estimate their corresponding TF values through matrix inverse.

The main contribution of this paper is proposing an algorithm to identify the active sources based on Euclidean distances for UBSS utilizing the continuity of communication signals in TF domain, which relaxes the sparsity condition comparing with subspace-based algorithm and diagonalization of covariance matrix algorithm, allowing the number of active sources at any TF point equals that of sensors. Furthermore, the proposed algorithm does not require special constraint condition about the mixing matrix unlike the method based on quadratic TF distribution. Another contribution of this paper is that the proposed method is performed at TF points without the partition of TF domain.

8.2 Problem Formulation

Consider the following delayed linear mixture model [17].

$$\mathbf{x}(t) = \mathbf{A}\mathbf{s}(t) + \mathbf{n}(t) \tag{8.1}$$

where $\mathbf{x}(t) = [x_1(t), \ldots, x_M(t)]^{\mathrm{T}}$ is the mixture vector, $\mathbf{s}(t) = [s_1(t), \ldots, s_N(t)]^{\mathrm{T}}$ is the source vectors, and $\mathbf{A} = [\mathbf{a}_1, \ldots, \mathbf{a}_N] \in \mathbb{C}^{M \times N}$ is complex valued mixing matrix. Applying STFT on both sides of Eq. (8.1), we can obtain the representations of the mixtures in the TF domain [5]:

$$\mathbf{X}(t,f) = \mathbf{A}\mathbf{S}(t,f) + \mathbf{N}(t,f) \tag{8.2}$$

where $\mathbf{X}(t,f) = [X_1(t,f), \cdots, X_M(t,f)]^{\mathrm{T}}$, $\mathbf{S}(t,f) = [S_1(t,f), \cdots, S_N(t,f)]^{\mathrm{T}}$ and $\mathbf{N}(t,f)$ are the STFT coefficients of the mixtures, sources and noise at the TF point $P(t,f)$ respectively.

Similar to other UBSS algorithms, the following assumption should be satisfied in this paper:

Assumption 1: Any $M \times M$ sub-matrix of mixing matrix A is of full rank.

Assumption 2: For each source, there exist some points in the TF domain where the source exists alone.

Different from the conventional algorithms, this paper assume that the sources satisfy the following assumptions in the TF domain.

Assumption 3: The number of active sources in any time-frequency neighborhood does not exceed the number of sensors.

This assumption relaxes the TF condition of the conventional algorithms which require the number of active sources is strictly less than the number of sensors.

Assumption 4: The signals are successive in the TF domain around the TF point where these signals occur simultaneously.

8.3 Source Signal Recovery

8.3.1 Detection of Single Source Points

The single source points denote those TF points where only one source is dominant. For a given support TF point $P(t,f)$, if it is single source point and the active signal is s_i, Eq. (8.2) can be written as

$$\mathbf{X}(t,f) = \mathbf{a}_i \mathbf{S}_i(t,f) \tag{8.3}$$

Let $\mathbf{Q}(i)$ represents the orthogonal project matrix onto the noise subspace of $\mathbf{a}_i$, $\mathbf{Q}(i)$ is expressed as

$$\mathbf{Q}(i) = \mathbf{I} - \mathbf{a}_i \left(\mathbf{a}_i^{\mathrm{H}} \mathbf{a}_i\right)^{-1} \mathbf{a}_i^{\mathrm{H}} \tag{8.4}$$

where $\mathbf{I}$ denotes a identity matrix. It is stated in [5] that

$$\begin{cases} \mathbf{Q}(i)\mathbf{a}_k = 0 & k = i \\ \mathbf{Q}(i)\mathbf{a}_k \neq 0 & k \neq i \end{cases} \tag{8.5}$$

We get

$$\min_{1 \le i \le N} \|\mathbf{Q}(i)\mathbf{X}(t,f)\|_2 = 0 \tag{8.6}$$

If it is single source point, consider two sources s_{i1} and s_{i2} occur at the TF point (t,f). Equation (8.2) can be written as

$$\mathbf{X}(t,f) = [\mathbf{a}_{i1}, \mathbf{a}_{i2}][\mathbf{S}_{i1}(t,f), \mathbf{S}_{i2}(t,f)]^T \tag{8.7}$$

Due to assumption 1, then

$$\min_{1 \le i \le N} \|\mathbf{Q}(i)\mathbf{X}(t,f)\|_2 > 0 \tag{8.8}$$

However, such condition too strict in practice. Hence, to take into account noise, we relax the condition as

$$\min_{1 \le i \le N} \|\mathbf{Q}(i)\mathbf{X}(t,f)\| > \varepsilon \tag{8.9}$$

where $\varepsilon > 0$ is a small threshold according to the noise. Then the TF points satisfying (8.9) are considered as single source points. The signal correspond to the single source point is gotten by Eq. (8.10)

$$i_{\min} = \arg\min_{i}\left(\|\mathbf{Q}(i)\mathbf{X}(t,f)\|_2\right) \tag{8.10}$$

In fact, if the estimating the mixing matrix is estimated according to literature [4], the detection of single source points can be completed simultaneously, without adding complexity of the algorithm.

8.3.2 *Source Signal Recovery*

A. Aïssa-El-Bey proposes a subspace-based linear TF-UBSS algorithm for TF non-disjoint sources using STFT in [5], assuming that there are K active sources at any given TF point, with $K<M$. But In this paper, we relax the sparse condition for $K = M$. Let $\mathbf{A}_M$ denote a $M \times M$ sub-matrix of mixing matrix $\mathbf{A}$. According to assumption 1, $\mathbf{A}_M$ is of full rank. We get any $\mathbf{A}_M$ can make

$$\left\|\mathbf{X}(t,f) - \mathbf{A}_M\mathbf{A}_M^{-1}\mathbf{X}(t,f)\right\|_2 = 0 \tag{8.11}$$

So the subspace-based algorithm can't adapt to this condition $K = M$.

Note the continuity of communication signals in TF domain, for any given TF support point $P(t,f)$, the active signals at the detected TF point can be estimated by calculating the Euclidean distances between the point and all single source points. The Euclidean distance between TF point $P(t,f)$ and other TF point $\widetilde{P}\left(\widetilde{t},\widetilde{f}\right)$ is defined as

$$d(P,\widetilde{P}) = \sqrt{(t-\widetilde{t})^2 + \left(f-\widetilde{f}\right)^2} \tag{8.12}$$

Let Ω_i denote the set of single source points for signal s_i. The Euclidean distance between detected TF point $P(t,f)$ and the set Ω_i is defined as

$$d_i = \min\left\{d(P,\widetilde{P})\middle|\widetilde{P} \in \Omega_i\right\} \tag{8.13}$$

Assume the number of active sources at $P(t,f)$ is M. We can estimate the $\mathbf{A}_M$ by sorting $\{d_i\}$

$$\mathbf{A}_M = \{\mathbf{a}_{i_1}, \ldots, \mathbf{a}_{i_M}\} = \arg\min_{i_1,\ldots,i_M}\left\{\sum_{m=1}^{M} d_{i_m}\right\} \tag{8.14}$$

Therefore whatever is the real number of active sources m, the TF coefficients of M active sources at detected TF point $P(t,f)$ are estimated by

$$\widehat{\mathbf{S}}_M(t,f) = \mathbf{A}_M^{-1}\mathbf{X}(t,f) \tag{8.15}$$

where $\widehat{\mathbf{S}}_M(t,f) = [\mathbf{S}_m(t,f), \mathbf{0}]$, especially when $m = M$, $\widehat{\mathbf{S}}_M(t,f) = \mathbf{S}_m(t,f)$.

8.4 Discussion

We discuss here certain points relative to the proposed EDTF-UBSS algorithms and their applications.

Discussion (1) Assumption 4: The signals are successive in the TF domain around the TF point where these signals occur simultaneously. In practical field, most communication signals are continuous in TF domain such as linear frequency modulation (LFM) signals, gaussian minimum shift keying (GMSK) modulated signals, direct sequence spread spectrum(DSSS) signals, and so on. Therefore, assumption 4 is easy to satisfy. The proposed method can recover all source signals as long as the active signals are included in the M signals with least Euclidean distance .

Discussion (2) Number of Overlapping Sources: In EDTF-UBSS method, the number of sources is considered as M. If the number of overlapping sources is less than M, the estimation error may be introduced. The estimated TF values of M sources at TF point (t,f) can be expressed as

$$\widehat{\mathbf{S}}(t,f) = \begin{cases} S_i(t,f) + \overline{N}_i(t,f) & i \in \{i_1, \cdots, i_m\} \\ \overline{N}_i(t,f) & i \in \{i_{m+1}, \cdots, i_M\} \\ 0 & i \notin \{i_1, \cdots, i_M\} \end{cases} \tag{8.16}$$

where $\overline{\mathbf{N}}(t,f) = \mathbf{A}_M^{-1}\mathbf{N}(t,f)$ is the noise projection. It is difficult and reliable to estimate the number of sources at only one TF point. Moreover, the estimate error brought by noise in (8.16) can be ignored at high SNRs [5], while the advantage of simplicity compared is achieved. Therefore in our simulation, we consider a maximum value to be equal the number of sensors M which is used for all support TF points.

8.5 Simulation Result

For performance evaluation of the proposed algorithm, simulation was performed on LFM signals. In the simulation, the number N of sources is four, the number M of sensors is two or three. The performance of the sources recovery is evaluated in terms of the average signal-to-interference ratio (SIR) [16]. Giver an original source $s_i(t)$ and its estimation $\hat{s}_i(t)$, SIR in decibel can be defined as

$$\mathrm{SIR} = 10\lg\left(\frac{\sum_{i=1}^{N} \mathrm{E}\{\mathrm{s}_i^2(t)\}}{\sum_{i=1}^{N} \mathrm{E}\left\{(\mathrm{s}_i(t) - \widehat{\mathrm{s}}_i(t))^2\right\}}\right) \tag{8.17}$$

In the simulation, the sources are 4 LFM signals $s_1(t), s_2(t), s_3(t)$ and $s_4(t)$, each of which has 10 k samples. The radio frequent (RF) signals have been transformed into intermediate frequent (IF) signals with frequency 1.0, 0.8, 0.6, 0.4 MHz and modulate slope –40, –100, 100, 80 MHz/s, respectively. The sample rate is 2.5 Msample/s. Due to the diverse directions of arrival (DOA) of source signals, the mixing matrix **A** for $M = 2$ is written as

$$\mathbf{A} = \begin{bmatrix} 0.054 - 0.705i & -0.130 - 0.695i & 0.676 - 0.209i & -0.584 - 0.399i \\ 0.054 + 0.705i & -0.130 + 0.695i & 0.676 + 0.209i & -0.584 + 0.399i \end{bmatrix} \tag{8.18}$$

For $M = 3$, the mixing matrix **A** is written as

$$\mathbf{A} = \begin{bmatrix} 0.044 - 0.576i & -0.106 - 0.568i & 0.552 - 0.171i & -0.477 - 0.326i \\ -0.472 + 0.334i & 0.114 + 0.566i & 0.366 + 0.470i & -0.556 + 0.155i \\ 0.295 - 0.496i & 0.535 + 0.216i & 0.460 - 0.349i & 0.547 - 0.186i \end{bmatrix} \tag{8.19}$$

It can be seen clearly from Fig. 8.1 the four sources are non-disjointed but not overlapped completely in the TF domain, and the proposed algorithm work well on the separation of the four LFM signals using two sensors at SNR = 20dB. However, the subspace-based algorithm is invalid at the overlapping TF points. The top two plots represent the TF representation of the two mixtures, the middle four plots represent the TF representation of the source estimates using the proposed algorithm, and the bottom four plots represent the TF representation of the source estimates using the subspace-based algorithm [5].

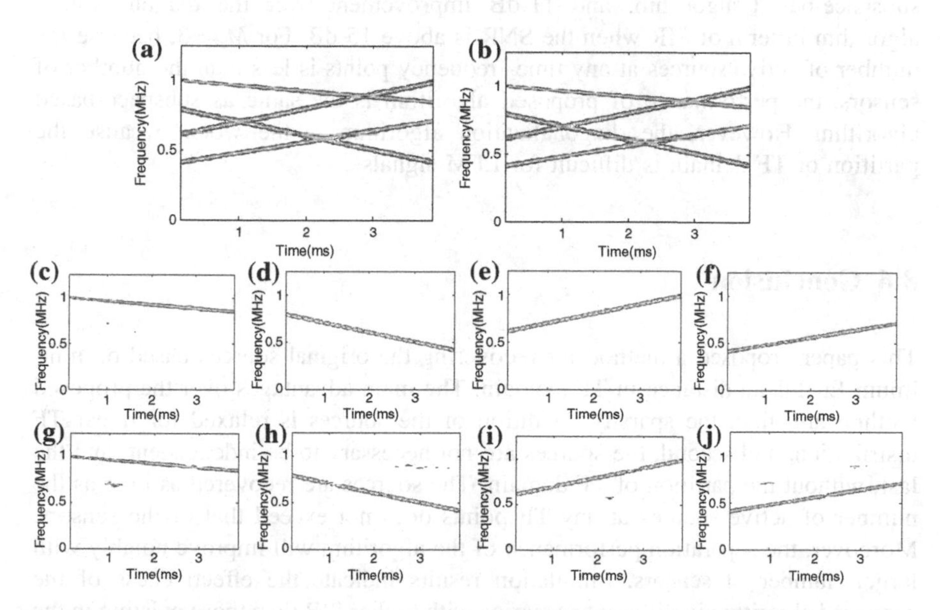

Fig. 8.1 Simulated example (viewed in TF domain) for the proposed algorithm and subspace-based algorithm in the case of four LFM signals and two sensors

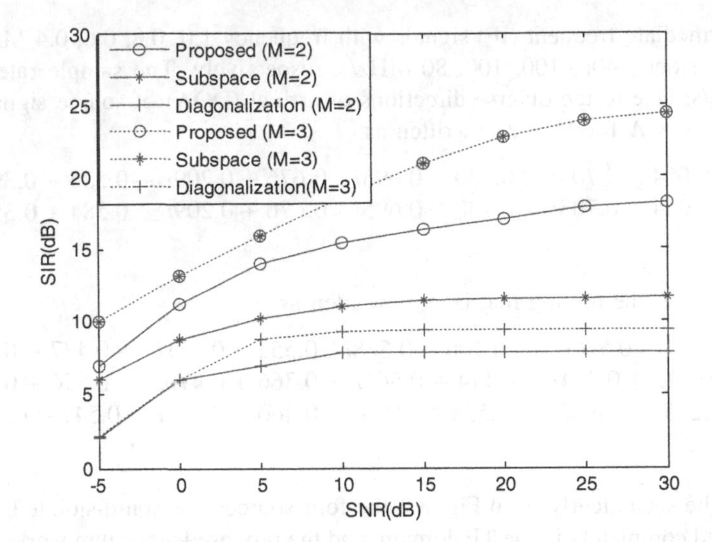

Fig. 8.2 Performance comparisons between the proposed, subspace-based and diagonalization algorithm: average SIR versus SNR for four LFM signals and two or three sensors

Figure 8.2 shows the SIR performance of the proposed algorithm according to the SNR from –5 to 30 dB for LFM signals, in comparison with the subspace-based algorithm and diagonalization algorithm [16]. As shown in the figure, for $M = 2$, the proposed algorithm achieves more than 7 dB improvement over subspace-based algorithm, and 11 dB improvement over the diagonalization algorithm in term of SIR when the SNR is above 15 dB. For $M = 3$, because the number of active sources at any time-frequency points is less than the number of sensors, the performance of proposed algorithm is the same as subspace-based algorithm. However, the diagonalization algorithm is the worst because the partition of TF domain is difficult for LFM signals.

8.6 Conclusion

This paper proposed a method for recovering the original sources based on minimum Euclidean distance in TF Domain. The main advantages over the proposed method are, first, the sparsity condition of the sources is relaxed for linear TF distribution, and second, the sources are not necessary to be independent, and the last, without the partition of TF domain. The sources are recovered as long as the number of active sources at any TF points does not exceed that of the sensors. Moreover, the separation performance of the algorithm will improve notably with larger number of sensors. Simulation results indicate the effectiveness of the proposed algorithm in different scenarios with higher SIR than those existing in the literature.

References

1. Xu R, Wunsch D (2005) Survey of clustering algorithms. IEEE Trans Neural Networks 16(3):645–678
2. Pham D (1999) An adaptive fuzzy C-means algorithm for image segmentation in the presence of intensity in homogeneities. Pattern Recognit Lett 20:57–68
3. Aharon M, Elad M, Bruckstein A (2006) K-SVD: an algorithm for designing over complete dictionaries for sparse representation. IEEE Trans Signal Process 54(11):4311–4322
4. Dong T, Lei Y, Yang J (2012) An algorithm for underdetermined mixing matrix estimation, Neurocomputing. http://dx.doi.org/10.1016/j.neucom.2012.09.018
5. Aissa-El-Bey A, Linh-Trung N, Abed-Meraim K et al (2007) Underdetermined blind separation of non disjoint sources in the time-frequency domain. IEEE Trans Signal Process 55(3):897–907
6. Bofill P, Zibulevsky M (2001) Underdetermined blind source separation using sparse representations. Signal Process 81(11):2353–2362
7. Donoho DL, Elad M (2003) Maximal sparsity representation via l1 minimization. Proc Nat Acad Sci 100:2197–2202
8. Zibulevsky M, Pearlmutter BA (2001) Blind source separation by sparse decomposition. Neural Comput 13(4):863–882
9. Snoussi H, Idier J (2006) Bayesian blind separation of generalized hyperbolic processes in noisy and underdetermined mixtures. IEEE Trans Signal Process 54(9):3257–3269
10. Zayyani H, Babaie-Zadeh M, Jutten C (2009) An iterative bayesian algorithm for sparse component analysis in presence of noise. IEEE Trans Signal Process 57(11):4378–4390
11. Ichir M, Mohammad-Djafari A (2006) Hidden Markov models for wavelet-based blind source separation. IEEE Trans Image Process 15(7):1887–1899
12. Peng D, Xiang Y (2009) Underdetermined blind sources separation based on relaxed sparsity condition of sources. IEEE Trans Signal Process 57(2):809–814
13. Peng D, Xiang Y (2010) Underdetermined blind separation of non-sparse sources using spatial time-frequency distributions. Dig Signal Process 20(2):581–596
14. Xie S, Yang L, Yang J et al (2012) Time-Frequency approach to underdetermined blind source separation. IEEE Trans Neural Networks Learn Syst 23(2):306–316
15. Lathauwer LD, Castaing J, Cardoso JF (2007) Fourth-order cumulant-based identification of underdetermined mixtures. IEEE Trans Signal Process 55(6):2965–2973
16. Lathauwer LD, Castaing J (2008) Blind identification of underdetermined mixtures by simultaneous matrix diagonalization. IEEE Trans Signal Process 56(3):1096–1105
17. Lu F, Huang Z, Jiang W (2011) Underdetermined blind separation of non-disjoint signals in time-frequency domain based on matrix diagonalization. Signal Process 91(7):1568–1577

Chapter 9
Analysis of Anisotropic Scattering of Aircraft Engine Blades in Micro-Doppler Signature

W. C. Zhang, B. Yuan and Z. P. Chen

Abstract Due to the fast rotational motion and skewed geometry of the blades, the Micro-Doppler (m-D) patterns of the aircraft rotating blades in measured data usually do not consistent with the expected m-D pattern via ideal point-scatterer model. In this paper, the backscatter field of the rotating blades is investigated based on the physical optic method, and anisotropic scattering behavior of the aircraft blades which is ignored in ideal point-scatterer model is predicted in the m-D signature. These conclusions in this paper will be useful for the m-D analysis and further high-quality m-D extraction techniques of aircraft rotating blades.

Keywords Micro-Doppler · Aircraft engine blades · Anisotropic scattering · Physical optic method

9.1 Introduction

In traditional inverse synthetic aperture radar (ISAR) imaging of aircraft, rigid-body motion is usually assumed and the classical range-Doppler imaging algorithm is used to obtain the focused image of targets after translational motion compensation. However, in real-world situations, non-rigid-bodies are often present in aircraft such as rotating blades in helicopters, propellers, or turbofans in airplanes. These mechanical rotating structures on the target will generate additional frequency modulation besides the Doppler frequency caused by the main body of the target in the ISAR image, which is known as "Micro-Doppler (m-D) Effect" [1]. On the one hand, the existence of m-D will contaminate the ISAR image of the target. But, on the other hand, the m-D signature can be regarded as

---

W. C. Zhang · B. Yuan (✉) · Z. P. Chen
ATR Key Lab, National University of Defense Technology, Changsha 410073, China
e-mail: yuanbin_163@163.com

A. A. Farag et al. (eds.), *Proceedings of the 3rd International Conference on Multimedia Technology (ICMT 2013)*, Lecture Notes in Electrical Engineering 278,
DOI: 10.1007/978-3-642-41407-7_9,

a unique characteristic that has special significance in the detection, classification, and recognition of the target.

Since the concept of the m-D was proposed, many available analysis and extraction techniques for the m-D signature of the aircraft engine have been proposed. Thayaparan et al. detect and extract the m-D signature from radar returns of helicopter utilizing the adaptive joint T-F analysis and wavelet transform theory [2]. Li and Ling utilize the chirplet basis decomposition algorithm to separate the component of the rotating parts returns from the echoes of an in-flight jet engine aircraft [3]. Bai et al. propound a m-D separation algorithm based on complex-valued empirical mode decomposition for a set of turboprop measured data [4]. All these techniques have good performance in the m-D signature extracting of the engine from the radar returns. However, the m-D modeling of the rotating parts in these literatures is based on an assumption of the ideal point-scatterer model. Due to the fast rotational motion and skewed geometry of the blades, the ideal point-scatterer model may be not appropriate in these situations [5], and the m-D patterns of the aircraft rotating blades in measured data usually do not consistent with the expected m-D pattern via ideal point-scatterer model. In order to present the m-D characteristic of the target more accurately, the radar cross-section prediction techniques, such as the hybrid finite element–boundary integral-multilevel fast multipole algorithm [6, 7] and the shooting and bouncing rays technique [8–12], could be utilized to obtain high fidelity simulation of the radar returns of the target.

In this paper, the backscatter field of the rotating blades is investigated based on the physical optic method, and the anisotropic scattering behavior of the blades which is ignored in ideal point-scatterer model is predicted in the m-D signature. The reminder of this paper is organized as follows. In Sect. 9.2, first, the m-D signature of target with rotating parts is analyzed with ideal point-scatterer model and, second, the m-D characteristic of the radar returns of rotating blades are modeled based on the physical optic method. In Sect. 9.3, experimental results indicate the good agreement with theoretical results, and the last part is the conclusion.

9.2 Micro-Doppler signature Analysis with Point-Scatterer Model

The ideal point-scatterer model is the simplest model in electromagnetic scattering mechanism and usually used in radar imaging to model the radar echoes of the target. Compared with other electromagnetic scattering models, the point-scatterer model can easily incorporate targets' motion in electromagnetic scattering. With the ideal point-scatterer model, aircraft with rotating device may be simplified to two parts by different motions, the main body and the rotating parts. Both of the

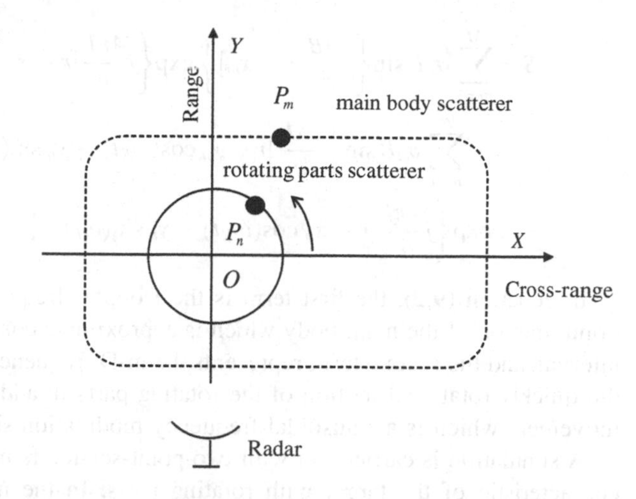

Fig. 9.1 Illustration of target with rotating parts

two parts move with respect to the radar, and the difference is the rotating parts have an additional rotation motions beside the motions of the main body.

Without loss of generality, this paper considers the imaging model of the target on a 2D imaging plane which is unchanged during the imaging time. The geometry of the target for radar imaging is depicted in Fig. 9.1. During the imaging interval, small angle approximation can be assumed for the main body. However, small angle approximation does not hold for the rotating parts, since its rotation rate is usually much larger than that of the main body. The rotating parts might undergo many cycles while the main body rotates only a few degrees during the imaging interval.

Then, the radar echo can be expressed as follow

$$\begin{aligned} S = & \sum_{m=1}^{M} \sigma_m \exp\left\{ -j\frac{4\pi f}{c} [x_m + y_m \omega_B t] \right\} \\ & + \sum_{n=1}^{N} \sigma_n \exp\left\{ -j\frac{4\pi f}{c} [x_n \cos(\omega_R t) + y_n \sin(\omega_R t)] \right\} \end{aligned} \tag{9.1}$$

where the radar echo is a two-dimensional function of transmitting frequency and pulse dwell time. σ_m and ω_B are the backscatterering coefficient and the effective rotation rate of the main body, respectively. σ_n and ω_R are the backscatterering coefficient and the effective rotation rate of the fast rotating blades, respectively.

Utilizing Fourier transform of the formula (9.1) with respect to f, the radar echo signal can be brought into the range dwell-time domain. The signal through a fixed range cell is given by

$$
\begin{aligned}
S = & \sum_{m=1}^{M} \sigma_n B \sin\left\{\frac{2\pi B}{c}[r - x_m]\right\} \exp\left\{j\frac{4\pi f}{c}[r - x_m - y_m \omega_B t]\right\} \\
& + \sum_{n=1}^{N} \sigma_n B \sin\left\{\frac{2\pi B}{c}[r - x_n \cos(\omega_R t) - y_n \sin(\omega_R t)]\right\} \\
& \exp\left\{j\frac{4\pi f}{c}[r - x_n \cos(\omega_R t) - y_n \sin(\omega_R t)]\right\}
\end{aligned} \tag{9.2}
$$

In equation (9.2), the first term is the Doppler frequency caused by the rotational motion of the main body which is approximate constant during the imaging interval, and the second term represents the m-D frequency component induced by the quickly rotational motion of the rotating parts in addition to the target's bulk movement which is a sinusoidal frequency modulation signal.

A simulation is carried out with two point-scatterers model to present the m-D characteristic of the target with rotating parts. In the model, the rotating parts scatterer rotates around the imaging center with a radius of 0.5 m, and the rotation rate of is 2 Hz. the main body scatterer rotates around the imaging center with a radius of 5 m, and the rotation rate of is 0.01 Hz.. The radar center frequency is 10 GHz, the bandwidth is 400 MHz, the time width of the chirp pulse is 25.6 μs, the pulse repetition frequency (PRF) is 400 Hz.

Fig. 9.2a shows the spectrogram of the range profile sequence of the target. It can be seen that the traces of the non-rotating scatter are presented as straight lines which are parallel to the time axis, whereas, the trace of the rotating scatter is a sinusoid line with respect to the pulse dwell time. The T-F spectrogram of the narrow bandwidth returns of the target with a 32 samples Hamming is shown in Fig. 9.2b. In the figure, it also can easily be identified that the stationary patterns as horizontal lines reflect scatterers from the main body, and the sinusoidal curve pattern reflects the frequency characteristic of the rotating parts.

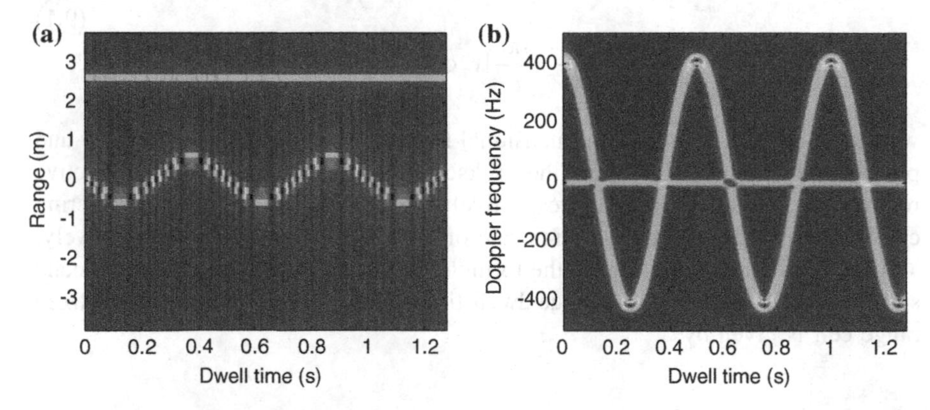

Fig. 9.2 Micro-Doppler signature analysis with point-scatterer model. **a** The range profile sequence of the echoes. **b** T-F spectrogram of the echoes micro-Doppler signature analysis with point-scatterer model

9.3 Anisotropic Behavior Analysis in Micro-Doppler Signature

The ideal point-scatterer model is the simplest model of electromagnetic scattering mechanism of the target. Compared with other electromagnetic scattering models, the point-scatterer model can easily incorporate the target's motion in electromagnetic scattering and isolate the electromagnetic scattering from each individual motion component. By theoretical and simulation analysis with ideal point-scatterer model, the m-D of the rotating parts scatterer can be expressed as a sinusoidal frequency modulation signal, whereas, each main body scatterer has a constant Doppler frequency with respect to the pulse dwell time. However, the m-D pattern of the rotating parts scatterer in real-world situation may usually not accord with the expected m-D from the ideal point-scatterer model. Especially, the characterization of the anisotropic behavior should be considered in the scattering model. Scattering models such as physical optics give functional forms for radar scattering from canonical scatterers which clearly conveys an azimuthally dependence on the orientation of a scatterer. For instance, a flat plate produces a strong specular response where the degree of specularity is directly related to the size of the plate in cross-range.

In order to present the m-D characteristic of the rotating blades of the aircraft more actually, the radar cross-section prediction method is used to calculate electromagnetic backscattering cross-section of the rotating blades. The commonly used methods for the RCS prediction are the physical optics method, ray tracing method, moments method, finite-difference methods, etc. The PO method is a high-frequency asymptotic algorithm to estimate the surface current induced on the target with a dimension much larger than the wavelength. According to the PO technique, the blade can be represented by arrays of triangular facets, the baseband signal in radar receiver is modified as [13]

$$s(t) = \sum_{k=1}^{n_p}\sum_{n=1}^{N_B}\sum_{m=1}^{N_F}\sqrt{\sigma_{u,v}(t)}rect\left\{t - kT_{pul} - \frac{2r_{u,v}(t)}{c}\right\}\exp\left\{-j4\pi f_c\frac{r_{u,v}(t)}{c}\right\} \quad (9.3)$$

where N_B is the number of blades, N_F is the total number of facets in each blade, n_p is the total number of pulses during the observation time interval, and $\sigma_{n,m}(t)$ is the RCS of each facet calculated by PO method, $r_{u,v}(t)$ is the distance between the radar and the mth facet in the nth blade at time.

As a simplified model of the rotating blades, we consider the plates position in a local coordinate system $X_kY_kZ_k$ around the Z axis of the reference coordinate system with an angular velocity ω, and the skew angle between the normal direction of the surface and the rotation axis is ϕ_k, the initial azimuth angle of the plate to the axis Y is φ_k. The elevation angle and azimuth angle of the direction of incidence in the reference coordinate system is α and β, respectively. The geometry of the rotating plates is shown in Fig. 9.3.

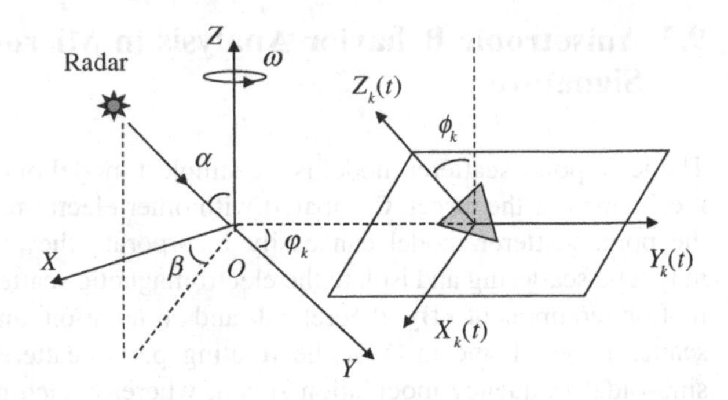

Fig. 9.3 Geometry of the rotating blades for radar cross-section prediction method

With the PO method, the backscattering cross-section of a perfectly conduction surface of the triangular facet on the plate can be approximated at high frequencies by the expression

$$\sigma(t) = \frac{4\pi}{\lambda^2} \left| \sum_{pixels} \sin c\left(K \frac{l}{\cos\theta} \sin\theta\right) \exp(2jKr) \right| \tag{9.4}$$

where θ is the angle between the normal direction of the surface and the direction of incidence, and r is the distance from the differential surface ds to the observer projected on the direction of incidence.

Due to the blades' rotation, the angle θ will change with the normal direction of the surface. The cosine of the angle can be expressed as

$$\begin{aligned} \cos\theta &= \cos\phi_k \cos(\omega t + \varphi_k) \cos\alpha \cos\beta \\ &\quad + \cos\phi_k \sin(\omega t + \varphi_k) \cos\alpha \sin\beta + \sin\phi_k \sin\alpha \end{aligned} \tag{9.5}$$

By analysis with formula (9.3), (9.4) and (9.5), the backscattering cross-section of a rotating blades will change with the dwell times.

A three-dimensional model of the rotating blades is designed and the PO method is applied to calculate the electromagnetic backscattering cross section. PO method is a high-frequency asymptotic algorithm to estimate the surface current induced on the target with a dimension much larger than the wavelength. The propellers 3D model of turboprop aircraft represented by the arrays of triangular facets in detail is shown in Fig. 9.4a. The T-F spectrogram of the echoes including the rotating blades radar returns with a 32 samples Hamming window is shown in Fig. 9.4b. Another 3D engine blades model of turbofan aircraft and the m-D T-F spectrogram are shown in Fig. 9.4c, d, respectively. The simulation parameters of blades are described in Table 9.1, and the simulation parameters of the PO method and Table 9.2.

In the figure, different from the T-F characterization with point-scatterer model, since the characterization of the anisotropic behavior in scattering model, only a section of each sinusoidal curve which reflects the m-D characteristic of the rotating parts can be observed in the T-F plane. However, the periodicity of the variation about the frequency characterization is still preserved.

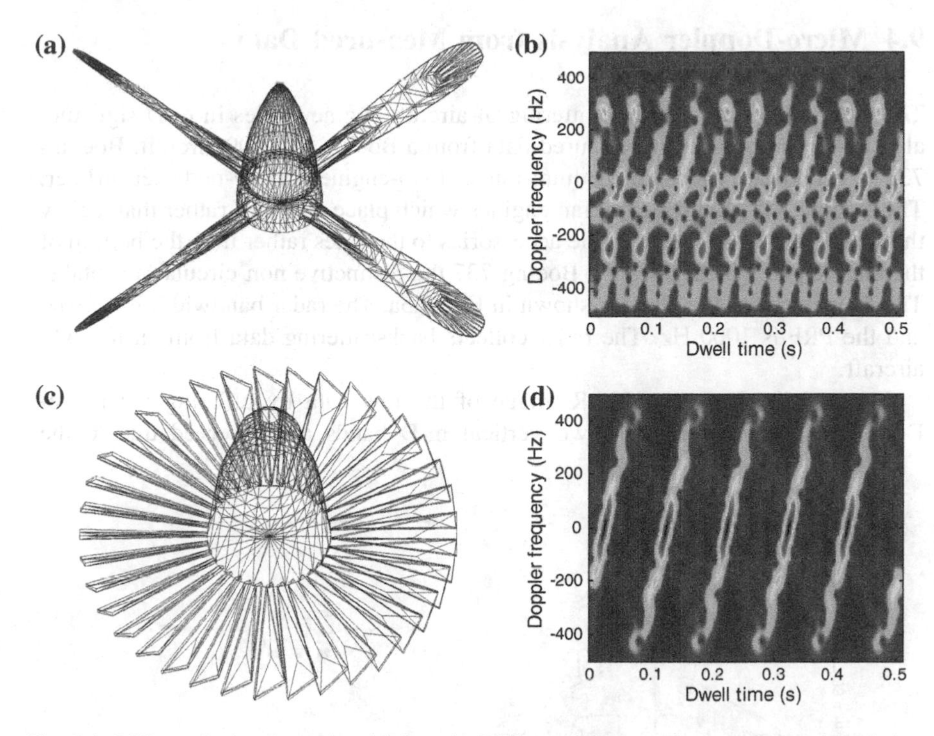

Fig. 9.4 M-D analysis of rotating blades via RCS prediction method. **a** 3D model of the turboprop aircraft blades represented by triangular facets **b** T-F spectrogram of the echoes including the turboprop engine **c** 3D model of the turbofan aircraft blades represented by triangular facets **d** T-F spectrogram of the echoes including the turbofan engine m-D analysis of rotating blades via RCS prediction method

Table 9.1 Simulation parameters of blades

Parameters	Turboprop	Turbofan
The numbers of blades	4	36
Diameter of blades	3900 mm	1524 mm
Revolutions per minute	300 rpm	3900 rpm

Table 9.2 Simulation parameters of the PO method

Central frequency	10 GHz
Bandwidth	1 GHz
Pulse repetition frequency	1 kHz
Interval of frequency	25 MHz
Number of frequency	40
Number of bursts	512
Incidence angle	10°

9.4 Micro-Doppler Analysis from Measured Data

The behavior of anisotropic scattering of aircraft engine blades in m-D signature also is verified in a set of measured data from a Boeing 737-300 aircraft. Boeing-737 is a midsize, short- to medium-range, twin-engine narrow-body jet airliner. The Boeing-737 featured turbofan engines which place ahead of rather than below the wings, and by moving engine accessories to the sides rather than the bottom of the engine pod which gives the Boeing-737 the distinctive non-circular air intake. The framework of the plane is shown in Fig. 9.5a. The radar bandwidth is 1 GHz, and the PRF is 1000 Hz. The radar collects backscattering data from an in-flight aircraft.

Figure 9.5b shows the ISAR image of the raw data by utilizing the range-Doppler algorithm directly. Two vertical m-D bands are observed due to the

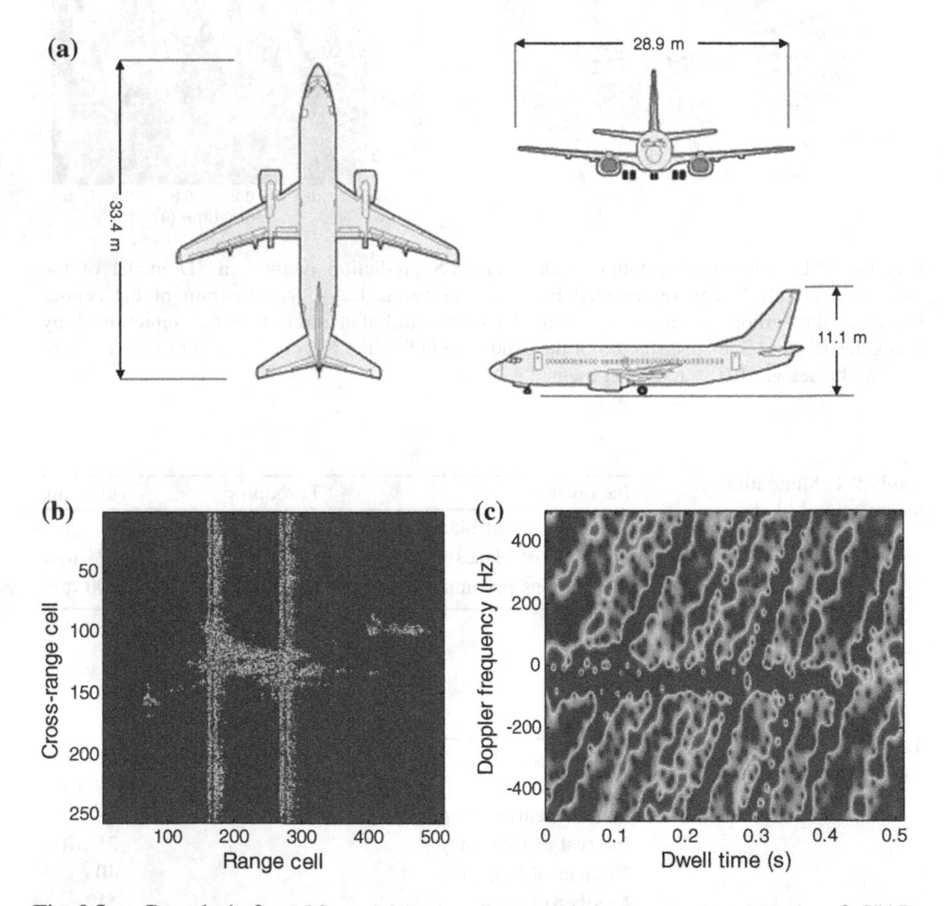

Fig. 9.5 m-D analysis from Measured Data. **a** Framework for Boeing 737-300 plane **b** ISAR image of the data by RD algorithm **c** T-F spectrogram of echo in range cell 296 m-D analysis from Measured Data

rotating engine blades. The geometry of the aircraft body is obscured due to the presence of the m-D interference. Figure 9.5c shows the T-F spectrogram of the echo in range cell 296. The Doppler of the main body around zero frequency, and the m-D signature of the rotating blades represent approximately as some bias which is wondrously similar to the m-D characteristic shown in Fig. 9.4d. Only a section of each sinusoidal curve that reflects the m-D signature of the rotating parts can be observed in the T-F plane because of the anisotropic behavior of the blades in scattering model.

9.5 Conclusion

We simulated the m-D signatures of aircraft engine blades and analyzed anisotropic behavior of the blades in scattering model. Two types of engine models were designed to be similar to the actual aircraft. And a detailed analysis of the modulation signals of aircraft engine blades models systematically is made. It is shown that the m-D patterns of the aircraft rotating blades in measured data usually do not consistent with the expected m-D pattern via ideal point-scatterer model, because the ideal point-scatterer model may not appropriate in real situations due to its fast rotational motion and skewed geometry. These results in this paper can be regarded as complementing and assisting of the traditional analysis and will be useful for the further high-quality m-D extraction techniques of the aircraft rotating blades.

References

1. Chen VC (2000) Analysis of radar micro-Doppler signature with time-frequency transform. In: Proceedings of 10th IEEE workshop statistical signal array process, Pocono Manor, PA, pp 463–466
2. Thayaparan T, Abrol S, Riseborough E (2007) Analysis of radar micro-Doppler signatures from experimental helicopter and human data. IET Radar Sonar Navig 1(4):289–299
3. Li J, Ling H (2003) Application of adaptive chirplet representation for ISAR feature extraction from targets with rotating parts. Proc Inst Elect Eng—Radar Sonar Navig 150(4):284–291
4. Bai X, Xing M, Zhou F, Lu G, Bao Z (2008) Imaging of micromotion targets with rotating parts based on empirical-mode decomposition. IEEE Trans Geosci Remote Sens 46(11):3514–3523
5. Yang SY, Yeh SM, (1991) Electromagnetic backscattering from aircraft propeller blades. IEEE Trans Magn 33(2):1432–1435
6. Guo KY, Sheng XQ (2009) A precise recognition approach of ballistic missile warhead and decoy. J Electromagn Waves Appl 23(14–15):1867–1875
7. Guo KY, Li Q, Sheng XQ, (2010) A precise recognition method of missile warhead and decoy in multi-target scene. J Electromagn Waves Appl 24(5–6):641–652
8. Gao PC, Tao YB, Lin H (2010) Fast RCS prediction using multiresolution shooting and bouncing ray method on the GPU. Prog Electromagn Res 107:187–202

9. Park JH, Yoo JH, Kim CH, Kwon KI, Myung NH (2011) Joint time-frequency analysis of radar micro-Doppler signatures from aircraft engine models. J Electromagn Waves Appl. 25(8–9):1069–1080
10. Lim H, Myung NH (2011) High resolution range profile-jet engine modulation analysis of aircraft models. J Electromagn Waves Appl 25(8–9):1092–1102
11. Park JH, Lim H, Myung NH (2012) Modified hilbert-huang transform and its application to measured micro doppler signatures from realistic jue engnie models. Prog Electromagn Res 126:255–268
12. Lim H, Yoo JH, Kim CH, Kwon KI, Myung NH (2011) Radar cross section measurement of a realistic jet engine structure with rotating parts. J Electromagn Waves Appl 25(7):999–1008
13. Chen VC (2011) The micro-Doppler effect in radar. Artech House, Boston, pp 113–117

Chapter 10
A Distributed Detection Scheme Based on Adaptive CUSUM and Weighted CAT Against DDoS Attacks

Zaihong Zhou, Xi Chen, Jiang Wang and Xueqiang Li

Abstract By designing a distributed hierarchical architecture, the detection task is distributed to the source end, the intermediate network, and the victim end over the Internet to implement the early detection against DDoS attacks. Based on the sensitivity of CUSUM algorithm to the slight change and the traffic characteristics at the source end and the intermediate network, the adaptive CUSUM on the estimation of both the mean value and the variance is adopted at the source end, which detects the outgoing traffic. And the adaptive CUSUM based on EWMA is adopted at the intermediate network, which detects the change and aggregation of the superflow. The detection at the victim end is based on the weighted CAT domain tree. Compared with DCD scheme, the detection rate of UDP attacks is raised from 72 % in DCD to 90 % in proposed scheme, and the detection rate of TCP attacks is improved too.

Keywords DDoS attacks · Distributed detection · Weighted CAT · Adaptive CUSUM

Z. Zhou (✉) · J. Wang · X. Li
School of Information Engineering, Guangdong Medical College, Dongguang, China
e-mail: zhou_zai_hong@126.com

Z. Zhou
Songshan Lake Science and Technology Industry Park, Dongguan, GuangDong, China

X. Chen
School of Information Engineering and Automation, Kunming University of Science and Technology, Kunming, China

A. A. Farag et al. (eds.), *Proceedings of the 3rd International Conference on Multimedia Technology (ICMT 2013)*, Lecture Notes in Electrical Engineering 278,
DOI: 10.1007/978-3-642-41407-7_10,

10.1 Introduction

DDoS attacks are easy to launch, but they are difficult to detect. Many tools are available on the Internet that helps attackers setup DDoS attacks. Thus, it makes DDoS attacks widely popular in the Internet. The distributed problem should be resolved by distributed scheme. There are some methods and theories in distributed detection scheme [1–7].

Among them, Y. Chen's schemes [6, 7] are very promising. A scheme for collaborative change detection of DDoS attacks on community and ISP networks was proposed at first [6]. But the scheme can only detect DDoS attacks in one AS collaboratively. To cope with the limitation of detection scope, the author proposed a new collaborative detection of DDoS attacks over multiple network domains-DCD (Distributed Change Detection) scheme [7]. The DCD scheme can collaboratively detect the DDoS attacks over multiple ISP domains. In addition, the superflow is considered as detection unit at the router. It enabled the detection against DDoS attacks effectively in real Internet environment. The hierarchical architecture and the detection implemented at the router and domain levels, respectively, simplified the alert correlation and global detection procedures and enabled the DCD system implementation in ISP networks. But, the DCD scheme has some disadvantages, such as high overhead of storage and computation at the victim-end CAT server, high false negative, and so on. Aiming at the disadvantages of the DCD scheme, we propose the distributed detection scheme against DDoS attacks based on adaptive CUSUM and weighted CAT (Hereinafter referred to AC_CAT).

10.2 DDos Attacks Detection Based on Adaptive CUSUM at the Source End and Routers

10.2.1 Estimation of Mean and Variance-Based Adaptive CUSUM Algorithm at the Source End

The change of traffic is slow at the source end and the abnormal message and attack features are not notable. A faster and more sensitive detection algorithm should be adopted. Nonparametric CUSUM algorithm can meet these requirements.

A source may access multiple destinations, so we improve the nonparametric CUSUM. In our scheme, let $X_{t,a}$ denote the traffic to the destination D_a in the tth monitoring cycle. One assumption to the nonparametric CUSUM is that the value of stochastic sequence is negative in normal and it will be positive when change happens. Therefore, make $X_{t,a}$ transformation as follows.

$$Y_{t,a} = X_{t,a} - XX_a \tag{10.1}$$

XX_a may be set to slightly larger than the expected value of the $X_{t,a}$ without attacks. Thus, the value of $Y_{t,a}(t = 1, 2, \ldots, a = 1, 2, \ldots)$ in normal is negative.

To judge whether suspicious event happened, we define cumulative sum of stochastic variable.

$$Z_{t,a} = (Z_{t-1,a} + Y_{t,a})^+ \tag{10.2}$$

where $Z_{0,a} = 0$

$$Z^+ = \begin{cases} Z, Z > 0 \\ 0, Z <= 0 \end{cases} \tag{10.3}$$

That is, $Z_{t,a} = \max\{0, Z_{t-1,a} + X_{t,a} - XX_a\}, t = 1, 2, \ldots, a = 1, 2, \ldots$.

Compared $Z_{t,a}$ with threshold T_a, if $Z_{t,a} > T_a$, an alert packet will be sent to the router. The alert packet is a 5-tuple: {source host, Destination host, source port, destination port, protocol type}.

For the parameter XX_a, its main role is to ensure that stochastic sequence $Y_{t,a}$ for detection in normal is negative. In general, $X_{t,a}$ will fluctuate around $\overline{X}_a$. For the stochastic variable X_n with mean μ and variance σ, there exists $P\{\mu - \sigma < X \le \mu + \sigma\} = 0.6826$, $P\{\mu - 2\sigma < X \le \mu + 2\sigma\} = 0.9544$, $P\{\mu - 3\sigma < X \le \mu + 3\sigma\} = 0.9974$. The value of stochastic variable X falls $[\mu - 3\sigma, \mu + 3\sigma]$ is almost certainly. That means, for 99.74 % of the cases, the value of $X - \mu - 3\sigma$ is less than 0. Here, the XX_a is set to $XX_a = \mu + 2\sigma = \overline{X}_a + 2\sigma$. At the source end, the network traffic is influenced greatly by user behavior, and the user behavior is arbitrary. So, we adopt Eqs. (10.4) and (10.5) to estimate $\overline{X}_a$ and σ online during each monitoring cycle.

$$\hat{\overline{X}}_{t,a} = \frac{1}{t}\sum_{i=1}^{t} X_{i,a} = \frac{1}{t}\left[(t-1)\hat{\overline{X}}_{t-1,a} + X_{t,a}\right] \tag{10.4}$$

$$\hat{\sigma}^2_{t,a} = \frac{1}{t-1}\sum_{i=1}^{t}(X_{i,a} - \overline{X}_{i-1,a})^2 = \frac{t-2}{t-1}\hat{\sigma}^2_{t-1,a} + \frac{1}{t}(X_{t,a} - X_{t-1,a})^2 \tag{10.5}$$

According to the estimation value of mean and variance online, parameter XX_a can be adjusted adaptively according to Eq. (10.6).

$$XX_a = \hat{\overline{X}}_{t,a} + 2\hat{\sigma}_{t,a} \tag{10.6}$$

For the threshold T_a, the initial value is zero. Having detected that $Y_{t,a}$ is greater than zero, the traffic in next c cycle will be monitored continuously. So, T_a is determined by the multiplication of the number of successive monitored cycle c and the average of $Y_{t,a}$ before abnormality. That is,

$$T_a = c * \text{avg}\,Y_{t-1,a} \tag{10.7}$$

Here, avg $Y_{t-1,a}$ is the average of traffic to destination D_a before time t. If $Z_{t,a} > T_a$, then an abnormality rises at the source host, and an alert packet will be sent to the connected router. If the number of alarms is approaching to a preset value, then the alarm stops. Call initialization module and start the next round of detection.

10.2.2 *The Adaptive CUSUM Detection Algorithm Based on EWMA at the Router*

10.2.2.1 The Detection to Attacks at the Router

We consider the traffic to port p_i of the router in the t_mth monitoring cycle as observation sequence, and denote it as $X(t_m, p_i)$. In order to apply nonparametric CUSUM, a transformation is made to $X(t_m, p_i)$ as follows.

$$XT(t_m, p_i) = X(t_m, p_i) - \overline{X}(t_m, p_i) \tag{10.8}$$

$\overline{X}(t_m, p_i)$ is the average of the traffic to port p_i of the router in the t_mth monitoring cycle.

The traffic passing through the router at the intermediate network per second is large, thus, using too complex algorithm to estimate the mean is not desirable. So, low complex EWMA is adopted to adaptively attain the value of $\overline{X}(t_m, p_i)$.

$$\overline{X}(t_m, p_i) = (1 - \lambda)\overline{X}(t_{m-1}, p_i) + \lambda X(t_m, p_i) \tag{10.9}$$

Here, λ is the smoothing factor, and its value is greater than zero and less than one. The greater the λ is, the greater the impact of current data to mean is. Because our goal is to get mean, λ is set from 0.1 to 0.5.

We define $Z_{\text{in}}(t_m, p_i)$ as the accumulation of traffic deviation between the incoming traffic and the average traffic at port p_i in the t_mth monitoring cycle.

$$Z_{\text{in}}(t_m, p_i) = \max\{0, Z_{\text{in}}(t_{m-1}, p_i) + X(t_m, p_i) - \overline{X}(t_m, p_i)\} \tag{10.10}$$

In order to reflect the change more accurately, we define $RD_{\text{in}}(t_m, p_i)$ (Relative Deviation) as an indicator of attack.

$$RD_{\text{in}}(t_m, p_i) = Z_{\text{in}}(t_m, p_i)/\overline{X}(t_m, p_i) \tag{10.11}$$

If $RD_{\text{in}}(t_m, p_i) > \beta$, here, β is the threshold of the router, then the DDoS attacks are declared at the router.

10.2.2.2 Detection of Aggregation at the Router

Whether attack traffic is aggregative or not, it can be determined by the ratio of the incoming traffic deviation and the outgoing traffic deviation at the port p_i. We define $Y(t_m, p_i)$ as the outgoing traffic from port p_i in the t_mth monitoring cycle and make a transformation to it as follows.

$$YT(t_m, p_i) = Y(t_m, p_i) - \overline{Y}(t_m, p_i) \tag{10.12}$$

$\overline{Y}(t_m, p_i)$ is the average of the outgoing traffic from port p_i in the t_mth monitoring cycle. It is also estimated using EWMA as follows.

$$\overline{Y}(t_m, p_i) = (1 - \lambda)\overline{Y}(t_{m-1}, p_i) + \lambda Y(t_m, p_i) \tag{10.13}$$

Define $Z_{\text{out}}(t_m, p_i)$ as the accumulation of traffic deviation between the outgoing traffic and the average traffic at the port p_i in the t_mth monitoring cycle.

$$Z_{\text{out}}(t_m, p_i) = \max\{0, Z_{\text{out}}(t_{m-1}, p_i) + Y(t_m, p_i) - \overline{Y}(t_m, p_i)\} \tag{10.14}$$

Then the ratio *DR* of I/O traffic deviation is,

$$\text{DR} = Z_{\text{out}}(t_m, p_i) / Z_{\text{in}}(t_m, p_i) \tag{10.15}$$

If $DR >= 1$, then the attack is aggregative at the router. So, an alert packet is sent to the local CAT server.

10.3 DDoS Detection Based on Weighted CAT at the Victim End

10.3.1 The Construction of the Weighted CAT Subtree

Weighted CAT is a CAT tree that consists of routers where exist abnormality and aggregate abnormality and each edge (R_i, R_j) has corresponding weight $w(R_i, R_j)$. The value of $w(R_i, R_j)$ is the number of attack sources that the router R_i connected. Only if the suspicious scenario is detected at the attack source in monitoring cycle, an alert packet is sent to the connected router. So, the router can determine the number of connected attack sources according to the number of alert packets.

The global weighted CAT domain tree is a CAT domain tree that the victim is the root, the domain with abnormality is the child, and the link between domains is the edge. Each edge (AS_i, AS_j) in CAT domain tree has corresponding weight $W(AS_i, AS_j)$, the value of $W(AS_i, AS_j)$ is the weight of CAT subtree in AS_i.

The weighted CAT subtree is constructed according to the maintained network topology and the alert packets from all routers at the local CAT server in a monitoring cycle. Denote *G* as the copy of the local network topology. The router is the node in *G*. It contains the following messages {*R_ID*, *Weight*, *UP_NUM*, UP_R_1, UP_R_2, …, $UP\_R_{\text{up\_num}}$, *DN_R*}, where *R_ID* is the router *ID*. *Weight* is the edge weight between current router and downstream router, *UP_NUM* is the number of upstream routers, $UP\_R_{\text{up\_num}}$ is the address of upstream routers, and

DN_R is the address of downstream routers, respectively. The alert packet contains {*R_ID*, *FLOW_ID*, *SOURCE_NUM*}. They are the router *ID-R_ID*, flow *ID-FLOW_ID*, the number of attack sources connected to the router, respectively. Denote the address of the root in G as R_0, the being processed node as R, the next being processed node as Q, alert packet as P.

10.3.2 Detection Based on Global Weighted CAT

Having constructed the weighted CAT subtree at the CAT server of the intermediate network, the weight W of all subtrees is calculated at the CAT server to reduce the storage and computation overhead of the victim end, and the weight $W = \sum w(Ri, Rj)$. An alert packet is sent to the CAT server at the victim end. The alert packet is a 4-tuple $\{AS\_ID, UP\_AS\_ID, DN\_AS\_ID, W\}$, where *AS_ID* is the *ID* of current AS, *UP_AS_ID* is the *ID* of upstream AS, *DN_AS_ID* is the *ID* of downstream AS, and W is the weight of CAT subtree.

The global weighted CAT domain tree is constructed according to the alert packets from upstream routers at the CAT server of the victim end. Starting from the AS of the victim end, if *DN_AS_ID* is equal to the victim's *ID*, adding the AS to the victim's children. Then, starting from one child, all ASs will be the child's children if *DN_AS_ID* is equal to the child's *ID*, and so on, till all ASs nodes are disposed. The weight $W(AS_i, AS_j)$ of the edge composed of domain nodes AS_i and AS_j is the weight of CAT subtree in AS_i domain. Having constructed global weighted CAT domain tree at the victim end, the weight of the domain tree is calculated, which is the sum of the weight W of all edges. Because the sum of weight W indicates the number of attack sources connecting to all routers. If $\sum W_i > \gamma$, it is determined that the victim end is suffered from DDoS attacks. Here, γ is the attack threshold that the victim can endure. Once DDoS attacks are detected, the victim end will respond and traceback distributedly according to the global CAT domain tree and all CAT subtrees.

10.4 Experimental Results

We use NS2 simulator and adopt real ISP topology downloaded from the Rocketfuel project website in University of Washington. In our experiments, there are 5–10 routers and about 10 source hosts in each domain. Bandwidth is set to 100 MB and network delay is set to 50 ms. Detection parameter c is set to 25 in our experiments. Some other threshold parameters λ is set to 0.2, β is set to 2 and the size of monitoring window is set to 500 ms by a series of experiments in a domain.

Having attained and set all parameters in AC_CAT scheme, we measure the detection rate of UDP attacks and TCP attacks in 1, 2, 4, 8, 16 ISPs when the

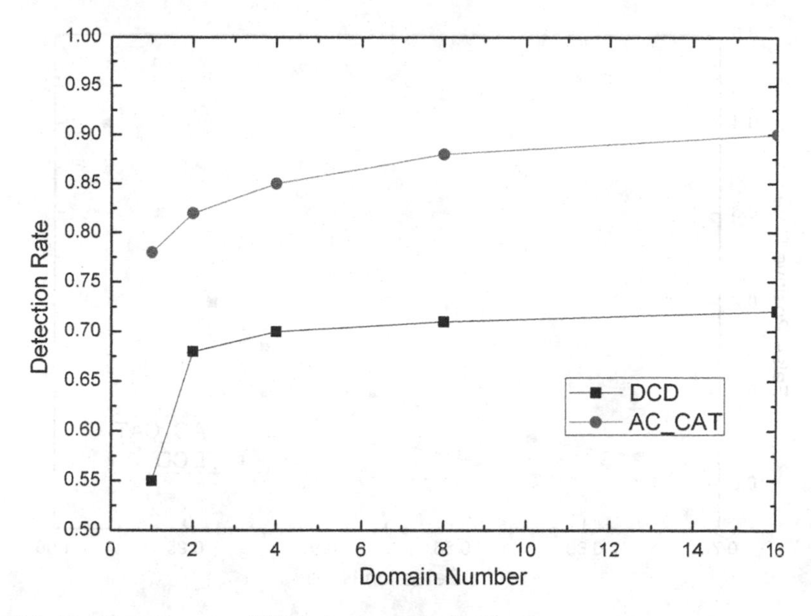

Fig. 10.1 The detection rate of UDP attacks in distribution

attack sources are distributed randomly. We study the trade-off between detection rate and false-positive rate of UDP attacks and compare the trade-off with DCD scheme. The detected results are shown as Figs. 10.1, 10.2 and 10.3.

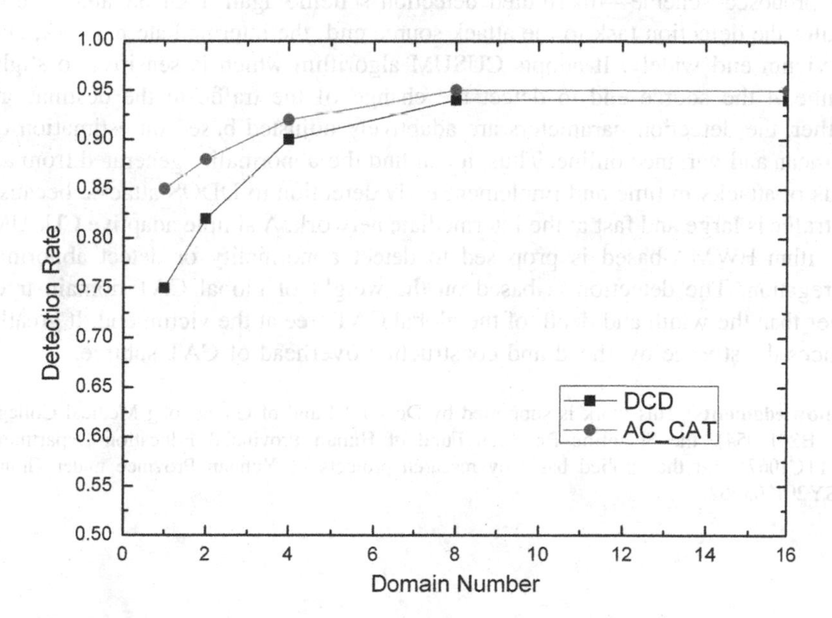

Fig. 10.2 The detection rate of TCP attacks in distribution

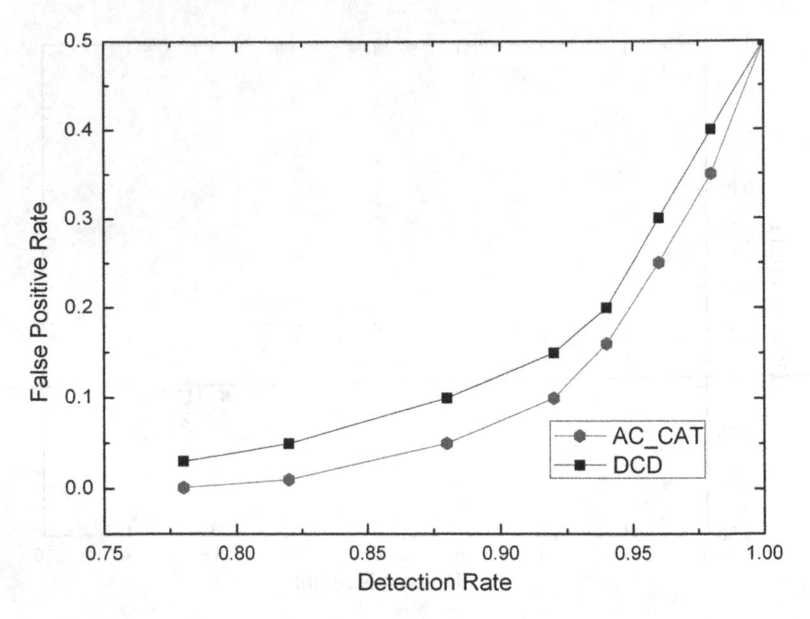

Fig. 10.3 The trade-off between detection rate and false positive rate of UDP Attacks in distribution

10.5 Conclusions

The proposed scheme—distributed detection scheme against DDoS attacks distributes the detection task to the attack source end, the intermediate network, and the victim end widely. It adopts CUSUM algorithm which is sensitive to slight change at the source end to detect the change of the traffic to the destination. Further, the detection parameters are adaptively adjusted based on estimation of the mean and variance online. Thus, it can find the abnormality generated from all kinds of attacks in time and implement early detection to DDOS attacks, because the traffic is large and fast at the intermediate network. A simple adaptive CUSUM algorithm EWMA-based is proposed to detect abnormality or detect abnormal aggregation. The detection is based on the weight of global CAT domain tree, rather than the width and depth of the global CAT tree at the victim end. It greatly reduces the storage overhead and construction overhead of CAT subtree.

Acknowledgments This work is supported by Doctoral Fund of Guangdong Medical College (No: B2012054), the Scientific Research Fund of Hunan Provincial Education Department (No:11C1067) and the applied basic by research projects of Yunnan Province under Grant: KKSY201203062.

References

1. Mirkovic J, Robinson M, Reiher P, Kuenning G (2003) Alliance formation for DDoS defense. In: Proceedings of the new security paradigms workshop, ACM SIGSAC, Aug 2003
2. Koutepas G, Stamatelopoulos F, Maglaris B (2004) Distributed management architecture for cooperative detection and reaction to DDoS attacks. J Netw Syst Manage 12:73–94
3. Lam HY, Li CP; Chanson ST, Yeung DY (2006) A coordinated detection and response scheme for distributed denial-of-service attacks. In: Proceedings of IEEE international conference on communications, vol 5, pp 2165–2170
4. Bouzida Y, Cuppens F, Gombault S (2006) Detecting and reacting against distributed denial of service attacks. In: Proceedings of IEEE international conference on communications, vol 5, pp 2394–2400
5. Xiao B, Chen W, He YX (2006) A novel approach to detecting DDoS attacks at an early stage. J Supercomput 3:235–248
6. Chen Y, Hwang K (2006) Collaborative change detection of DDoS attacks on community and ISP networks. In: Proceedings of international symposium on collaborative technologies and systems, pp 401–410
7. Chen Y, Hwang K, Ku WS (2007) Collaborative detection of DDoS attacks over multiple network domains. IEEE Trans Parallel Distrib Syst 18(12):1649–1662

Chapter 11
A Research on Speech Enhancement Based on Hybrid Parallel Subbands HMM and Neural Network Model

Zhao lv, Li Ni, Shiyu Chen and Xiaopei Wu

Abstract Robustness is a very important issue in the field of automatic speech recognition (ASR) research, especially to provide high recognition accuracy in practical applications. However, the recognition rate based on the traditional whole frequency band HMM will decrease when only partial frequency bands are corrupted by noise. In order to solve this problem, a speech enhancement algorithm based on hybrid parallel subbands HMM and neural network model was proposed. The whole frequency band HMM was split into a few subbands HMM and extract some new feature parameters according to all subbands HMM outputs, and then merged them by BP neural network to yield a global recognition decision. The experimental results show that the hybrid parallel subbands HMM and neural network (PSHMM/NN) model can improve the robustness of speech recognition system in noisy environments.

Keywords Parallel subbands HMM (PSHMM) · Neural network (NN) · Speech enhancement · Noise

Z. lv · L. Ni · S. Chen · X. Wu (✉)
The Key Laboratory of Intelligent Computing and Signal Processing,
Anhui University, Hefei 230039, China
e-mail: wxp2001@ahu.edu.cn

Z. lv
e-mail: kjlz@163.com

L. Ni
e-mail: 2294335475@qq.com

S. Chen
e-mail: 514251756@qq.com

A. A. Farag et al. (eds.), *Proceedings of the 3rd International Conference on Multimedia Technology (ICMT 2013)*, Lecture Notes in Electrical Engineering 278,
DOI: 10.1007/978-3-642-41407-7_11,

11.1 Introduction

As we known, HMM model has a strong ability of modeling the dynamic time series, but it ignores the correlation between each model for only considering the variation within class. Moreover, the cumulative probability of the biggest state is used in recognition and ignores the similarity between each mode, as a result, the performance of speech recognition system based on HMM would be affected [1–3]. On the other hand, most feature parameters of speech recognition systems are extracted from the whole frequency band, but the mechanism of human auditory perception reveals that the human hearing decoding system will acquire different subband information first, and then analyses them synthetically [4, 5]. When the training environment and test environment are not same, the frequency response of inconsistencies in the different subbands have some differences; consequently, the recognition accuracy will decrease.

In order to solve this problem, a speech enhancement algorithm based on hybrid parallel subbands HMM and neural network model was proposed. A parallel structure of the subband multisystem framework was adopted in the proposed algorithm, and each subband HMM model information were extracted and reintegrated by neural network, as a result, the performance of speech recognition system will be improved in noisy environments.

11.2 Principles of Parallel Subbands HMM

Typically, a series of features vectors of the speech recognition system will be extracted from the whole frequency band, and then output them to the recognizer for analysis. Obviously, all features vectors will be corrupted when partial frequency band was corrupted by noise, as shown in Fig. 11.1a. To address this situation, the algorithm of feature extraction and recognition in different subbands have been studied [6]; the experimental results show that the multiband method can improve the robustness of speech recognition system.

From Fig. 11.1a we can see that, for the whole-band HMM, all feature parameters are disturbed when the low frequency band of the input original speech signal was corrupted by noise interference. However, only some feature vectors corresponding to different subbands were corrupted in the parallel subbands HMM, and other subbands are unchanged, as shown in Fig. 11.1b.

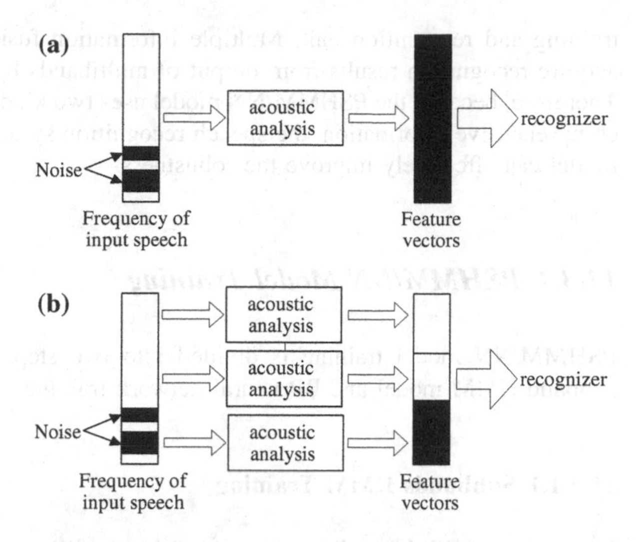

Fig. 11.1 Model and motivations of the parallel subbands HMM

11.3 Hybrid Parallel Subbands HMM and Neural Network Model

The architecture of hybrid parallel subbands HMM and neural network model is shown in Fig. 11.2.

From Fig. 11.2, we can see that the PSHMM/NN model is composed of speech recognition subsystem and multiple information fusion subsystem. Speech recognition subsystem consists of different subbands speech recognition subsystem, which it uses to recognize the input speech in different subbands and includes the extraction of feature parameters of subbands unit, the subband HMM model

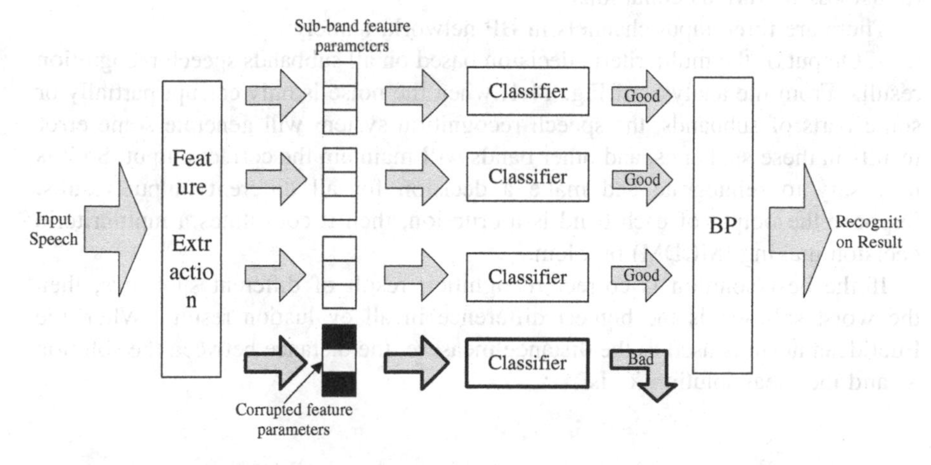

Fig. 11.2 PSHMM/NN model architecture

training and recognition unit. Multiple information fusion subsystem is used to acquire recognition results from output of multibands by BP neural network [7]. Therefore, because the PSHMM/NN model uses two kinds of identification pattern comprehensive information, the speech recognition system based on the proposed model can effectively improve the robustness.

11.3.1 PSHMM/NN Model Training

PSHMM/NN model training is divided into two steps, namely the training of subband HMM model and BP neural network training.

11.3.1.1 Subbands HMM Training

In subbands HMM training stage, a band-pass filter was used to split the input speech signal into different subband signals, and then each subband signal will be trained to a subbands HMM model. As the training process is independent, the trained subbands HMM models are independent too.

11.3.1.2 BP Network Training

Though the subband HMM models are different for different pattern classes, the BP neural network is a common model. Hence, multivoice model was used in training stage. This training mode not only ensures the relevance and the interference between different modes of training samples, but also effectively enhances the classification capability of the whole system and improves the system robustness in various conditions.

There are three input channels in BP network, namely:

1. Output of the multicriteria decision based on all subbands speech recognition results. From the analysis of Fig. 11.1, when the noise is only corrupt partially or some parts of subbands, the speech recognition system will generate some error results in these subbands, and other bands will maintain the correct output. So it is necessary to reintegrate and make a decision for all different output results. Suppose the output of each band is a criterion, then it constitutes a multicriteria decision making (MCDM) problem.

If the best solution is correct recognition result of different subbands, then the worst solution is the biggest difference in all evaluation results. When the Euclidean norm is used as the distance measure, the distance between the solution X_i and the ideal solution X^* is:

$$S_i^* = \sqrt{\sum_{j=1}^{m} (X_{ij} - X_j^*)^2} \quad i = 1, 2, \cdots, n \tag{11.1}$$

In Eq. 11.1, X_{ij} includes X_i independent, namely, the speech recognition results of the j_{th} subband. X_j^* is the j_{th} independent of the ideal solution X^*. And then a measure of the relative proximity in ideal solution is used to judge the merits of solutions:

$$C_i^* = S_i^- / (S_i^* + S_i^-) \quad i = 1, 2, \cdots, n \tag{11.2}$$

If X_i is the ideal solution X^*, then C_i^* is 1, otherwise, C_i^* is 0. Usually, C_i^* is between 0 and 1. If the value of C_i^* is closer to 1, the corresponding criteria will locate more top surface. In PSHMM/NN model, output results of all the subbands closest to the ideal solution will be used as a neural network input value.

2. The recognition results corresponding to the highest SNR in all subbands. The highest SNR means that the band with minimal impact in all subbands, in other words, the subband output result corresponding to the highest SNR has the highest reliability. In addition, the model added a whole-band subsystem because the whole frequency band can reflect a variety of frequency information correlation.

3. The average energy of the frame. Suppose the input speech signal includes N frames which consist of K sample points, then the average energy of the frame is calculated as follows:

$$E = (1/N) \sum_{i=1}^{N} \sum_{j=0}^{K-1} x_i^2(j) \tag{11.3}$$

11.3.1.3 PSHMM/NN Model Training

The training process of the PSHMM/NN model is shown as follows:

Step 1. Split input speech signal into different subband signals by the band-pass filter, then estimate SNR of each subband and extract their feature parameters;

Step 2. Establish HMM model of each subband signals;

Step 3. Calculate multicriteria decision result (X_i), the highest SNR result (Y_i) and the average energy of the frame (E_i); then combine these values to form a new feature vector $S_i = [X_i, Y_i, E_i]$;

Step 4. Normalize the feature vectors as BP neural network input vectors, and output a matrix $R = [r_1, r_2, \cdots, r_i, \cdots, r_N]$ corresponding to the input speech data. In matrix R, r_i represents the i_{th} output vector of the neural network. If $r_i = 1$ (i is an order number in speech library), then it means that the corresponding data element is the output result of the neural network;

Step 5. Train the BP neural network until the network converges.

11.3.2 PSHMM/NN Model Recognition

The recognization process of the PSHMM/NN model is similar to training process, some difference are described as follows:

Step 1. Calculate output results of the test speech in different subbands by the Viterbi algorithm;

Step 2. Use the spectral subtraction to estimate SNR in different subbands;

Step 3. Compute X_i, E_i and Y_i of the test speech, then combine these values to form a new feature vector $T_i = [X_i, Y_i, E_i]$;

Step 4. Normalize the feature vectors T_i and using as an output vector of the neural network, a output vector matrix R can be acquired. Then search the element of $r_i = 1$ and output final recognition result by the predefined rule.

11.4 Experiments Results

The speech database used in the experiments contains the isolated digits in English produced by 500 speakers. Here, 150 speakers among these 500 speakers were taken as the testing speakers, while the speech data produced by other speakers are used to train a speaker-independent model. The recognition accuracy was evaluated by the average of the 150 testing speakers. BP network uses three layer networks, namely, the input layer includes 3 nodes, the middle layer includes 20 nodes, and the output layer includes 10 nodes.

11.4.1 Selection the Number of Subband

Experiments were carried out to compare the different number of subbands and different division styles, respectively, namely:

- 3 subbands: 60–948 Hz, 867–1,935 Hz, 1,790–4,000 Hz;
- 4 subbands: 60–901 Hz, 797–1,661 Hz, 1493–2,547 Hz, 2298–4,000 Hz;

Table 11.1 Speech recognition error rate

Recognizer	Error ratio (%)	Recognizer	Error ratio (%)	Recognizer	Error ratio (%)
Base-line	3.85	Base-line	3.85	Base-line	3.85
Subband 1	12.5	Subband 1	13.4	Subband 1	30.6
Subband 2	24.6	Subband 2	30.6	Subband 2	24.3
Subband 3	18.7	Subband 3	36.7	Subband 3	27.8
		Subband 4	32.1	Subband 4	32.5
				Subband 5	31.1
				Subband 6	37.2
PSHMM/NN	2.92	PSHMM/NN	2.73	PSHMM/NN	3.77

- 6 subbands: 60–495 Hz, 438–778 Hz, 707–1,144 Hz, 1,051–1,631 Hz, 1,506–2,292 Hz, 2,121–4,000 Hz.

The error ratio of speech recognition is shown in Table 11.1.

Table 11.1 compares the speech recognition error ratio between different number subband and the whole-band of clean speech. The "base-line" means the recognition rate of speech recognition systems of using all-pass HMM model, "sub-band *i*" (*i* is the serial number of different subband models) means that the recognition rate of using different subband speech recognition systems. "PSHMM/NN" means that the recognition rate of the proposed algorithm. From Table 11.1, we can draw some conclusions:

1. Main characteristics of the speech signal concentrated in the low frequency band (below 1,000 *Hz*), the speech recognition system in this band has a relatively higher recognition rate. Observation of all HMM models identification is that relatively high recognition rate in low frequency subband rate, and the recognition ratio of other higher frequency subbands will decrease.
2. The four subbands HMM model acquires the best affection, recognition error rate is only 2.73 %. When the number of subband reduces, such as three subbands, the recognition accuracy rate dropped 0.19 %. The reason is the less fusion information in the BP network input terminal. When the number of subband increases to six, the recognition rate dropped 1.04 % due to the subband divides too fine to reduce information of each subband relatively.
3. Consider the computing load, the less number subband can reduce the amount of calculation. The reasons include: (a) the number of HMM models decrease, as a result, the time of training and recognition HMM model will reduce greatly; (b) the amount of calculation of the SNR estimation is also correspondingly reduced in the noise environment; and (c) the parameter calculation of neural network will decrease for subband quantity reduction.

In summary, if the system requires a higher recognition accuracy, four subband HMM model can be adopted; if the system requires higher speed and reduced computation time, then three subband HMM model can be used in case of sacrificing recognition accuracy.

11.4.2 Speech Recognition Experiments in Noisy Environments

A variety of noise types were collected from the NOISEX-92 noise-in-speech database including white noise, babble noise, volvo noise (car noise), F-16 noise, machine gun noise and so on. The clean speech signals and various noise signals are mixed at four styles different SNR (0, 5, 10, 20 db) to simulate the real noise environments. The accurate rate of speech recognition system in all-pass HMM model (A) and the PSHMM/NN(P) model are shown in Table 11.2.

Observing the speech recognition accuracy in different noisy environments from Table 11.2, some conclusions can be drawn:

1. For most types of noise, the performance of the PSHMM/NN model is better than the traditional all-pass HMM model in the same SNR and different noisy environments, and the performance improvement depends on the type of noise. For example, for babble, factory, machinegun, destroyer engine noise and so on, the recognition accuracy ratio of the PBHMM/NN model is higher than the all-pass HMM model; especially in the machinegun noise environment, the PBHMM/NN model obtains the relative increasing of 10 % compared with the all-pass HMM model in the weak SNR (20 dB) case. The reason is that these types of noise energy are concentrated in the partial spectrum, that is to say, the individual subband was corrupted and the other subband can maintain the correct output. On the other hand, some of the energy covering the whole spectrum of noise, such as: buccaneer1, buccaneer2, volvo, and white, the correct rate of the PSHMM/NN model is lower than the all-pass HMM model; especially for the white noise, the correct recognition rate has decreased by about 15 %, which is caused by almost all subband recognition error for the interference of covering the whole spectrum; as a result, the PBHMM/NN model outputs some error recognition results.

Table 11.2 Comparison between all-pass HMM and the PSHMM/NN in noisy environments

SNR	20 dB		10 dB		5 dB		0 dB		Avg.	
	A	P	A	P	A	P	A	P	A	P
White	55.8	40.4	38.9	21.6	5.3	2.8	0.0	0.0	25.0	16.2
Babble	66.5	74.1	47.8	52.4	5.7	8.9	0.4	2.1	30.1	34.4
F-16	69.1	73.2	51.3	53.3	10.4	11.8	0.0	0.2	32.7	34.6
Machine-G	68.5	78.2	61.4	68.9	39.3	43.0	13.4	13.1	45.7	50.8
Bucca1	62.7	58.7	58.0	52.1	21.1	15.6	2.9	1.7	36.2	32.0
Bucca2	62.1	44.9	42.5	28.8	10.7	4.2	1.4	0.1	29.2	19.5
Factory1	70.7	78.3	47.6	53.5	18.2	20.0	3.2	3.8	34.9	38.9
Factory2	84.5	90.0	60.7	64.3	38.6	40.4	14.5	13.9	49.6	52.2
Volvo	89.3	89.2	69.0	67.2	35.5	31.7	9.8	4.5	50.9	48.2
Des-engine	69.4	78.7	57.6	62.5	23.8	23.2	5.2	4.9	39.0	42.3
Des-ops	80.0	84.3	68.8	70.6	36.4	37.0	12.9	12.4	49.5	51.1

2. The recognition accuracy of the PSHMM/NN model is higher than that of the traditional all-pass HMM model in different SNRs and the same noise environments. The reason is each PSHMM/NN model is independent of each other and the neural network has better classification ability, so the output result will not change when the partial subband is corrupted by noise. However, the recognition performance of PSHMM/NN model will decrease when the whole frequency bands are corrupted in a low SNR (0–5 dB) environment.

11.5 Conclusions

In order to improve the performance of HMM model in noise environments, a PSHMM/NN model was proposed. The algorithm first splits the whole frequency band HMM into a few subbands HMM, in which different speech recognizers can be independently applied. And then, some new feature parameters can be extracted according to all subbands HMM outputs. Finally, these new feature parameters are merged by the neural network in order to yield a global recognition decision. The results show that the proposed model can provide better robustness in the case of noisy speech.

Acknowledgments The research work described in this paper was supported by Anhui University Academic and Technical Leaders Introduce Engineering Foundation (02303203), National Nature Science Foundation (61271352), and Training Program on Anhui University College Students Innovation Experiment (KYXL2012058).

References

1. San-Segundo R, Martínez-Hinarejos CD, Ortega A (2012) Review of research on speech technology: main contributions from Spanish research groups. J Speech Sci 1(1):31–53
2. Ishizuka K, Miyazaki N (2004) Speech feature extraction method representing periodicity and aperiodicity in sub bands for robust speech recognition. Acoust Speech Sig Processing, 2004. Proceedings of (ICASSP'04). IEEE international conference on, 2004, I- 141-4 vol. 1
3. Hu Y, Loizou PC (2007) A comparative intelligibility study of single-microphone noise reduction algorithms. J Acoust Soc Am 122:1777
4. Fan H, Tsai Y, Hung J (2012) Enhancing the sub-band modulation spectra of speech features via nonnegative matrix factorization for robust speech recognition. System science and engineering (ICSSE), 2012 international conference on IEEE, pp 179–182
5. Bourlard H, Dupont S (1996) A new ASR approach based on independent processing and recombination of partial frequency bands. Spoken language, 1996, vol 1. Proceedings of fourth International Conference on IEEE, 1996, ICSLP 96, pp 426–429
6. Hennansky H, Tibrewala S, Pavel M (1996) Towards ASR on partially corrupted speech. Spoken language, 1996, vol 1. Proceedings of fourth international conference on IEEE, 1996, ICSLP 96, pp 462–465
7. Wang S S, Hung J, Tsao Y (2012) A study on cepstral sub-band normalization for robust ASR. Chinese spoken language Processing (ISCSLP), 2012 8th International Symposium on IEEE, 2012: 141–145

Chapter 12
Surface Texture Detection of Double-Feature Apple Based on Computer Vision

Hui Guo, Yuzhi Tan and Wei Li

Abstract The apple surface texture feature is one of the important indicators of its quality grading. We took Fuji apples as the object of our study and obtained feature information, including the color distribution and textural properties by using computer vision technique, and then, we conducted surface texture detection through back-propagation (BP) neural network. We calculated the RGB means of collected images in RGB color model, and took the *R/G* mean ratio as the feature parameters of color distribution; meanwhile, we converted the collected images into gray level images and by using gray level co-occurrence matrix, we extracted the angular second moment, contrast, correlation, and entropy of apple surface texture as the texture feature parameters. We planned to detect the apple surface texture with BP neural network; took the obtained two kinds of features as the input unit of BP neural network classifier and applied one unit for output layer to output the detection result of apple surface. The test result showed that the recognition rate of the proposed method was 91.25 %, which has laid foundation for further realization of apple quality grading in accordance with surface texture.

Keywords Surface texture · Computer vision · Feature extraction · BP neural network · Fuji apple

H. Guo · Y. Tan (✉) · W. Li
College of Engineering, China Agricultural University, Beijing, China
e-mail: yztan@cau.edu.cn

H. Guo
e-mail: foxgh@163.com

W. Li
e-mail: liww@cau.edu.cn

H. Guo
Institute of Mechanical and Electrical Engineering, Beijing Vocational College of Agriculture, Beijing, China

A. A. Farag et al. (eds.), *Proceedings of the 3rd International Conference on Multimedia Technology (ICMT 2013)*, Lecture Notes in Electrical Engineering 278,
DOI: 10.1007/978-3-642-41407-7_12, © Springer-Verlag Berlin Heidelberg 2014

12.1 Introduction

Apples are inspected and graded mainly based on their size, color, and defect in class production, while their internal qualities are reflected by surface texture feature. Therefore, the surface texture feature of apple is one of the important indicators of its quality grading.

Non-destructive and accurate detection can be realized by applying image processing techniques to acquire and extract apple surface texture features, which provides a feasible way for the non-destructive quality grading of apple. At present, it has been widely used to extract texture feature information with computer vision techniques. Li et al. [1] proposed an image hierarchy algorithm based on surface texture, which achieved effective classification of apple surface texture by extracting gradient information of images and building texture classification model. Zhoa and Hou [2] described the apple color features with HIS model and based on the color histogram features of apples in different colors, they applied chroma mean instead of histogram to achieve classification through improved BP neural network. Castleman [3] analyzed the texture defects of sand paper and leather using the gray level of Fourier power spectrum. Li and Yu [4] separated different types of textures effectively using different metrics based on Choquet content. Liu and Zhang [5] extracted the textural features with gray level co-occurrence matrix. In the past few years, it has not yet been reported that image processing techniques and BP neural network were used together to conduct apple surface texture features detection through color distribution and texture properties.

In the paper, we used comprehensive image processing techniques and artificial neural network technology to extract the two kinds of features of surface textures by taking Fuji apple as the object of study; we built BP neural network and studied the detection method of apple surface texture features.

12.2 Introduction of Image Acquisition System

Image acquisition system contains mainly an illuminating system, digital camera, digital image input interface, and computer, as shown in Fig. 12.1. A quadrate closed light box with white inner walls is used for illuminating system; the light source consists of 12 high-frequency fluorescent lamps rated at 20 W, which are evenly distributed on both sides inside the box. The bottom part of the box is the image acquisition area where the background layout is black rolling transmission device. This ensures that the apple rolls in the process of image acquisition and the three continuous different surface images are collected of each apple which covers 90 % of the whole surface. Right above the acquisition area digital camera is installed, which applies the OK_AC1300 type CCD camera from Beijing Joinhope Image Technology Co., Ltd, whose max resolution is 1300 × 1024 (pixels).

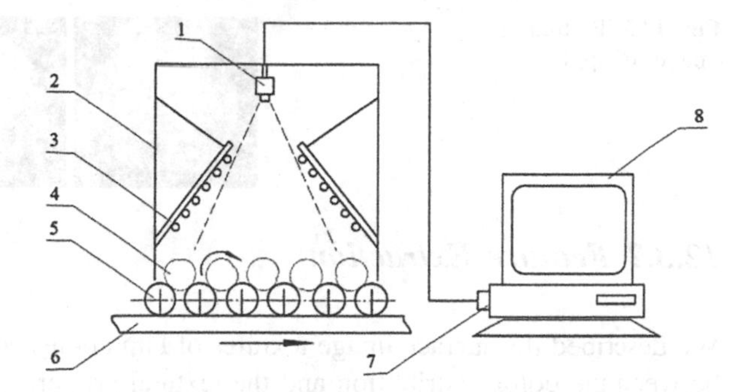

Fig. 12.1 Detection system of image acquisition. *1* CCD camera, *2* light box. *3* light source, *4* apple, *5* roller, *6* transmission band, *7* image acquisition card, *8* computer

12.3 Surface Texture Detection Feature Extraction of Fuji Apple

Crystal Fuji apples produced in Shandong province are used in the test. According to surface texture features, they are divided into all red, partial red, and strip red, as shown in Fig. 12.2. Apples 1 and 2 are strip red ones with red strips spreading evenly over the yellowish. The surface texture is distinct and the apples taste sweet. The surface color of apple 3 presents a flaky distribution without distinct textures. Apple 4 is all red since the surface color is basically red without distinct textures. It has poor taste. Therefore, we acquired mainly the color distribution and texture properties of the apple surface to conduct detection.

12.3.1 Image Preprocessing

We conducted graying and noise reduction process for the collected images. Since noises existing commonly in collected images because of certain interference, we applied median filter to reduce noise considering the digital images and apple shape features [6, 7], and 3 × 3 rectangular window was selected to be the filtering window, as shown in Fig. 12.3.

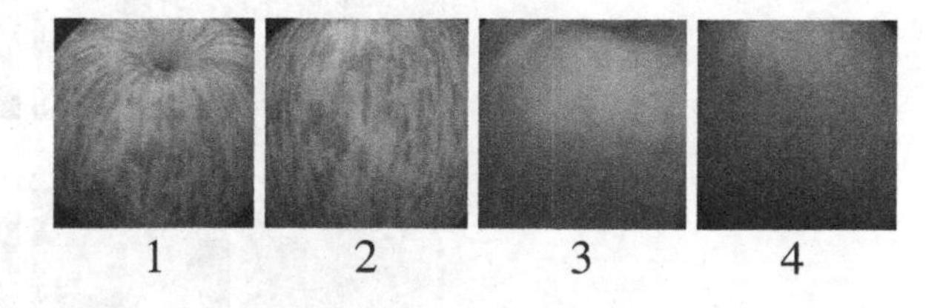

Fig. 12.2 Fuji apple surface texture

Fig. 12.3 Pretreatment images of apple

12.3.2 Feature Extraction

We described the surface image textures of Fuji apples collected using the relation between the color distribution and the textural properties of the images. Through the analysis of the color distribution of the apple surface, we extracted the color distribution features; through the analysis of textural distribution rules of apple surface, we extracted the texture properties features. Then, the feature combination of the two kinds was taken as the input parameters of BP neural network classifier to conduct recognition and detection of the Fuji apple surface textures.

12.3.2.1 Color Distribution Feature Extraction

For the analysis and characterization of the feature value of Fuji apple surface color distribution, we selected 12 Fuji apple samples with different proportions of red parts and conducted image acquisition. The collected images were shown in Fig. 12.4.

We analyzed the differences of apple color distribution and calculated the *R*, *G*, and *B* mean of all the apples and analyzed the results. Then, we used VC programming to calculate the *R*, *G*, and *B* mean of No. 1 to No. 12 apple and the results were shown in Table 12.1.

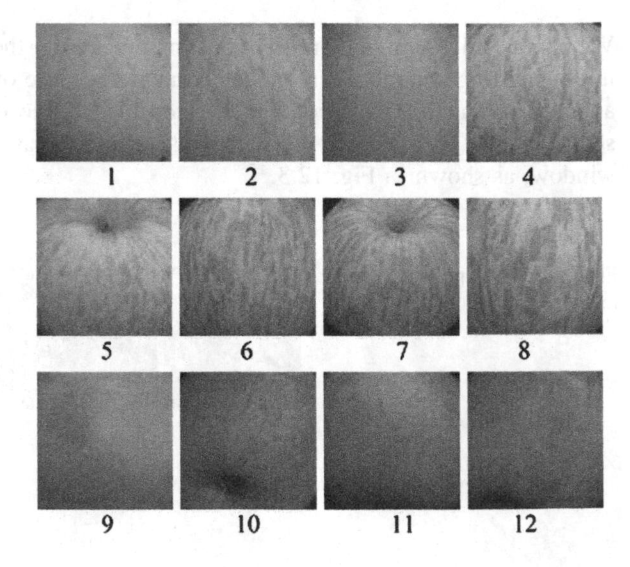

Fig. 12.4 Red Fuji apple samples image

Table 12.1 RGB mean of apple images

No.	*R*	*G*	*B*	*R/G*
1	155.86	126.44	93.04	1.23
2	162.28	132.37	98.34	1.23
3	153.48	111.19	88.48	1.38
4	179.56	127.07	105.10	1.41
5	171.24	124.44	99.85	1.38
6	171.99	92.35	71.68	1.86
7	168.94	92.78	71.90	1.82
8	179.97	102.23	85.02	1.76
9	146.59	76.16	71.43	1.92
10	130.87	47.80	41.50	2.74
11	134.13	46.69	48.91	2.87
12	130.09	39.20	50.69	3.32

It was easy to discover from Table 12.1 that with the increasing of apple surface coloring proportion, the mean value of Channel *R* varied little and the mean value of Channel *G* tended to decrease, while the mean value of Channel *B* varied irregularly. It was, therefore, clear that the Fuji apple surface coloring proportion was mainly controlled by Channels *R* and *G*.

Based on the comprehensive analysis of the data, the Fuji apple surface coloring proportion was basically in monotone increasing relation with the *R/G* mean ratio, as shown in Fig. 12.5. Therefore, we selected feature value *R/G* mean ratio as the color distribution feature parameter of Fuji apple surface textures.

12.3.2.2 Texture Feature Extraction

Generally, the analytic statistics methods of image texture feature include: gray co-occurrence matrix, gradient method, and the method to analyze texture using Fourier Transform of images and varying frequency of feature images. Fuji apple

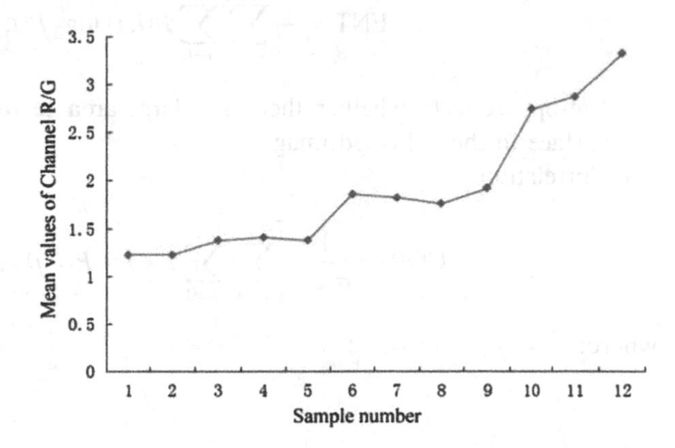

Fig. 12.5 Coloring proportional distribution of apple images

surface textures are produced in the process of growth by the distribution of different colors, so the texture shows overall regularity rather than local regularity. Large areas of apples that do not have distinct textures have the same color and gray level co-occurrence matrix can be selected to characterize the texture properties features of Fuji apple surface images collected.

Assuming that the gray level of images is N, the gray level co-occurrence matrix is N × N matrix, which could be expressed as $P_\delta(i,j)$. Where, the value of unit $P(i,j)$ at point (*i*, *j*) represents the occurrence probability of pixel pair, the gray level of which are, respectively, *i* and *j* and the distance between which is $\delta = (\Delta x, \Delta y)$ [8–10].

The four feature quantities of gray level co-occurrence matrix include:

(1) Contrast

$$\mathrm{CON} = \sum_{i=0}^{n}\sum_{j=0}^{n}(i-j)^2 P(i,j) \tag{12.1}$$

Contrast reflects the definition of images and the depth of texture grooves. If the grooves of rough textures of apple surface are deep and the CON value is large, the texture display is relatively distinct; if the grooves of fine textures of apple surface are shallow and the CON value is small, the texture display is relatively vague.

(2) Angular second moment

$$\mathrm{ASM} = \sum_{i=0}^{n}\sum_{j=0}^{n} P(i,j)^2 \tag{12.2}$$

Angular second moment reflects the uniformity of gray level distribution of collected images. When the gray level distribution of apple surface in the image is uniform, the textures are rough and the corresponding ASM value is small; in contrast, the ASM value is large.

(3) Entropy

$$\mathrm{ENT} = -\sum_{i=0}^{n}\sum_{j=0}^{n} P(i,j)\log_2 P(i,j) \tag{12.3}$$

Entropy reflects whether there are large-area texture features at the apple surface in the collected image.

(4) Correlation

$$\mathrm{COR} = \frac{1}{\sigma_x\sigma_y}\left[\sum_{i=0}^{n}\sum_{j=0}^{n} i \times j \times P(i,j) - \mu_x\mu_y\right] \tag{12.4}$$

where:

$$\mu_x = \sum_{i=0}^{n}\sum_{j=0}^{n} iP(i,j) \tag{12.5}$$

$$\mu_y = \sum_{i=0}^{n}\sum_{j=0}^{n} jP(i,j) \tag{12.6}$$

$$\sigma_x = \frac{1}{n}\sum_{i=0}^{n}(i-\mu_x)^2\sum_{j=0}^{n} P(i,j) \tag{12.7}$$

$$\sigma_y = \frac{1}{n}\sum_{j=0}^{n}(j-\mu_y)^2\sum_{i=0}^{n} P(i,j) \tag{12.8}$$

Correlation reflects the local variation of textures in the collected image. A large CON value indicates a uniform distribution of apple surface textures in a certain scope.

12.4 Design of BP Neural Network Classifier

BP neural network is a multi-layer mapping artificial neural network with the ability of back propagation and error correcting. We applied BP neural network as the classifier to conduct the recognition of apple surface texture features. The classifier contained one input layer, one hidden layer, and one output layer. In the input layer, there was the same quantity of input neurons as the collected features; in the hidden layer, the quantity of neurons was variable and tangent Sigmoid function that could continuously value in (0, 1) was applied as the transfer function [11]; there was only one neuron in the output layer, the result of which was expressed with 1 or 0, and 1 represented that there were textures and 0 represented the opposite. Before inputting the training samples, it was necessary to conduct normalization processing. Then, we conducted BP neural network training with texture features and color features and saved the obtained network weights and threshold values to files [12].

12.4.1 BPNN Classifier Based on Double Features

We conducted image preprocessing over the collected Fuji apple surface texture images and divided them into Group A (apples with distinct textures) and Group B (apples without distinct textures), and each group is composed of 24 images. Three images were collected from each apple and were joined together to form a complete image which could show 90 % of the apple surface, as shown in Fig. 12.6. Therefore, the detection accuracy was improved.

We extracted eight complete images from both Groups A and B and took their *R/G* man value as the color feature, as shown in Table 12.2.

After data analysis, we found there were differences in *R/G* mean values of sample images in Group A and Group B as color feature value. The color distribution was even in Group A and the *R/G* mean value ranged from 1.4 to 1.9. Comparing with the samples, we found that with the increase of red coloring proportion of apple surface, *R/G* mean value also increased. In Group B, if *R/G* mean values were larger than 1.9, the apples tended to be all red; otherwise, if the *R/G* mean values were less than 1.4, the apples tended to be yellow. At the same time, it was easy to discover that the *R/G* mean vale of Sample 2 and 5 in Table 12.3 were not in corresponding ranges, which was because there was a calyx in the image of Sample 2 and there was a fruit stem in the image of Sample 5, both of which affected the *R/G* mean value. Through analysis, we knew that what the *R/G* mean value could reflect was the coloring proportion of apple surface, and what it could not reflect were the texture distribution properties and uniformity. Therefore, we had to combine with the following texture property features of apple surface to conduct recognition, so as to improve the accuracy of recognition.

We determined that the step size of gray level co-occurrence matrix was 4, and the generating direction was 45° based on the tests and comparisons. On this condition, we extracted the contrast, angular second moment, entropy, and correlation of the apple surface complete image as the texture properties features, as shown in Table 12.3.

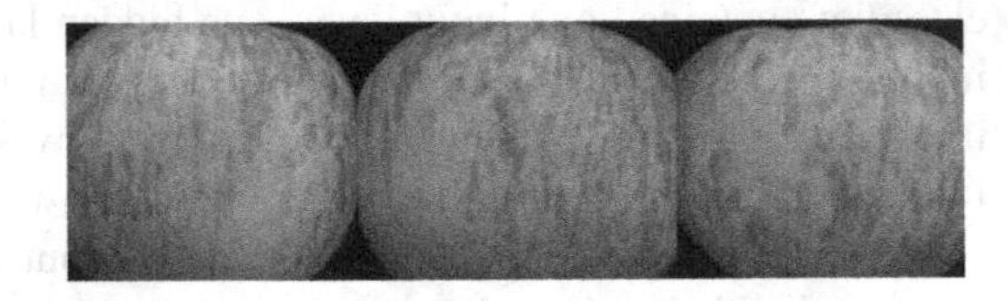

Fig. 12.6 Three images stitching effects

Table 12.2 Color features parameters of Fuji apple images

Color feature parameters			
R/G			
	Group A (even color distribution)		Group B (uneven color distribution)
1	1.53	9	1.23
2	1.35	10	1.32
3	1.40	11	1.22
4	1.41	12	1.38
5	1.38	13	1.36
6	1.86	14	2.87
7	1.82	15	1.92
8	1.76	16	2.73

Table 12.3 Texture features Parameters of Fuji apple images

	Texture featurc parameters				
		CON	ASM	ENT	COR
1	Group A (distinct texture)	0.1119	0.2877	2.3216	0.921
2		0.0933	0.2804	2.3655	0.947
3		0.1105	0.2643	2.4023	0.9354
4		0.1174	0.2389	2.4709	0.9338
5		0.0941	0.2377	2.4895	0.9527
6		0.1322	0.216	2.6924	0.9354
7		0.1211	0.1824	2.8995	0.9593
8		0.1165	0.1913	2.7709	0.9532
9	Group B (indistinct texture)	0.1014	0.3375	2.227	0.9331
10		0.0839	0.5056	1.5056	0.8436
11		0.1045	0.2989	2.179	0.9086
12		0.0929	0.3378	1.9236	0.8965
13		0.1231	0.3207	2.1159	0.8714
14		0.0817	0.4979	1.6088	0.8752
15		0.0936	0.3578	2.0164	0.8879
16		0.0901	0.4733	1.5788	0.8642

Through analysis of the data in thc table, we found that the contrasts in the images of Group A were larger than those of Group B, which indicated that the textures were distinct and the grooves of textures were deep; the angular second moments in the images of Group A were obviously smaller than those of Group B, which indicated that when the textures were obviously rough, the gray levels of images were distributed evenly; the entropies in the images of Group A were obviously larger than those of Group B, which indicated that when the textures were distinct, there were large amount of texture information of apple surface images; the correlations in the images of Group A were larger than those of Group B, which indicated that the textures were locally even. From this, it could be seen that there were differences among the four texture feature values extracted from the gray level co-occurrence matrix of the two groups of apple sample images. This could be used for the recognition of apple surface textures. At the same time, since sometimes there were similarities in the four texture feature values of sample images in the table, erroneous judge may be caused.

In conclusion, it was necessary to combine the color distribution features and texture properties features when recognizing Fuji apple surface textures. We divided the input layer of BP classifier into five neuron parameters, namely, contrast, angular second moment, entropy, correlation, and *R/G* mean ratio; there was only one unit in output layer, corresponding to, respectively, apples with distinct textures or without distinct textures; the number of neurons in hidden layer was chosen based on the empirical formula (12.9) [13].

$$h = (m + n)^{\frac{1}{2}} + 1 \tag{12.9}$$

Table 12.4 Recognition results of BP neural network classifier based on both texture and color features

Apple image	Quantity of samples	Quantity of correct recognition	Recognition accuracy (%)
With distinct textures	80	74	92.5
Without distinct textures	80	72	90

where, h was the number of neurons in hidden layer; m was the number of neurons in input layer; n was the number of neurons in output layer; 1 represented an integer selected from 1 to 10. In the optional range of 1, we built BP neural network for different values of 1 and compared the training results. Finally, we decided that the number of hidden layer was 6 [14, 15].

12.4.2 Recognition Experiment of Sample Set

We inputted 80 images of the two kinds of samples to conduct recognition and at the same time tested the recognition accuracy of BP neural network classifier. The result was shown in Table 12.4.

From analysis of the data in the table, we knew that it was possible to conduct recognition of Fuji apple surface textures using BP neural network built by double features as color distribution and texture properties; the recognition accuracy reached, respectively, 92.5 % and 90 % and the average accuracy reached 91.25 %. Among this, erroneous judge was caused since some collected images of apple surface contained the calyx and fruit stem and lead to an intensive color distribution and lack of distinct textures.

12.5 Conclusions

After conducting Fuji apple surface image preprocessing, we used both the image color distribution and the texture property features to recognize Fuji apple surface texture. We applied BP neural network and selected color distribution *R/G* mean ratio as the color features; then, we calculated the texture property features with gray level co-occurrence matrix and conducted the recognition of Fuji apple surface textures with double feature values. The experiment results showed that it was possible to recognize correctly the apple surface textures with complex information, and the average recognition accuracy reached 91.25 %. However, since the calyx and fruit stem were not eliminated when collecting images, errors occurred in the texture recognition. This restricted the detection of apple surface textures by this method. We should further study how to eliminate the impact of calyx and fruit stem on the image recognition and improve the accuracy of this method in the detection.

References

1. Li W, Kang Q, Zhang J, Xun Y (2008) Detecting technique for surface texture on apples based on machine vision. J Jilin Univ: Eng Technol Ed 38(5):1110–1113
2. Zhao M, Hou W (2009) Method of apple automatic grading based on neural network. J Nanjing For Univ: Nat Sci Ed 33(1):136–138
3. Castleman KR (1996) Digital image processing. Prentice, HallUpper Saddle River
4. Li H, Yu B (2003) Image segmentation approach based on new multifractal feature vectors. Opt Precis Eng 11(6):627–631
5. Liu Z, Zhang Y (1999) Image retrieval using both color and texture features. J China Inst Commun 20(5):36–40
6. Sun H, Shi W, Ju Y (2003) Image processing with medium value filter. J Chang'an Univ: Nat Sci Ed 23(2):104–106
7. Li H, Zhang Z, Yi Z (2004) The application of median filtering on image processing. Inf Technol 28(7):26–27
8. Wang H, Shi P (2006) Methods to extract images texture features. J Commun Univ China: Sci Technol Ed 13(1):49–52
9. Gao C, Hui X (2010) GLCM-based texture feature extraction. Comput Syst Appl 19(6):195–198
10. Guo B, Zhou C, Zhao Y, Wang G (2010) Texture of continuous image based on gray Co-Occurrence matrix. Laser Optoelectron Prog 47(051002):1–5
11. Zhang H, Dong H, Long F, Guo S (2008) Research on image recognition based on BP neural network. Comput Modern (5):17–19
12. Wan L, Chen J, Wang W, Li J (2009) Image recognition based on BP neural network. J Wuhan Technol Univ: Sci Technol Ed 29(3):277–279, 292
13. Zhang L (1994) Models and applications of artificial neural networks. Fudan University Press, Shanghai
14. Shen Y, Wang Z, Gao C et al (2008) Determining the number of BP neural network hidden layer units. J Tianjin Univ Technol 24(5):13–15
15. Yan P, Zhang C (2005) Artificial neural networks evolutionary computation and simulation. Tsinghua University Press, Beijing

Chapter 13
A Fast Search Algorithm of Multi-View Video Coding

Mei-leng Yuan and Yang Zhang

Abstract Multi-view video coding is the key technology in Multi-view video technology field. The motion estimation is the most time consuming segments of multi-view video coding. This paper analysis the EPZS algorithm applied in JMVC. For EPZS algorithm's disadvantages, a fast hybrid search algorithm based on multi-view video coding is proposed. The optimization is suggested mainly from the following four aspects: predictive vector sets, search model, threshold setting as well as search strategy. Experimental results on JMVC8.3 show that the performance of modified algorithm is so good that coding efficiency is also improved, on the promise of right code rate and reconstruction video quality comparing to EPZS in JMVC. The average encoding speed increased from 55.66 to 69.62 percent.

Keywords JMVC · Multi-view video coding · EPZS · Coding efficiency

13.1 Introduction

Multi-view video is a set of video signals caught on the same scene by multiple cameras from different angles. Compared with single viewpoint or 2D video, multi-view video is more comply with person's view habit, contains more information, stereoscopic, high quality, multimedia experience, and interactive, which can be widely used in free viewpoint video, 3D television, teleconference, remote medical diagnosis, virtual reality, and video surveillance system. Multi-view video coding is the key technology and hot spot which will bring new opportunity and prospects for development in multimedia application [1, 2]. Mode decision, multi

M. Yuan (✉) · Y. Zhang
Shenzhen Polytechnic, Shenzhen 518055, China
e-mail: mlyuan@szpt.edu.cn

A. A. Farag et al. (eds.), *Proceedings of the 3rd International Conference on Multimedia Technology (ICMT 2013)*, Lecture Notes in Electrical Engineering 278,
DOI: 10.1007/978-3-642-41407-7_13,

reference frames selective and motion estimation are the most time consuming segments of multi-view video coding, their complexity covers over 96 % of the whole coding. Fast search algorithm is the core process to influence these three parts coding complexity, and plays very important role in decreasing whole coding complexity [3]. Therefore, researching high efficient fast search algorithm has great significance in decreasing coding complexity and improving coding real-time performance.

The earlier fast search algorithms mainly include TSS, TDL, CSA, BBGDS, DS, and HEXBS, the biggest features of these algorithms are searching mode unitary, rules simple and low complexity, searching can easily fall to partial minimum or false direction [4]. The newly mixed fast search algorithm such as ARPS, PMVFAST, UMHexagonS, and EPZS adopt multi-searching models, large step and small step combination, coarse-grained, and refined search complement with each other, draw into early termination rules and predict search starting point to fasten speed of finding the best matching point. Reference [5] proposed a unidirectional and bidirectional searching algorithm based on 3D constrained model; Reference [6] proposed a fast motion estimation algorithm based on motion vector interview and space correlation, compared with whole search of JMVC and EPZS algorithm, the coding time decrease 99.56 and 22.2 %, the decreased range still needs to improve; Reference [7] proposed a self-adaptive early termination strategy; Reference [8] proposed a fast algorithm combined with motion and parallax estimation, which saved 87.69 % coding time compared with whole search algorithm; Based on researching fast searching algorithm of JMVC, this article proposed a motion estimation fast search algorithm point at 8 parallel view points, while guarantee video coding rate distortion properties, this algorithm decreases coding time and improves coding real-time performance.

13.2 EPZS Algorithm Analysis

EPZS is the core motion estimation algorithm based on H.264 standard and JMVC8.3 (multi-view video coding reference software).

EPZS is a mixed algorithm which combines multi-search models, the basic process is : at first confirm the best search starting point, second eight points diamond search successively with step length in 1, 2, 4, 8, 16, 32, and 64, third grating search, and the last is refined search. The shortages of EPZS are as below:

1. In the first step selection of search starting point, predicted vector collection only consider median prediction vector, upper right prediction vector, upper prediction, and null vector.
2. In the second step of model search, eight points diamond search with step length in 1, 2, 4, 8, 16, 32, and 64 proceed successively which totally need seven rounds matching search, not using threshold early termination, according to [9], using threshold early termination could decrease some complexity.

3. The third step adopts dense gratings search which is more complicated and time consumption, not combining video contents in the searching strategy, video contents motion direction are mainly in horizontal and vertical, thus, flexible horizontal and vertical gratings search are adoptable.
4. In the last step refined search does not draw into early termination, at the round of the best point there is no further search of possible modification of the best point.

Based on above analysis, this article proposes an improved mixed fast search algorithm.

13.3 Motion Estimation Fast Search Algorithm Design

Multi-view video coding modes mainly include Skip, Direct, B16*16, B16*8, B8*16, B8*8, B8*4, B4*8, B4*4, IPCM, Intra4*4, Intra8*8, and Intra16*16. From experimental analysis we get that time consumption of mode decision, multi reference frames selective, and motion/parallax estimation covers over 97 % in the whole video coding, in consideration of the best mode and reference frames, motion estimation search algorithm is used to calculate rate distortion figure to supply basis of selecting best mode and reference frames. Although using whole search algorithm can find the minimum, the calculation is so complicated which is not beneficial for video coding in real-time application. Usually natural shoot multi videos have some relativity, static background covers large portion, target in the video moves more in horizontal and vertical than other directions, moving vector has center-biased property, in the past fast search algorithm which was aimed at single point motion usually use center-biased property to search from null vector, but video contents have very strong space relativity, the best search starting point will use prediction vector collection to confirm; To avoid search fall into partial minimum, fast search algorithm combines widespread rough search and small range refined search together, adopts decomposed cross gratings search according to video array motion conditions and extended hexagon matching search; finally proceed refined hexagon and small diamond search and modification of best matching point, the algorithm processing flow is as Fig. 13.1.

13.3.1 Enlarge Prediction Vector Collection, Select Best Starting Point

1. Combine median prediction vector $MV_{\text{pred\_}MP}$, left prediction vector MV_{Left}, upper prediction vector MV_{Up}, right upper prediction vector MV_{RU} and null vector (0,0) to motion vector prediction collection S_1, as shown in formula (13.1):

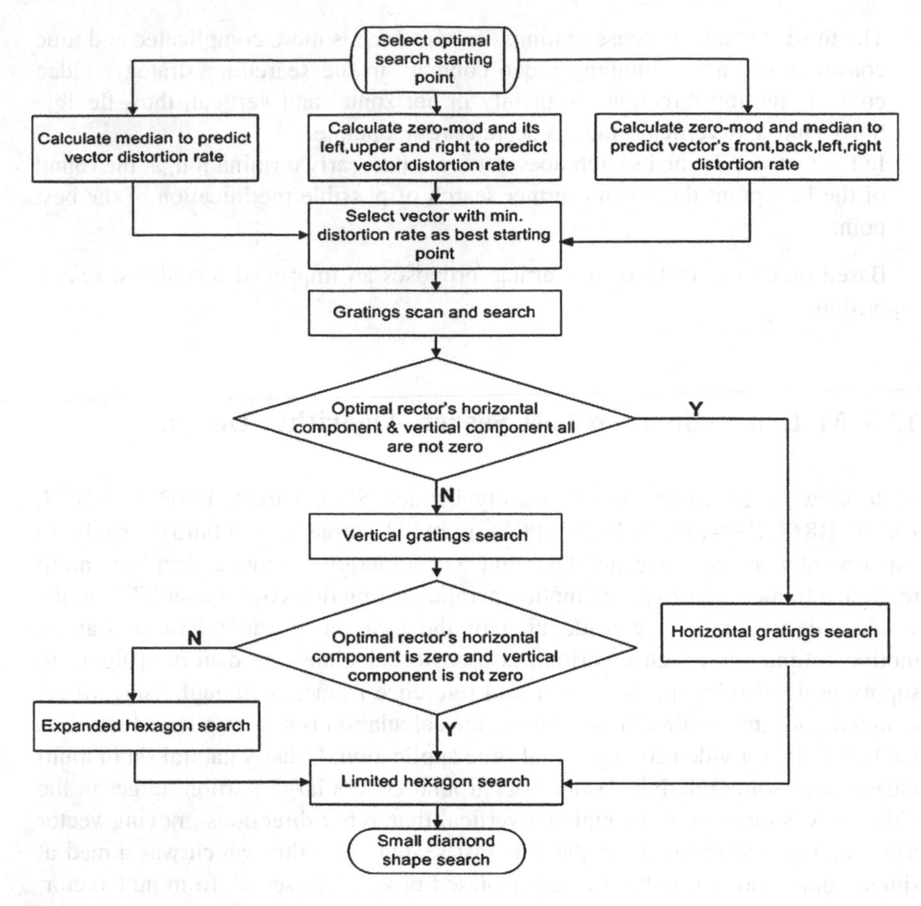

Fig. 13.1 Algorithm flowchart

$$S_1 = \left\{ \overrightarrow{M_i} \middle| \overrightarrow{M_i} = \overrightarrow{MV_{\text{pred\_}MP}}, \overrightarrow{MV_{\text{Left}}}, \overrightarrow{MV_{Up}}, \overrightarrow{MV_{RU}}, \left(\overrightarrow{0,0}\right) \right\} \tag{13.1}$$

2. As the surrounding points of motion vector S_1 also might be the best motion vector, thus, median vector and null vector's four points of right, left, upper and below are added in S_1 formula (13.2):

$$\psi\left(\overrightarrow{\text{CMV}}\right) = \left\{ \overrightarrow{MV_i} \middle| \overrightarrow{MV_i} = \left(\text{CMV}_x \pm 1, \text{CMV}_y\right), \left(\text{CMV}_x, \text{CMV}_y \pm 1\right) \right\} \tag{13.2}$$

3. Prediction vector collection S_2, is union by median prediction vector $MV_{\text{pred\_}MP}$ and null vector (0,0), as shown in formula (13.3):

$$S_2 = \psi\left(\overrightarrow{MP_{\text{pred\_}MP}}\right) \cup \psi\left(\overrightarrow{0,0}\right) \tag{13.3}$$

4. The total prediction vector collection S is union by S_1 and S_2, as shown in below formula:

$$S = S_1 \cup S_2. \tag{13.4}$$

5. The best search starting point vector $\vec{MV}_{\min}$ is calculated according to formula (13.5):

$$\overrightarrow{MV_{\min}} = \arg\left[\min_{\overrightarrow{MV_i}} \mathrm{Cost}(\overrightarrow{MV_i}, \lambda_{\mathrm{MOTION}})\right], \overrightarrow{MV_i} \in S. \tag{13.5}$$

13.3.2 Grating Search According to Motion Conditions

Taken the best vector $MV_{\min}$ as search starting point from step 3.1, the best matching point horizontal and vertical vectors are CMV_x and CMV_y, T_1 and T_2 are horizontal and vertical motion threshold, T_3 is the step length of gratings search, iNegNor, iPosNor, iNegVer, iPosVer means left, right, upper, and below range, respectively, Best Distance is distance from best matching point to starting point, Search Times is the biggest circulating search times, T_4 is searching round threshold (when selecting the best point, if after sequent T_4 rounds as shown in Fig. 4.1, all the SAD value of matching points are over the current best point, then the current best point is the best matching point), in order to get searching times vector, proceeding motion judge search are as follows:

1. Defined below equation is to select the best motion vector with minimum rate distortion figure from motion vector collection; the following process will also use this formula to calculate.

$$\overrightarrow{MV_{\min}} = \arg\left[\min_{\overrightarrow{MV_i} \in \left(\overrightarrow{MV} \cup \overrightarrow{MV_{\min}}\right)} \mathrm{Cost}(\overrightarrow{MV_i}, \lambda_{\mathrm{MOTION}})\right]\Bigg\} \tag{13.6}$$

2. If the best vector's vertical component is 0, horizontal component is not 0, the initial judge is video content has trend to move in horizontal direction, so proceeding horizontal grating search can confirm the step length is 4 or 3 according to relation between motion range and setting threshold.
3. If the best vector's horizontal component is 0, vertical component is not 0, we can judge initially that video content has trend to move in vertical direction, so proceeding vertical grating search can confirm the step length is 3 or 4 according to relation between motion range and setting threshold.

13.3.3 Expand Search Range to Extended Hexagon

After starting point and grating search of step 3.1 and 3.2, the starting point and direction are basically confirmed, but they belong to partial search. According to rules of combining small and wide range search together, they needs further expanding search to increase search range, in some related articles it is confirmed hexagon mode has better performance than diamond mode, but less complexity. So in this algorithm, according to the video motion is more in horizontal than vertical direction, the original hexagon mode adds search point at horizontal direction to the extended one.

13.3.4 Draw into Early Termination Strategy to Limited Times Search

After small and wide range search, the last is proceeding small range best vector confirmation and limited times hexagon search; in this process the early termination strategy is drawn into to avoid unlimited circulating search and consume more coding time. This strategy is mainly draw in early termination threshold. According to experiment analysis, in hexagon search step length is gradually extended, when circulating times threshold is set to 2 or 3, it can probably get the best matching point, which can early terminate the extended step length hexagon search. Thus, drawing into threshold early termination circulating searching strategy can help decreasing coding complexity.

13.3.5 Small Diamond Search to be Refined

At the round of final best motion vector, the last step is refined search to increase the property of finding the best motion vector.

13.4 Experimental Simulations

13.4.1 Experimental Description

1. JMVC basic parameter configuration (Table 13.1).
2. Formulas of average peak signal to noise ratio (dB) decrease value PSNR, coding time (s) decrease range Time and output code rate (Kbit/s) increase range Bitrate are shown in (13.7), (13.8) and (13.9).

Table 13.1 JMVC 8.3.1 basic parameter configuration

Name	Configuration
Quantization parameter	24, 28, 32, 36
GOP length	12
Coding frame number	61
Uni-direction range	96
Max.refer frame number	2
Bidirectional iterative times and range	4 and 8

$$\begin{aligned} \mathrm{PSNR}_P &= \mathrm{PAV}_j[(Y*4+U+V)/6] \\ \mathrm{PSNR}_J &= \mathrm{PAV}_i[(Y*4+U+V)/6] \\ \Delta\mathrm{PSNR} &= \sum_{j=0}^{7}\mathrm{PSNR}_P - \sum_{i=0}^{7}\mathrm{PSNR}_J \end{aligned} \tag{13.7}$$

$$\nabla\mathrm{Time} = \frac{\sum_{i=0}^{7}\mathrm{OVTime}_i - \sum_{j=0}^{7}\mathrm{NVTime}_j}{\sum_{i=0}^{7}\mathrm{OVTime}_i} \times 100\ \% \tag{13.8}$$

$$\Delta\mathrm{BitRate} = \frac{\sum_{J=0}^{7}\mathrm{NVBR}_j - \sum_{i=0}^{7}\mathrm{OVBR}_i}{\sum_{i=0}^{7}\mathrm{OVBR}_i} \times 100\ \% \tag{13.9}$$

PSNR, Bitrate, and ∇Time means dB decreasing value, code rate increasing range, and coding time decreasing range, respectively, comparing this algorithm to JMVC; $\mathrm{PSNR_P}$, $\mathrm{NVBR_j}$, NVTime means average dB, output coding rate, coding time at j video point of this algorithm; $\mathrm{PSNR_J}$, $\mathrm{OVBR_i}$, OVTime means average dB, output coding rate, coding time at i view point of JMVC. The experiment result is shown in Table 13.2.

Comparison of rate distortion and complexity among JMVC, [10] and this paper are shown in Figs. 13.2 and 13.3.

13.4.2 Experiment Data Analysis

From Table 13.2, when Qp value is 22, 27, 32, 37, and 42, the algorithm ballroom sequence coding time decrease range is [68.88, 76.58 %], the average is 69.62 %, peak signal to noise ratio decrease $0 \sim 0.1$ dB, the average decrease is 0.04 dB,

Table 13.2 Result comparison

Array	Quantization value	Way	Times (s)	∇Time	Bitrate (Kbit/s)	ΔBitRate	PSNR (dB)	∇PSNR
Ballroom	22	JMVC	8,382		1526.08		40.32	
		Ref. [10]	5,711	31.86	1529.68	0.24	40.32	0
		This paper	1,963	76.58	1538.05	0.78	40.31	−0.01
	27	JMVC	7,577		746.04		38.22	
		Ref. [10]	5,158	31.93	749.94	0.50	38.21	−0.01
		This paper	1,909	74.80	756.04	1.31	38.20	−0.02
	32	JMVC	6,879		385.84		35.92	
		Ref. [10]	4,438	35.48	388.74	0.75	35.91	−0.01
		This paper	1,842	73.22	393.01	1.88	35.89	−0.03
	37	JMVC	6,335		214.61		33.66	
		Ref. [10]	3,698	41.62	218.18	1.63	33.62	−0.04
		This paper	1,761	72.20	220.56	2.77	33.60	−0.06
	42	JMVC	5,439		125.84		31.27	
		Ref. [10]	2,785	48.80	128.16	1.81	31.19	−0.07
		This paper	1,692	68.88	129.58	2.97	31.17	−0.10
Average value		Ref. [10]		37.93		0.99		−0.02
		This paper		69.62		1.94		−0.04

(continued)

Table 13.2 (continued)

Array	Quantization value	Way	Times (s)	∇Time	Bitrate (Kbit/s)	ΔBitRate	PSNR (dB)	∇PSNR
Exit	22	JMVC	7,386		864.08		41.25	
		Ref. [10]	4,623	37.41	866.52	0.28	41.24	−0.01
		This paper	1,929	73.88	872.47	0.97	41.24	−0.01
	27	JMVC	6,649		352.74		39.78	
		Ref. [10]	4,053	39.03	354.90	0.61	39.77	−0.01
		This paper	1,886	71.62	357.15	1.25	39.76	−0.02
	32	JMVC	6,063		177.40		38.08	
		Ref. [10]	3,500	42.27	179.14	0.97	38.05	−0.03
		This paper	1,807	70.18	180.22	1.58	38.05	−0.03
	37	JMVC	5,491		101.95		36.10	
		Ref. [10]	2,825	48.54	103.63	1.63	36.06	−0.04
		This paper	1,762	67.90	104.49	2.49	36.05	−0.05
	42	JMVC	4,747		60.32		33.76	
		Ref. [10]	2,164	54.41	62.07	2.81	33.69	−0.07
		This paper	1,683	64.54	62.26	3.21	33.70	−0.06
Average value		Ref. [10]		44.33		1.26		−0.03
		This paper		69.62		1.9		−0.03

(continued)

Table 13.2 (continued)

Array	Quantization value	Way	Times (s)	∇Time	Bitrate (Kbit/s)	ΔBitRate	PSNR (dB)	∇PSNR
Vassar	22	JMVC	5,670		1071.86		40.11	
		Ref. [10]	3,489	38.47	1070.77	–0.10	40.11	0
		This paper	1,885	66.75	1072.23	0.03	40.11	0
	27	JMVC	4,723		370.55		38.47	
		Ref. [10]	2,803	40.66	370.55	0.48	38.47	0
		This paper	1,784	62.23	372.98	0.66	38.47	0
	32	JMVC	4,023		148.45		36.83	
		Ref. [10]	2,370	41.08	149.17	0.48	36.82	–0.01
		This paper	1,741	56.70	149.20	0.50	36.82	–0.01
	37	JMVC	3,423		66.37		35.07	
		Ref. [10]	1993.38	41.77	66.99	0.93	35.06	–0.01
		This paper	1,707	50.10	66.79	0.63	35.05	–0.02
	42	JMVC	2,897		31.83		33.08	
		Ref. [10]	1,693	41.55	32.77	2.95	33.06	–0.02
		This paper	1,664	42.54	32.22	1.22	33.06	–0.02
Average value		Ref. [10]		40.70		0.95		–0.008
		This paper		55.66		0.6		–0.01

bit rate increase range is [0.78, 2.97 %], the average increase is 1.94 %, in [10] coding time decrease range is [31.86, 48.80 %], the average decrease is 37.93 %, peak signal to noise ratio decrease 0 $\sim$ 0.07 dB, the average decrease is 0.02 dB, bit rate increase range is [0.24, 1.81 %], the average increase is 0.99 %. In this article, exit sequence coding time decrease range is [64.54, 73.88 %], the average is 69.62 %, peak signal to noise ratio decrease 0.01 $\sim$ 0.06 dB, the average decrease is 0.03 dB, bit rate increase range is [0.97, 3.21 %], the average increase is 1.9 % [10] coding time decrease range is [37.41, 54.41 %], the average decrease is 44.33 %, peak signal to noise ratio decrease 0.01 $\sim$ 0.07 dB, the average decrease is 0.03 dB, bit rate increase range is [0.28, 2.81 %], the average increase is 1.26 %. In this article algorithm vassar sequence coding time decrease range is [42.54, 66.75 %], the average is 55.66 %, peak signal to noise ratio decrease 0 $\sim$ 0.02 dB, the average decrease is 0.01 dB, bit rate increase range is [0.03, 1.22 %], the average increase is 0.06 %, [10] coding time decrease range is [38.47, 41.55 %], the average decrease is 40.70 %, peak signal to noise ratio decrease 0 $\sim$ 0.02 dB, the average decrease is 0.008 dB, bit rate increase range is [$-$0.1 $\sim$ 2.95 %], the average increase is 0.95 %.

From Fig. 13.2, the rate distortion curve graph of these three algorithms in this article, [10] and JMVC are almost coincide, thus, algorithm of this article has good rate distortion performance which guarantee rebuild video quality and output code rate. From complexity comparison in Fig. 13.3, the complexity of this article declines along with Qp decreasing, the decrease ranges are all over 50 %, some even higher than 80 %, which largely decrease coding complexity, and each video sequence shows to be unanimous. Thus, algorithm of this article has lower complexity.

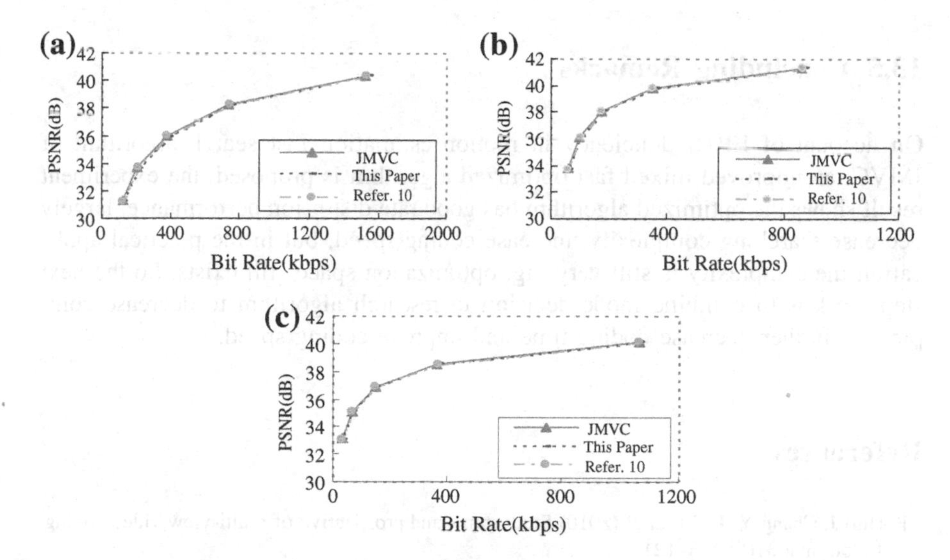

Fig. 13.2 Distortion rate curve graph. **a** Ballroom, **b** exit, **c** vassar

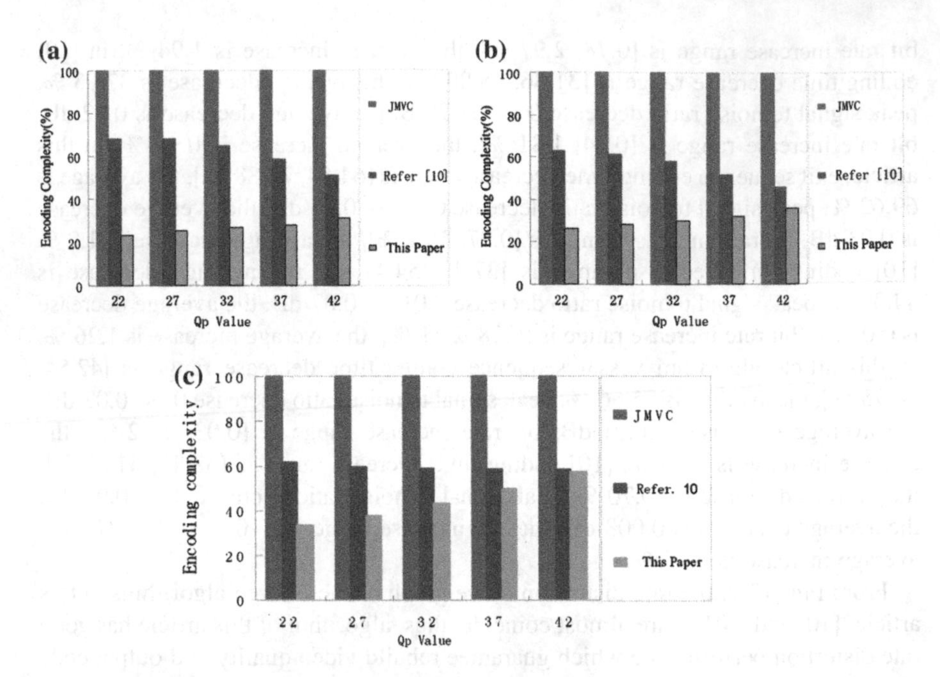

Fig. 13.3 Complexity comparison. **a** Ballroom, **b** exit, **c** vassar

From above analysis, we can get that on the premise of guarantee video rebuild quality, this algorithm largely decreased coding complexity and increase coding speed.

13.5 Concluding Remarks

On account of EPZS deficiency in motion estimation fast search algorithm of JMVC, an improved mixed fast optimized algorithm is proposed, the experiment result shows the optimized algorithm has good rate distortion performance, largely decrease searching complexity, increase coding speed, but in the practical application the complexity is still very big, optimization space still exists. So the next step work is to combine mode decision to research algorithm to decrease complexity, further decrease coding time and improve coding speed.

References

1. Huo J, Chang Y, Li M, et al (2010) Studystatus and prospective of multi-view video coding. J Commun 31(5):113–121

2. Anthony V, Thomas W, Garyjs (2011) Overview of the stereo and multi-view video coding extensions of the H.264/MPEG-4 AVC standard 99(4):626–642
3. Ding X, Fan H (2011) Mix-pattern motion estimation search algorithm based on direction adaptation. J Image Graph 16(1):14–20
4. Gu H, Chen S, Sun S (2011) Fast motion estimation algorithm based on multi-search centers. Acta Electronica Sinica 39(3):695–699
5. Deng ZP, Jia KB, Chan YL (2011) A fast algorithm for MVC using HBP prediction structure. J Electron Inf Technol 33(8):1955–1962
6. Zhu W, Chen Y (2011) A fast motion estimation algorithm for multi-view video coding. J S China Univ Technol (Nat Sci Ed) 39(2):39–45
7. Zhang P, Jiang G, Yang S (2010) Anadaptive early termination algorithm for motion estimation in multi-view video coding. In: Proceedings of the 2010 3rd international congress on image and signal processing (CISP), IEEE, Yantai, pp 72–75
8. Deng ZP, Jia KB, Chan YL (2011) Fast motion and disparity estimation of multi-view video coding. J Beijing Univ Technol 37(5):683–690
9. Yang ZH, Dai SK (2011) New optimization algorithm for multi-view video coding. J Comput Appl 31(9):2461–2464
10. Tang XL, Dai SK, Cai CH (2010) An analysis of TZ search algorithm in JMVC. In: Proceedings of the 1st international conference on green circuits and systems (ICGCS), IEEE, Shanghai, pp 516–520

Chapter 14
A New Texture Direction Feature Descriptor and Its Application in Content-Based Image Retrieval

Yu Xia, Shouhong Wan and Lihua Yue

Abstract Local Binary Pattern (LBP) has been widely used in texture analysis and content-based image retrieval (CBIR). LBP encodes the relationship between the referenced pixel and its surrounding neighbors by computing gray-level variation. However, LBP is unable to reflect the spatial distribution information of gray variation direction in the whole image. Therefore, in this paper, we propose a new texture direction feature descriptor to extract the spatial distribution information of gray-level variation between pixels. After the calculation of the gray variation pattern on different directions, we construct the statistic histograms of pattern pairs between the referenced pixel and its neighbor pixels. The performance of the proposed feature descriptor is compared with different methods using two benchmark image databases. Performance analysis shows that the proposed feature descriptor improves the retrieval precision rate, as well as the recall rate both in texture and natural scene images.

Keywords CBIR · Local binary pattern · Texture analysis · Feature direction

14.1 Introduction

With the increasing demand of image retrieval on large image database, there exists a strong need for developing an efficient technology that can automatically search the desired image from the huge database. Content-based image retrieval

Y. Xia · S. Wan (✉) · L. Yue
University of Science and Technology of China (West Campus), Hefei, China
e-mail: wansh@ustc.edu.cn

Y. Xia
e-mail: xiay1989@mail.ustc.edu.cn

L. Yue
e-mail: llyue@ustc.edu.cn

A. A. Farag et al. (eds.), *Proceedings of the 3rd International Conference on Multimedia Technology (ICMT 2013)*, Lecture Notes in Electrical Engineering 278,
DOI: 10.1007/978-3-642-41407-7_14,

(CBIR) is one of the most useful solutions for such applications. CBIR refers to using the visual contents of images such as color, texture, shape, space layout, semantics to describe the image, and designing proper similarity measurement of image features to retrieval images which a user may be interested in.

Local Binary Pattern (LBP) has emerged as a shining spot in the field of texture classification and retrieval. Ojala et al. [1] proposed LBP, which has been improved to rotational invariant version for texture classification (Rotational invariant LBP, RILBP) [5]. Various extensions of the LBP, such as uniform LBP (ULBP) [4], completed LBP (CLBP) [3], dominant LBP (DLBP) [2], local ternary pattern (LTP) [6] are proposed for texture classification. ULBP re-encoded the 256-dimensional LBP value and normalized the code result as the new feature vector. Using the same calculation method of LBP, CLBP encoded the gray value variation magnitude as well. CLBP obtained the texture feature which can reflect gray value variation information. Subrahmanyam et al. [7]. proposed the local maximum edge binary pattern (LMEBP) feature, which changed the coding method of center pixel and its neighbors. Through sorting the local difference of pixel and neighbor pixels to obtain eight maximum edges. LMEBP achieved better retrieval results in texture image database. In general, LBP, CLBP, LTP, LMEBP only calculate gray value relation of all pixels in the image, but ignore the spatial distribution information of variation direction, which leads to the deficiency in feature extraction.

To solve the above problem, this paper proposes a new texture direction feature descriptor: local gray-scale variation co-occurrence matrix (LGVCM). First, we calculated the gray value variation mode between each pixel and its neighbor pixels on the directions of 0, 45, 90, and 135°. In addition, the pattern pairs of center pixel and its neighbor pixels were counted to construct the pattern pair histogram. Finally, the variation directional feature vectors were extracted from the normalization of pattern pair histogram. Combining with the LBP features, not only the spatial distribution information of variation direction, but also magnitude information were well described. Therefore, local gray-scale co-occurrence matrix can reflect much more texture detail information for feature extraction.

14.2 Feature Extraction

14.2.1 Local Gray-Level Variation Co-Occurrence Matrix

Given an image G, the width and height of G is N_x and N_y. The definition of local gray-scale variation co-occurrence matrix (LGVCM) is Formula (14.1).

$$LGVCM^{\alpha}(m,n) = \{(x_1,y_1),(x_2,y_2)_x \times N_y | CM^{\alpha}_{(x1,y1)} = m, \\ CM^{\alpha}_{(x_2,y_2)} = n\}, m,n \in M = \{1,2,3,4,5,6\}, \alpha \in A = \{0^{\circ},45^{\circ},90^{\circ},135^{\circ}\} \quad (14.1)$$

where α represents the direction, *m, n* is the gray variation pattern value, $CM^{\alpha}_{(x1,y1)}$ and $CM^{\alpha}_{(x2,y2)}$, respectively, represents the gray variation pattern in direction α at pixel (x_1, y_1) and (x_2, y_2). Using a 3 × 3 window to traverse the whole image, as the pixels' distribution is shown in Fig. 14.1. Supposing the three pixels' values in direction α are (P_x, P, P_y), CM^{α}_{p} can be calculated as Formula (14.2).

$$CM^{\alpha}_{p} = \begin{cases} 1, & if\,(P = P_x \;\&\; P = P_y \;\&\; P < Thres) \\ 2, & if\,(P = P_x \;\&\; P = P_y \;\&\; P \geq Thres) \\ 3, & if\,(P \leq P_x \;\&\; P \leq P_y) \;\&\; (!(P = P_x \,\&\, P = P_y)) \\ 4, & if\,(P \geq P_x \;\&\; P_y) \;\&\; (!(P = P_x \,\&\, P = P_y)) \\ 5, & if\,(P \geq P_x \;\&\; P \leq P_y) \;\&\; (!(P = P_x \,\&\, P = P_y)) \\ 6, & if\,(P \leq P_x \;\&\; P \geq P_y) \;\&\; (!(P = P_x \,\&\, P = P_y)) \end{cases} \tag{14.2}$$

where *Thres* is the threshold we choose to distinguish the situation of same texture with different gray value. In our experiments, we choose *Thres* as the average gray value of the whole image.

Based on four pattern images in four directions, LGVCM extracts the pattern pair of pixel $P(x, y)$ and its neighbor pixel $Q(x + \Delta_x,\, y + \Delta_y)$. Suppose, $(\Delta_x,\, \Delta_y)$ is $(1, 0)$ and the direction is 0°, so we need to calculate $CM^{0°}_{p}$ and $CM^{0°}_{Q}$, while these two patterns both need to calculate the gray value variation relation between *P* and *Q* (according to Fig. 14.1). To avoid the redundant information, we do not count the pattern pairs in 0°when $(\Delta_x,\, \Delta_y)$ is (1,0). Similarly, when $(\Delta_x,\, \Delta_y)$ equal to (1,1), (0,1), (−1,1), we do not count the pattern pairs in direction 45, 90, and 135°. Therefore, through this process, we can finally obtain all 12 LGVCMs, with three LGVCMs on each direction. For each LGVCM, CM belong to the set $M = \{1, 2, 3, 4, 5, 6\}$, so the amount of pattern pairs is $6 \times 6 = 36$. As LGVCM pattern pair histograms and feature vectors are calculated as Formula (14.3).

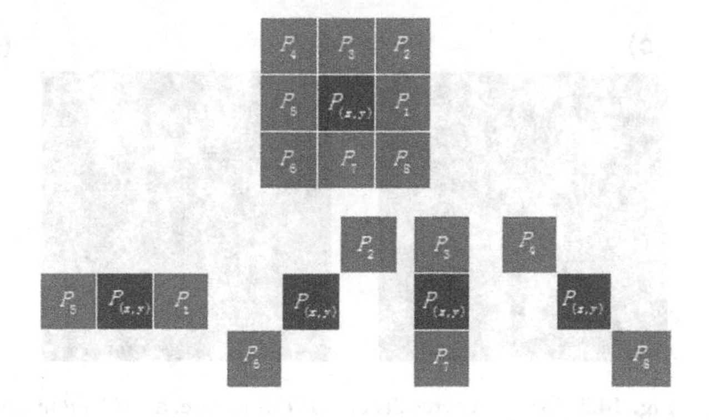

Fig. 14.1 Pixels distribution in 3 × 3 window

$$H^{\alpha}_{LGVCM}(m,n) = \sum_{i=2}^{N_y-1}\sum_{j=2}^{N_x-1} f_{LGVCM}\left(CM^{\alpha}_{(i,j)}, CM^{\alpha}_{(i+\Delta_i,j+\Delta_j)}, m, n\right) \quad (14.3)$$

$$f_{LGVCM}(x,y,m,n) = \begin{cases} 1, \mathit{if}\ x = m\ \&\ y = n \\ 0, \mathit{else} \end{cases} \quad (14.4)$$

$$F^{\alpha}_{LGVCM}(m,n) = \frac{H^{\alpha}_{LGVCM}(m,n)}{(N_y-2)\times(N_x-2)};\ m,n \in M \quad (14.5)$$

While the pixels at the edge of image have no neighbor pixels, we do not take these patterns into consideration. As there are 36 kinds of different pattern pairs, so the level of H is 36, while F is a normalized 36-dimensional vector with every value ranging from 0 to 1. Finally, we can get a 12×36 dimensional vector to describe the spatial distribution information of gray value variation direction.

To verify the effectiveness of this feature, we choose an image with four direction intersecting lines, as shown in Fig 14.2. First, we calculate the 12 LGVCMs in four directions of the original image ((Δ_x, Δ_y) chose as $(1,0)$), then the pixel's gray value is set as 255 in result image where its pattern pair is (1,1), otherwise 0. The feature extraction result is shown in Fig. 14.2, which proves that spatial distribution information can be well reflected using LGVCM.

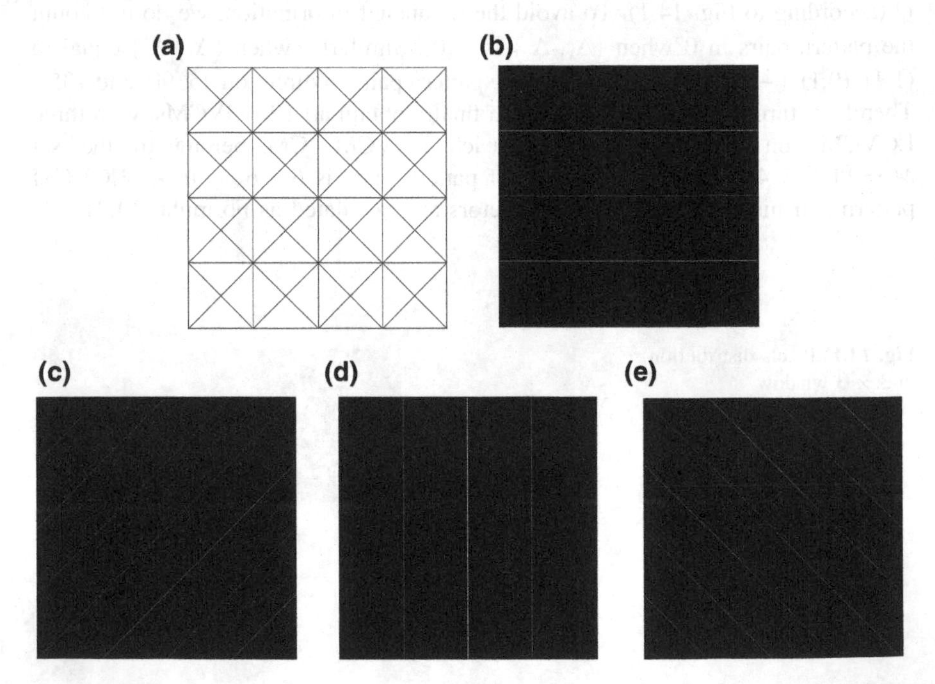

Fig. 14.2 Experiment results of *LGVCM* feature. **a** *Oridinal* image. **b** LVGCM in 0 °. **c** LVGCM in 45 °. **d** LVGCM in 90 °. **e** LVGCM in 135 °

14.2.2 Gray-Scale Variation on Different Patterns

Assuming the center pixel is P and the whole three pixels' values in direction α are (P_x, P, P_y). The variation Var_P^α is calculated by

$$Var_P^\alpha = |P_x - P| + |P - P_y| \tag{14.6}$$

For the pixels whose pattern value in direction $\alpha$is m, the mean ($\overline{V}_m^\alpha$) and standard deviation (σ_m^α) are calculated by

$$\overline{V}_m^\alpha = \sum_{i=1}^{K} Var_{P_i}^\alpha / K, P_i \in \Phi_{MP}, \sigma_m^\alpha = \sqrt{\sum_{i=1}^{K} \left(Var_{P_i}^\alpha - \overline{V}_m^\alpha\right)^2 / K, P_i \in \Phi_{MP}} \tag{14.7}$$

where Φ_{MP} is the pixels set whose pattern value is $m$in direction α, K is the number of Φ_{MP}. Through this step, we can calculate the mean and standard deviation of the six different patterns in 0, 45, 90, and 135° directions, and a $4 \times 6 \times 2 = 48$ dimensional feature vector was obtained to reflect the gray variation magnitude information.

After combining the LBP with LGVCM and GVDP, we can extract more texture detail information and give LBP more effective complementary information. In general, our new texture feature descriptor can fully extract spatial distribution information of variation direction and variation magnitude information between pixels. In other words, it can comprehensively express the texture characteristics of the image.

14.3 Image Retrieval Algorithm

To verify the effectiveness of the proposed texture feature descriptor, we apply the texture features to image retrieval. The framework of the proposed image retrieval algorithm is shown in Fig. 14.3. First, the query image and the images in database are converted into gray image. Then the LGVCM feature vectors of 0, 45, 90, and 135°, the LBP feature vectors and the GVDP feature vectors are extracted from gray images. Finally, we use feature similarity to measure the similarity between

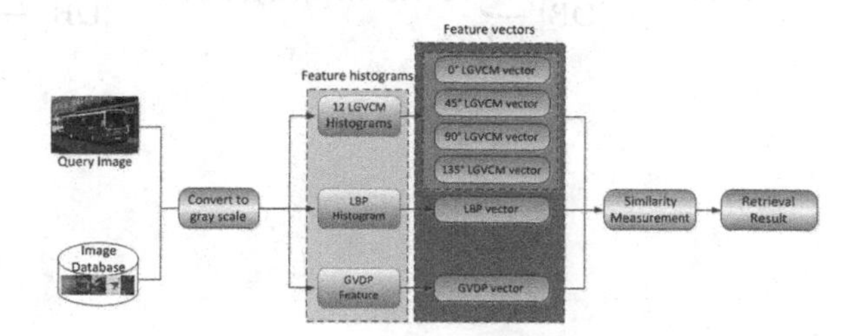

Fig. 14.3 Framework of the proposed image retrieval algorithm

the query image and the database images, and get retrieval result based on similarity calculation results.

The feature vectors for the query image Q is represented as $f_Q(f_{Q_1}, f_{Q_2}, \cdots, f_{Q_L})$, each image in the database is represented as $f_{DB}(f_{DB_1}, f_{DB_2}, \cdots, f_{DB_L})$. L is the dimension of feature vector. The similarity to match the images is computed using

$$d = \sum_{i=1}^{L} \left| \frac{f_{DB_i} - f_{Q_i}}{1 + f_{DB_i} + f_{Q_i}} \right| \tag{14.8}$$

where f_{Q_i} and f_{DB_i} is the ith feature of the query image and the database image.

For each query, we save the top N images with highest similarity as the retrieval result. Considering the retrieval result as $X = \{x_1, x_2, \ldots, x_n\}$, if the category of x_i is the same with query image, we define image x_i as a correct retrieval. The performance of the proposed method is measured in terms of average precision, average recall, all average precision (*AAP*), and all average recall (*AAR*). For the query image Q, the precision, recall, average precision, and average recall is defined as Formula (14.9), (14.10), and (14.11).

$$Precision(I_q, n) = \frac{1}{n} \sum_{i=1}^{|DB|} \left| \delta(\phi(I_i), \phi(I_q)) | Rank(I_i, I_q) \leq n \right|, \delta(x, y) = \begin{cases} 1, x = y \\ 0, else \end{cases} \tag{14.9}$$

$$\text{Recall}(I_q, n) = \frac{1}{N_G} \sum_{i=1}^{|DB|} \left| \delta(\phi(I_i), \phi(I_q)) | Rank(I_i, I_q) \leq n \right| \tag{14.10}$$

$$\begin{aligned} Avg\ Precision^k(n) &= \frac{1}{N_G} \sum_{i \in N_G} Precision(I_i, n), Avg\ \text{Recall}^k(n) \\ &= \frac{1}{N_G} \sum_{i \in N_G} \text{Recall}(I_i, n) \end{aligned} \tag{14.11}$$

where n indicates the number of retrieved images, and $|DB|$ is the size of the image database. $\phi(x)$ is the category of x, $Rank(I_i, I_q)$ return the rank of image I_i(for the query image I_q) among all images of $|DB|$. Finally, the AAP and ARR are defined by Formula (14.12).

$$\text{AAP(n)} = \frac{1}{|\text{DB}|} \sum_{\text{i}=1}^{|\text{DB}|} \text{Precision}(\text{I}_\text{i}, \text{n}),\ \text{AAR} = \frac{1}{|\text{DB}|} \sum_{\text{i}=1}^{|\text{DB}|} \text{Recall}(\text{I}_\text{i}, \text{n})| \tag{14.12}$$

14.4 Experiment and Discussions

14.4.1 Experiment on DB1

DB1 is consists of 111 different textures, which are chosen from Brodatz texture photographic album [8]. The size of each texture is 512 × 512. Each 512 × 512 image is divided into 16 128 × 128 nonoverlapping subimages, thus creating a database of 1,776 (16 × 111) images. In this experiment, each image in the database is considered as the query image, and some examples of DB1 are shown in Fig. 14.4.

Figure 14.5 illustrates the retrieval performance of the proposed method and other existing methods. It is evident that our method outperforms the other existing methods.

14.4.2 Experiment on DB2

In this experiment, images from the Corel database [9] have been used. This database consists of a large number of images of various contents ranging from animal images to outdoor sports and natural images. These images have been preclassified into different categories. Each category has 100 images. For our experiment, we have chosen 1,000 images to form DB2, the size of each image is

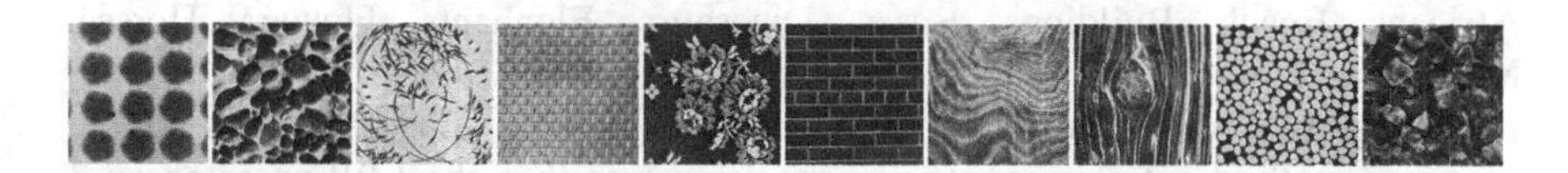

Fig. 14.4 Sample texture images in DB1

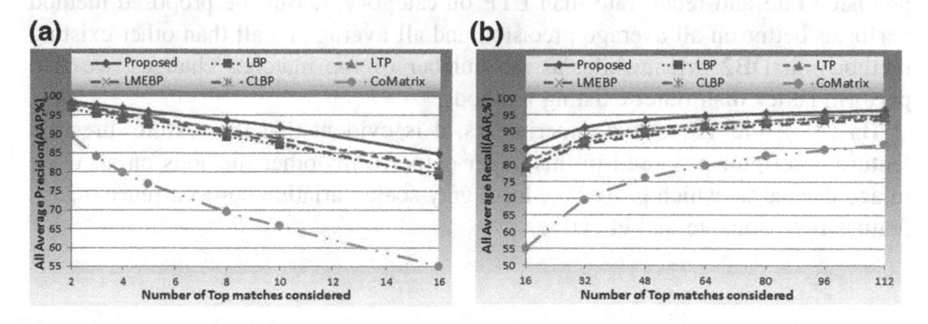

Fig. 14.5 Comparison of the proposed method with other existing methods in terms of AAP and AAR on Database DB1. **a** AAP results of existing methods on DB1. **b** AAR results of existing methods on DB1

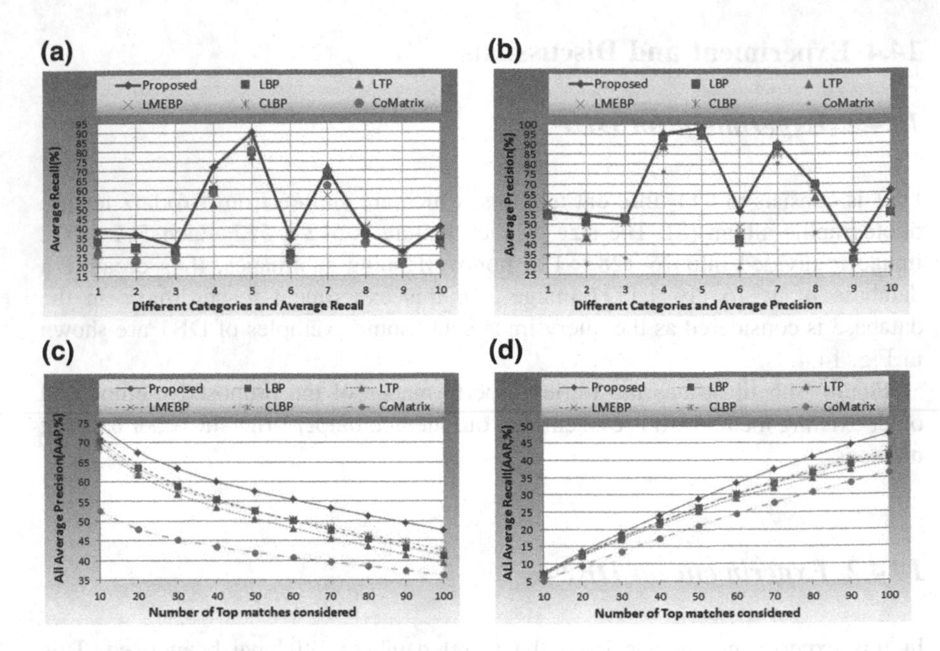

Fig. 14.6 Retrieval performance comparison of the proposed method with other existing methods on DB2. **a** AP of existing methods when $N = 20$ on DB2. **b** AR of existing methods when $N = 100$ on DB2. **c** AAP results of existing methods on DB2. **d** AAR results of existing methods on DB2

256 × 384 or 384 × 256. These images are collected from ten different domains: Africans, Beach, Buildings, Buses, Dinosaurs, Elephants, Flowers, Horses, Mountains, Food.

As Fig. 14.6 shows, while the number of top matches considered N is 20, the proposed method gets lower retrieval precision rate than the LBP on category 3 and 8, lower retrieval precision rate than the LMEBP on category 1 and 3, lower retrieval recall rate than the LMEBP and CLBP on category 8, lower retrieval precision rate and recall rate than LTP on category 7. But the proposed method performs better on all average precision and all average recall than other existing methods on DB2. Meanwhile, as the number of top matches changes, we also perform better than other existing methods.

Based on the above two experiments, it is evident that the texture direction feature descriptor proposed in this paper outperforms other methods on all these image databases, which proves the local gray-scale variation co-occurrence matrix feature is reasonable and effective.

14.5 Conclusions and Future Work

In this paper, we have presented a new texture direction feature descriptor referred to as local gray-scale variation co-occurrence matrix (LGVCM). LGVCM can reflect the spatial distribution information of gray value variation direction between center pixels with its surrounding neighbors. However, local gray-scale co-occurrence matrix does not have rotation invariance, which leads to the poor retrieval result of the texture images after rotation. In the future, we will focus our research on how to construct the local gray-scale co-occurrence matrix which has the character of rotation invariance.

Acknowledgments This work is supported by the National Natural Science Foundation of China (Grant No. 61272317) and the General Program of Natural Science Foundation of AnHui of China (Grant No. 1208085MF90).

References

1. Ojala T, Pietikäinen M, Harwood D (1996) A comparative study of texture measures with classification based on featured distributions. Pattern Recogn 29(1):51–59
2. Liao S, Law MWK, Chung ACS (2009) Dominant local binary patterns for texture classification. IEEE Trans Image Process 18(5):1107–1118
3. Guo Z, Zhang L, Zhang D (2010) A completed modeling of local binary pattern operator for texture classification. IEEE Trans Image Process 19(6):1657–1663
4. Ojala T, Pietikainen M, Maenpaa T (2002) Multiresolution gray-scale and rotation invariant texture classification with local binary patterns. IEEE Trans Pattern Anal Mach Intell 24(7):971–987
5. Guo Z, Zhang L, Zhang D, Zhang S (2010) Rotation invariant texture classification using adaptive LBP with directional statistical features. The 17th IEEE international conference on image processing (ICIP), IEEE
6. Tan X, Triggs B (2010) Enhanced local texture feature sets for face recognition under difficult lighting conditions. IEEE Trans Image Process 19(6):1635–1650, June
7. Subrahmanyam M, Maheshwari RP, Balasubramanian R (2012) Local maximum edge binary patterns: a new descriptor for image retrieval and object tracking. Signal Process 92(6):1467–1479
8. Brodatz P (1996) Textures: a photographic album for artists and designers. Dover, New York
9. Corel 1000 and Corel 10000 image database. Available via http://wang.ist.psu.edu/docs/related.shtml

Chapter 15
An Efficient Fast CU Depth and PU Mode Decision Algorithm for HEVC

Zong-Yi Chen, He-Yan Chen and Pao-Chi Chang

Abstract High Efficiency Video Coding (HEVC) is a new video coding standard, which improves the coding efficiency significantly. To achieve the best performance, HEVC encoder evaluates all possible candidates to determine the best depth of coding unit (CU) and mode of prediction unit (PU). This increases substantial computational complexity that might become an obstacle for practical applications. This paper proposes a fast algorithm for CU and PU to reduce the encoding time of HEVC. By referring spatial and temporal depth information of CU and motion/texture characteristics of PU, the proposed algorithm skips rarely used depths and modes in certain situations. The experimental results show that our proposed method averagely achieves 57 % time saving in high efficiency configuration and 61 % in low complexity configuration with negligible rate-distortion loss compared with the reference software.

Keywords HEVC · Mode decision · Fast algorithm

15.1 Introduction

With increasing popularity of high resolution video format, video coding technologies which can provide a substantially higher compression capability than the existing H.264/AVC [1] standard have received increased attention. In 2010, the Joint Collaborative Team on Video Coding (JCT-VC) jointed by ITU-T VCEG and ISO/IEC MPEG started to develop the next generation video coding standard called High Efficiency Video Coding (HEVC) [2]. In HEVC, the basic encoding unit which plays a similar role as a macroblock in H.264/AVC is separated into

Z.-Y. Chen · H.-Y. Chen · P.-C. Chang (✉)
Department of Communication Engineering, National Central University, Jhongli, Taiwan
e-mail: pcchang@ce.ncu.edu.tw

A. A. Farag et al. (eds.), *Proceedings of the 3rd International Conference on Multimedia Technology (ICMT 2013)*, Lecture Notes in Electrical Engineering 278,
DOI: 10.1007/978-3-642-41407-7_15, © Springer-Verlag Berlin Heidelberg 2014

coding unit (CU), prediction unit (PU), and transform unit (TU), as shown in Fig. 15.1. To achieve efficient and flexible compression of video content with various resolutions, the size of CU can be ranged from 64 × 64 to 8 × 8 and a quadtree-based block partition is adopted. Figure 15.2 shows the quadtree-structured CU and the size of CU is corresponding to its depth. Each CU contains one to four PUs and the size of PU is limited to that of CU. PU is the basic unit for prediction, and all information related to prediction, e.g., motion vector, is signaled on a PU basis. The partition type of a PU depends on the prediction mode as shown in Fig. 15.3.

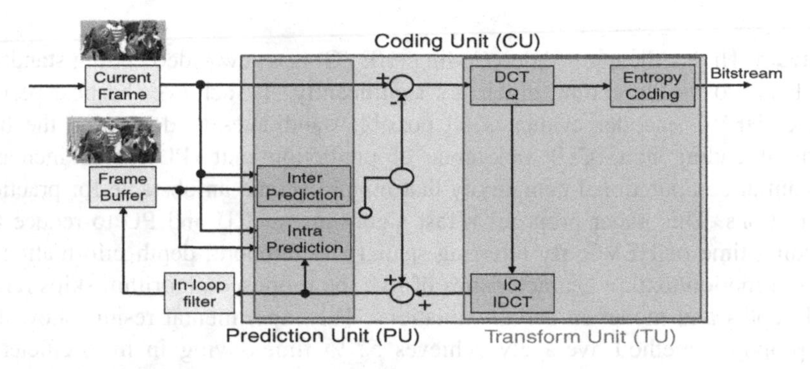

Fig. 15.1 Basic unit in HEVC

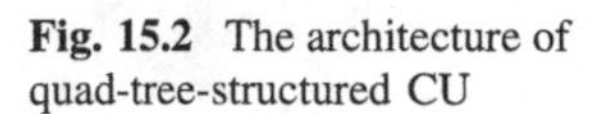

Fig. 15.2 The architecture of quad-tree-structured CU

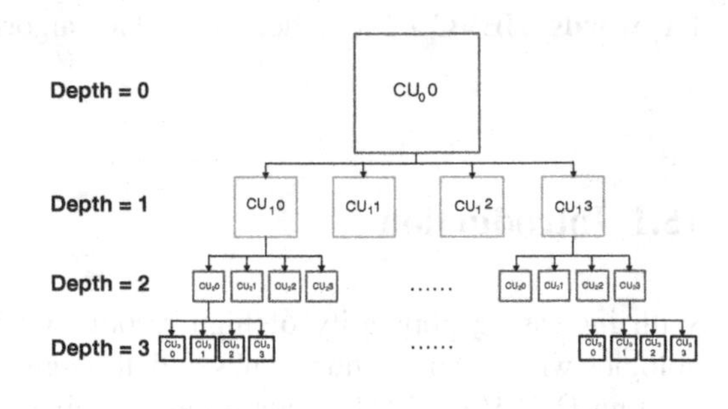

Fig. 15.3 Partition modes of a PU in HEVC

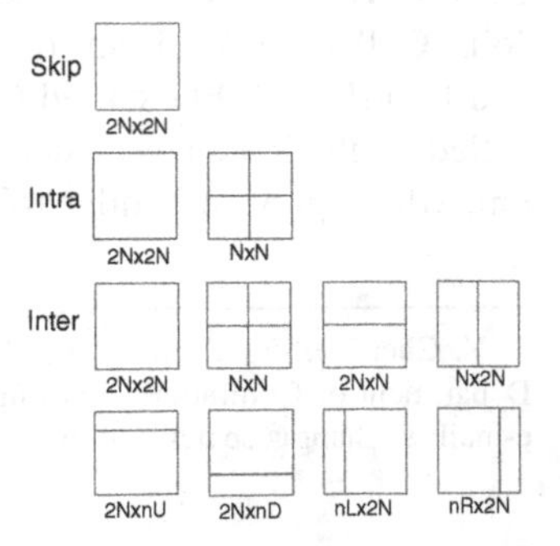

Generally, large size of CU is tend to be chosen as the best one in homogeneous regions to reduce the side information. On the other hand, small size of CU is selected to preserve the texture details in nonhomogeneous regions.

In the reference software HM5.0, the encoder has to test all CU depths recursively to find the best one, and all modes of PU must be further searched in each CU. This procedure leads to significant complexity increase in the encoder. Thus, it is critical to reduce the encoding complexity to make the real-time applications practicable. From our analyses, all encoding processes except in-loop filtering are computing on a CU basis which occupies about 95 % total encoding time in the encoder, and PU including motion estimation, motion compensation also occupies 57 % encoding time in CU. Therefore, we can efficiently reduce the complexity of encoder by decreasing the candidate depths of CU and modes of PU.

Several early termination methods [3, 4] have been proposed during HEVC standardization process. In [3], if the best prediction mode in current CU depth is Skip mode, the computation of subdepths will be terminated. In [4], if coded block flag (cbf) values of luma and two chromas are zero after encoding current PU mode, the remaining modes will be skipped. A fast CU decision algorithm which contains frame level and CU level is proposed in [5]. And some other fast encoding algorithms [6, 7] are proposed continuously. But the time savings of most of the previous works can be achieved only about 40 %. In this work, we provide an efficient method to skip the rarely used CU depths and PU modes under some conditions to reduce the encoding time by 60 % approximately.

The rest of this paper is organized as follows. Section 15.2 presents the proposed fast CU depth decision. Section 15.3 describes the fast PU mode decision algorithm. Experimental results are shown in Sect. 15.4 and the conclusion is remarked in Sect. 15.5.

15.2 Fast CU Depth Decision Algorithm

In the proposed CU depth decision algorithm, the temporal correlation of CUs is first considered. Two continuous frames in a video sequence are usually highly correlated. Therefore, the depth in previous frame can be utilized to predict the depth of current frame. Table 15.1 shows the average depth distribution between

Table 15.1 Depth distribution between current and co-located CUs

Co-located CU depth	Current CU depth				
	0 (%)	1 (%)	2 (%)	3 (%)	Total (%)
0	66	23	9	2	100
1	33	41	19	7	100
2	22	28	33	17	100
3	9	18	27	46	100

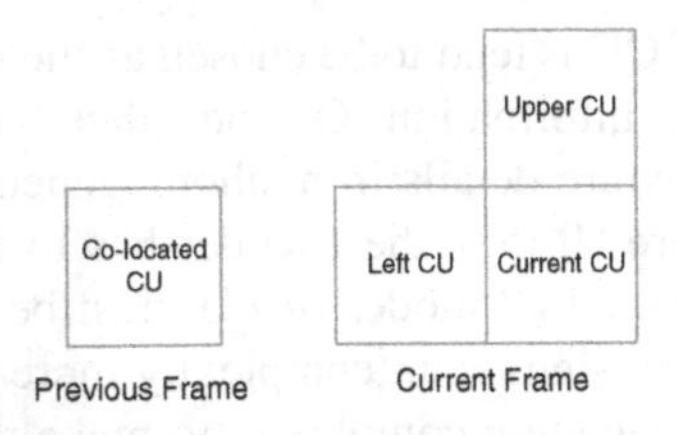

Fig. 15.4 CUs referred in the proposed algorithm

current and co-located CUs by running four sequences in different classes (B, C, D, and E) and five QPs (22, 27, 32, 37, and 42). The reference software that we adopted in our analyses is HM5.0.

We can observe that when co-located CU depth equals to zero or one, small CU size is rarely used in current CU. Similarly, large size is seldom used in current CU if co-located CU depth is three. This motivates us to skip the current CU depths when the conditional probability in Table 15.1 is below 10 % (the cases colored red with boldface). For example, when the co-located CU depth is 1, the search of depth = 3 will be skipped.

Not only object motion but details in video sequences will affect the depth determination of current CU. Thus, the spatial correlation of CUs is also taken into account for our algorithm. When an object is moving into a homogeneous area, the edge of object tends to be encoded with small CU size. Therefore, we refer to upper and left CU depths to determine whether a moving object is in the neighborhood of current CU. We propose to do full search on the current CU if any neighboring depth of CU is three. Figure 15.4 shows the CUs referred in our algorithm. And Fig. 15.5 is the flowchart of the proposed fast CU depth decision algorithm.

15.3 Fast PU Mode Decision Algorithm

In H.264, the encoder exists high probability to choose Skip mode as the best mode when the current coding mode is 16 × 16 and the following conditions are all satisfied [8]: (a) reference frame is previous one, (b) motion vector difference (MVD) is zero, and (c) transform coefficients are all quantized to zero. Similarly, we analyze CU depth and PU mode distribution under these three conditions in HEVC. The search of Inter 2N × 2N mode in each depth is performed first. We observe that more than 92 % candidate PU modes will be selected as Skip mode when previous three conditions are met as shown in Table 15.2. Moreover, there is 99 % probability that the current CU depth will be the best one if PU satisfies these conditions as Table 15.3 shows. Hence we propose to select the best CU depth as current depth and the best PU mode as Skip mode once the conditions (a), (b), and (c) are all satisfied.

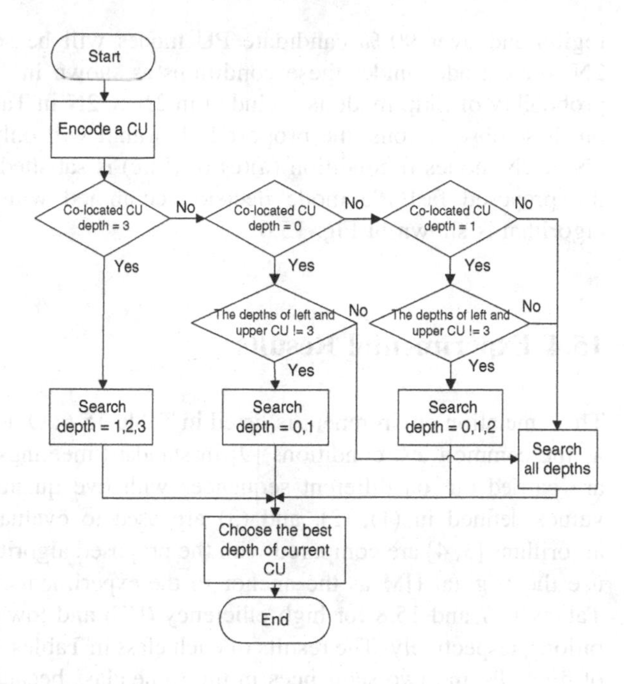

Fig. 15.5 Flowchart of the proposed fast CU depth decision algorithm

Table 15.2 PU mode distribution when (a), (b), and (c) are met

	Skip mode (%)	Inter mode (%)
ClassB_Cactus	93	7
ClassC_BQMall	91	9
ClassD_BQSquare	89	11
ClassE_vidyo3	96	4
Average	92	8

Table 15.3 CU depth distribution when (a), (b), and (c) are met

	Depth = k (%)	Depth = k + 1 (%)
ClassB_Cactus	99	1
ClassC_BQMall	99	1
ClassD_BQSquare	98	2
ClassE_vidyo3	100	0
Average	99	1

The encoder tends to choose Skip and Inter 2N × 2N modes in a stationary region. We utilize (a), (b), and (c) to represent the degree of motion and texture in a block. When (a) is satisfied, the current PU tends to be simple motion. And the motion characteristic of current PU is similar to neighboring PU if (b) is satisfied. Both conditions (a) and (b) combine with (c) can be used to represent a stationary region and over 90 % candidate PU modes will be selected as Skip and Inter 2N × 2N modes under these conditions as shown in Tables 15.4 and 15.5. The probability of Skip mode is included in 2N × 2N in Tables 15.4 and 15.5. Based on these observations, the proposed algorithm will only examine Skip and Inter 2N × 2N modes if condition (a)(c) or (b)(c) is satisfied. Finally, the flowchart of the proposed fast PU mode decision combined with fast CU depth decision algorithm is shown in Fig. 15.6.

15.4 Experimental Results

The simulation environment is listed in Table 15.6. Other settings are set the same as the common test conditions [9] in standard meetings. A group of experiments are carried out on different sequences with five quantization parameters. Three values defined in (1), (2), and (3) are used to evaluate the performance. Two algorithms [3, 4] are compared with the proposed algorithm and all fast algorithms use the original HM as the anchor in the experiments. The results are shown in Tables 15.7 and 15.8 for high efficiency (HE) and low complexity (LC) configurations, respectively. The results of each class in Tables 15.7 and 15.8 are averages of five QPs and two sequences in the same class because of limitations of space.

$$\Delta\text{PSNR (dB)} = \text{PSNR}_{\text{proposed}} - \text{PSNR}_{\text{HM}} \tag{15.1}$$

$$\Delta\text{Bitrate (\%)} = \frac{\text{Bitrate}_{\text{proposed}} - \text{Bitrate}_{\text{HM}}}{\text{Bitrate}_{\text{HM}}} \times 100 \tag{15.2}$$

$$\Delta\text{EncTime (\%)} = \frac{\text{EncTime}_{\text{proposed}} - \text{EncTime}_{\text{HM}}}{\text{EncTime}_{\text{HM}}} \times 100 \tag{15.3}$$

From Tables 15.7 and 15.8, we can observe that the proposed fast CU depth decision algorithm achieves about 40 % time saving on average while maintaining very good RD performance. We can also observe that the proposed algorithm performs worse for videos with smaller resolutions (classC and classD). For a smaller resolution video, there exists more textures or objects in a fixed-size CU block than that in a larger resolution video, and large CUs are used less frequently in these classes. This is the reason why the time reduction of class C and class D are fewer than that of others. When comparing with ECU [3], the proposed fast CU algorithm can provide almost the same PSNR and time saving performance with just a little less rate reduction.

For the proposed fast PU mode decision algorithm, about 45 % time saving on average and little rate decrease with PSNR degradation is achieved. For videos with slow motion and texture such as class E, the proposed algorithm saves more encoding time due to more PUs satisfy our proposed conditions. And the proposed fast PU algorithm performs significantly better than CFM [4] in rate reduction and time saving with almost the same video quality.

Table 15.4 Inter mode distribution under (a) and (c)

	2N × 2N (%)	2N × N (%)	N × 2N (%)	N × N (%)	2N × nU (%)	2N × nD (%)	nL × 2N (%)	nR × 2N (%)
ClassB_Cactus	93	3	4	0	0	0	0	0
ClassC_BQMall	94	2	4	0	0	0	0	0
ClassD_BQSquare	89	4	7	0	0	0	0	0
ClassE_vidyo3	94	2	4	0	0	0	0	0
Average	92	3	5	0	0	0	0	0

Table 15.5 Intermode distribution under (b) and (c)

	2N × 2N (%)	2N × N (%)	N × 2N (%)	N × N (%)	2N × nU (%)	2N × nD (%)	nL × 2N (%)	nR × 2N (%)
ClassB_Cactus	97	1	2	0	0	0	0	0
ClassC_BQMall	96	1	3	0	0	0	0	0
ClassD_BQSquare	92	3	5	0	0	0	0	0
ClassE_vidyo3	96	1	3	0	0	0	0	0
Average	95	2	3	0	0	0	0	0

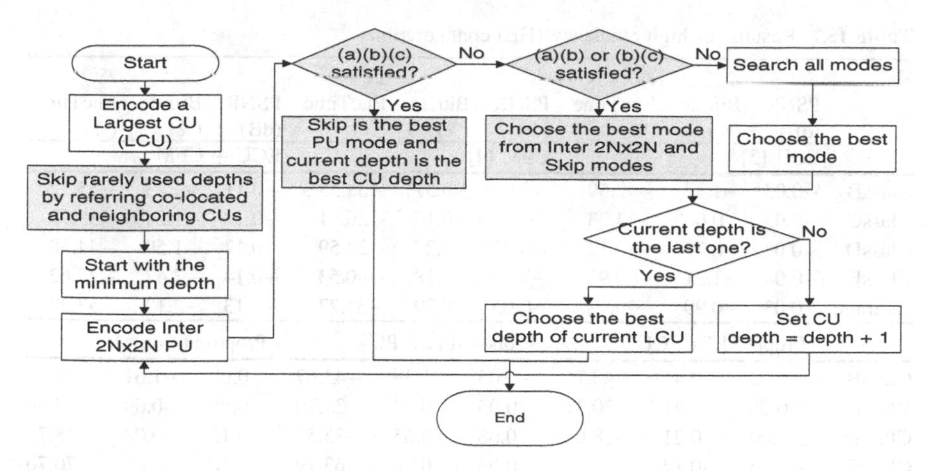

Fig. 15.6 Flowchart of the proposed overall algorithm

Table 15.6 Simulation environment

HM5.0		
Sequence	ClassB (1920 × 1080)	BasketballDrive, BQTerrace
	ClassC (832 × 480)	BasketballDrill, RaceHorses
	ClassD (416 × 240)	BlowingBubbles, BasketballPass
	ClassE (1280 × 720)	Vidyo1, Vidyo4
Configuration file	Low delay P (IPPP...)−High Efficiency (HE)	
	Low delay P (IPPP...)−Low Complexity (LC)	
MotionSearch	EPZS	
SearchRange	64	
FramesToBeEnclosed	100	
QP	22, 27, 32, 37, 42	
GOPSize	4	

Finally, the overall performance of the proposed algorithm is compared with the combination of ECU [3] and CFM [4]. The proposed algorithm achieves better PSNR and time saving performance than the reference algorithm (ECU + CFM) but with less rate reduction. We can observe that the proposed algorithm saves much more encoding time than the reference algorithm for ClassC (9.59 % in LC case). The two sequences in ClassC exist complex textures and objects such that CU depth is less early terminated at large CU stage in ECU and zero blocks occurs less in CFM. Compared with the original HM encoder, the average PSNR degradation of the proposed algorithm is no more than 0.12 dB and the rate reduction is up to 1.08 % on average. The encoding time can be saved up to 75 %. On average, 57 and 61 % encoding time reduction are achieved respectively for HE

Table 15.7 Results for high efficiency (HE) configuration

HE	△ PSNR (dB)	△ Bitrate (%)	△ EncTime (%)	△ PSNR (dB)	△ Bitrate (%)	△ EncTime (%)	△ PSNR (dB)	△ Bitrate (%)	△ EncTime (%)
	ECU [3]			CFM [4]			ECU + CFM		
ClassB	−0.03	−1.37	−42.59	−0.04	−0.57	−33.53	−0.11	−3.23	−55.72
ClassC	−0.03	−0.66	−24.23	−0.04	−0.16	−26.41	−0.12	−1.24	−39.95
ClassD	−0.04	−0.71	−30.12	−0.07	−0.27	−24.59	−0.17	−1.59	−44.58
ClassE	−0.04	−1.21	−60.97	−0.03	−0.18	−40.53	−0.14	−2.42	−72.62
Average	−0.03	−0.99	−39.48	−0.05	−0.29	−31.27	−0.13	−2.12	−53.22
	Proposed Fast CU			Proposed Fast PU			Proposed Overall		
ClassB	−0.03	−0.41	−43.71	−0.05	−1.39	−43.67	−0.08	−1.61	−59.35
ClassC	−0.03	0.31	−30.31	−0.05	−0.44	−31.70	−0.08	−0.09	−47.36
ClassD	−0.04	0.21	−28.67	−0.08	−0.65	−35.52	−0.12	−0.08	−48.71
ClassE	−0.03	−0.17	−52.02	−0.06	−0.99	−63.79	−0.10	−0.61	−70.70
Average	−0.03	−0.02	−38.68	−0.06	−0.87	−43.67	−0.10	−0.60	-56.53

Table 15.8 Results for low complexity (LC) configuration

LC	PSNR (dB)	△ Bitrate (%)	△ EncTime (%)	△ PSNR (dB)	△ Bitrate (%)	△ EncTime (%)	△ PSNR (dB)	△ Bitrate (%)	△ EncTime (%)
	ECU [3]			CFM [4]			ECU + CFM		
ClassB	−0.04	−1.89	−44.13	−0.04	−0.69	−33.75	−0.13	−3.74	−57.83
ClassC	−0.04	−0.81	−31.02	−0.05	−0.40	−31.11	−0.15	−1.90	−44.80
ClassD	−0.06	−0.88	−31.58	−0.07	−0.61	−30.44	−0.21	−2.16	−49.03
ClassE	−0.05	−1.70	−62.83	−0.03	−0.08	−42.68	−0.14	−2.88	−75.07
Average	−0.05	−1.32	−42.39	−0.05	−0.44	−34.49	−0.16	−2.67	−56.68
	Proposed Fast CU			Proposed Fast PU			Proposed Overall		
ClassB	−0.06	−0.97	−42.80	−0.05	−1.48	−44.13	−0.11	−2.42	−60.23
ClassC	−0.05	0.24	−35.95	−0.06	−0.64	−39.26	−0.10	−0.27	−54.39
ClassD	−0.07	0.23	−30.82	−0.09	−0.85	−38.48	−0.16	−0.44	−53.33
ClassE	−0.07	−0.31	−52.17	−0.06	−1.32	−67.27	−0.13	−1.18	−75.39
Average	−0.06	−0.20	−40.44	−0.07	−1.08	−47.29	−0.12	−1.08	−60.84

and LC cases. Although the rate-distortion performance of the proposed overall algorithm is slightly worse than the reference algorithm, higher encoding time can be saved.

15.5 Conclusion

In this paper, we propose a fast algorithm for HEVC to skip rarely used depths of CU and modes of PU in certain situations. In CU, the depth information of co-located block is employed to reduce the depth candidates and the upper and left blocks are referred to determine whether to search all depths or not. In PU, the combination of motion and texture characteristics including reference previous one, zero block, and zero MVD are utilized. On average, approximately 60 % reduction in encoding time compared to HM5.0 encoder can be yielded with only negligible rate-distortion losses.

References

1. H.264: International Telecommunication Union (2003) Recommendation ITU-T H.264: advanced video coding for generic audiovisual services. ITU-T
2. JCT-VC (2012) High efficiency video coding (HEVC) test model 8 encoder description. JCTVC-J1002
3. JCT-VC (2011) Coding tree pruning based CU early termination. JCTVC-F092. 6th JCT-VC meeting, Torino
4. JCT-VC (2011) Early termination of CU encoding to reduce HEVC complexity. JCTVC-F045. 6th JCT-VC meeting, Torino
5. Leng J, Sun L, Ikenaga T, Sakaida S (2011) Content based hierarchical fast coding unit decision algorithm for HEVC. IEEE CMSP 1:56–59
6. Tan LH, Liu F, Tan YH, Yeo C (2012) On fast coding tree block and mode decision for high-efficiency video coding (HEVC). IEEE ICASSP 825–828
7. Shen L, Liu Z, Zhang X, Zhao W, Zhang Z (2013) An effective CU size decision method for HEVC encoders. IEEE Trans Multimedia 15(2):465–470
8. Lee J, Jeon B (2004) Fast mode decision for H.264. IEEE ICME 2:1131–1134
9. JCT-VC (2011) Common HM test conditions and software reference configurations. JCTVC-G1200. 7th JCTVC meeting. Geneva, CH

Chapter 16
Digital Audio Watermarking Technique Exploiting the Properties of the Psychoacoustic Model 2 of the MPEG Standard

Maha Bellaaj and Kais Ouni

Abstract In this paper, we propose a watermarking technique for digital audio data operates in the frequency domain. Time–frequency mapping is often done using the Modified Discrete Cosine Transform (MDCT) [1]. It is based on the design of the psychoacoustic model 2 (MPH2) of the MPEG standard [2] layer 3 but specific to audio watermarking. To ensure more inaudibility, the insertion of the mark bits will be in the least significant bit (LSB). In this technique, we duplicate the bits of the mark in order to have a maximum capacity of insertion and robustness of growing against different types of attacks. In order to increase the detection rates, we used Hamming [3] code as error correction code. We studied the robustness of this technique against compression/decompression MP3 attack and we evaluated the inaudibility by calculating the Signal-to-Noise Ratio (SNR) and the objective difference grade (ODG) notes given by PEAQ. To highlight our results, we compared the proposed technique with three other existing techniques.

Keywords Watermarking · Digital audio · MDCT · Psychoacoustic model 2 (MPH2) · Inaudibility · Robustness · SNR · ODG · PEAQ

16.1 Introduction

Through the development of Internet and the emergence of new communications media digital takes a place increasingly important, which poses serious problems since it is easy to copy and deal with these computer documents. As a result,

M. Bellaaj · K. Ouni (✉)
U. R. Signals and Mechatronic Systems, Higher School of Technology and Computer Science, Carthage University, Tunis, Tunisia
e-mail: kais.ouni@esti.rnu.tn

M. Bellaaj
e-mail: maha_bellaaj@yahoo.fr

A. A. Farag et al. (eds.), *Proceedings of the 3rd International Conference on Multimedia Technology (ICMT 2013)*, Lecture Notes in Electrical Engineering 278,
DOI: 10.1007/978-3-642-41407-7_16,

the copyrights become increasingly unprotected and we also suffer from illegal redistribution of data. As effective solution to these problems comes the digital watermarking [4], whose basic idea is to insert into the digital document (image, sound, video…) a signature in a way robust and imperceptible. Since the 1990s, the articles continue to multiply in order to find a watermarking technique which satisfies the following characteristics: robustness, large insertion capacity, and imperceptibility of the mark [5].

In this paper we propose a watermarking technique for digital audio based on a spectral approach of insertion of the mark combined with modeling of psychoacoustic phenomena to improve the robustness of the technique.

This paper is organized as follows: In Sect. 16.2, we detail the process of insertion and detection for the proposed technique. Section 16.3 presents the experimental results.

In Sect. 16.4, we compare the results obtained by the proposed technique with three other existing techniques. In the last section, we give a conclusion and perspective to this work.

16.2 Presentation of the Proposed Watermarking Technique

A detailed bibliographic study on digital watermarking [6–8] showed that the frequency domain space is a good point of view robustness and inaudibility hence the idea of using Modified Discrete Cosine Transform (MDCT) to move from time domain to the frequency domain [9, 10]. In addition, MDCT allows a finer frequency resolution.

16.2.1 Insertion Schema of the Mark

In the first step, the original audio signal (.wav) will be divided into block of 1024 samples. Thereafter, we will apply the MDCT to move to the frequency domain. This transformation will break the frame into low frequency (LF) and high frequency (HF). To separate these two frequency bands, we will use a frequency separation module. At the end of this step, we get all the low frequencies where we will insert bits of the brand. The choice of the LF band is due to the fact that the latter is much less sensitive against the attacks than the HF band (especially against MP3 compression). In parallel and to search for the places of insertion them less audible to the human ear, we will apply the psychoacoustic model 2 (MPH2) of the MPEG standard on the temporal samples of each sub-block of 1024 samples. Insertion places are located under the final threshold of energy hearing generated by this model for each block. This approach provides a good compromise between robustness and inaudibility.

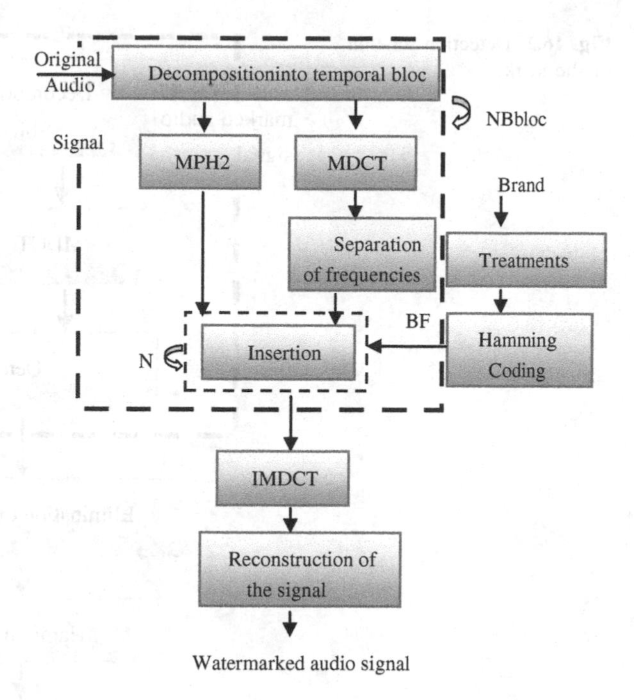

Fig. 16.1 Insertion schema of the mark

After the application of several treatments (binarization of the brand, decomposition into portions of 8 bits each) and Hamming coding (12, 8) to ensure the correction bits if necessary, since the bits of the signature can undergo changes during the insertion and detection, each bit is duplicated N times where N is calculated based on number of components that are below the final threshold of energy hearing and the size of the brand. Next, we will make a substitutive insertion of each bit of the mark in the least significant bit (LSB) of the components searched by the MPH2. All the previous steps will be repeated NB block times (number of blocks in the audio signal) and the insertion is done on all the blocks of the audio signal. Thereafter, we apply the IMDCT on the frequency-watermarked blocks of 1024 samples to obtain watermarked blocks in the time domain. The last step is to reconstruct the watermarked audio signal.

Figure 16.1 will give the general scheme and the different steps necessary for the insertion of the brand.

16.2.2 Detection Schema of the Mark

According to the Fig. 16.2, we note that the detection scheme of the brand is the inverse of the insertion. It is a blind detection that does not require the original audio signal or the presence of the mark originally inserted. Only the secret key

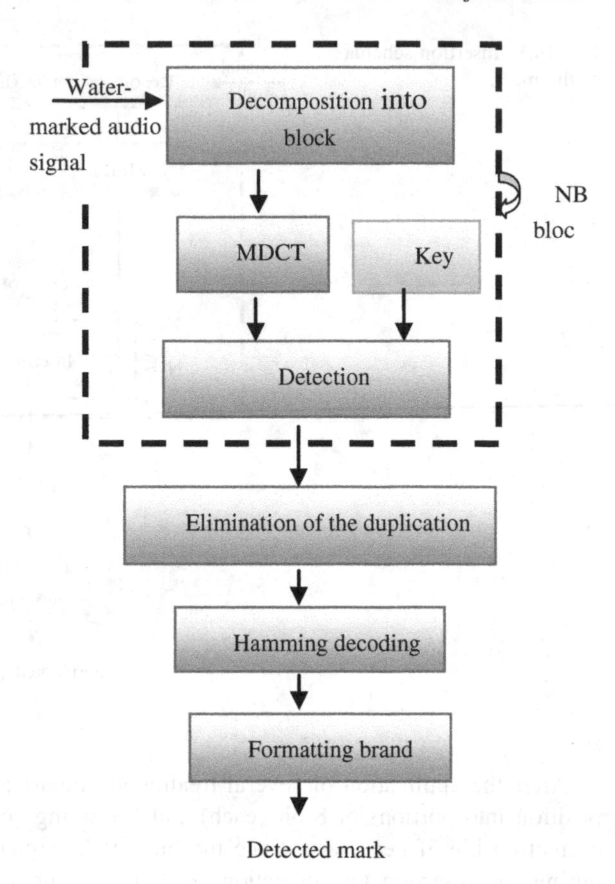

Fig. 16.2 Detection schema of the mark

(all the positions of the less sensitive components sought by the MPH2 in the insertion phase and the number of duplication N) is required. The output of the detection process is the final mark decoded and formatting.

16.3 Test Result

This section will present the different experimental results obtained by this technique. These results were focused on an experimental corpus composed of 12 audio signals. These signals are sampled at CD quality (at a sampling frequency Fe = 44.1 kHz), duration 20 s on average and different style: symphony orchestras, spoken voices (male and female), jazz, rock, singing voice…

16.3.1 Inaudibility

16.3.1.1 Spectrogram

For testing the watermarking system presented above, we inserted the text mark "audiowatermarking" of length 136 bits and after the hamming coding its length reaches 204 bits (after that each bit will be duplicated N times). From the tests, we were able to detect correctly and without error the mark which is identical to the original brand.

The Figs. 16.3 and 16.4 shows the spectrograms of the original audio signal and the watermarked audio signal. We will use an extract to the comparison:

Jazz.wav: extract of jazz

- Interpretation:

If we compare the spectrogram of the watermarked signal with the spectrogram of to the originals signals (by comparing Fig. 16.3 with Fig. 16.4), we notice that they are very similar.

Also, while listening to the original signal and the watermarked signal we do not perceive a difference. Despite the large number of bits already inserted, we do not perceive the existence of the signature in the watermarked signal which remains faithful to the original signal.

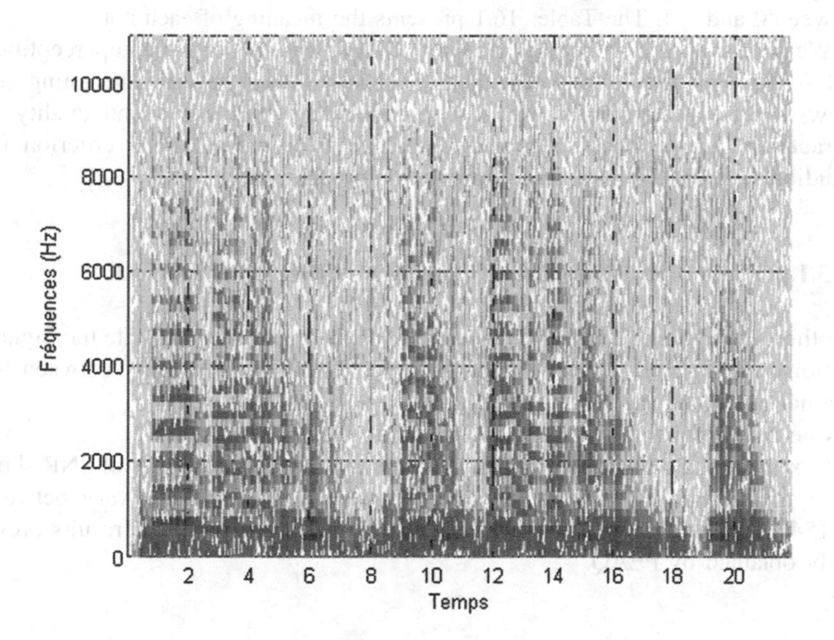

Fig. 16.3 Spectrogram of the original audio (jazz.wav)

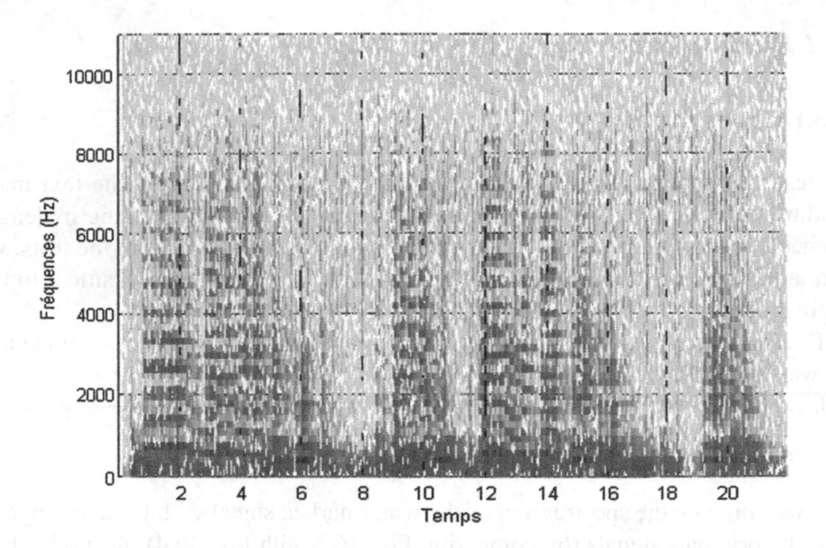

Fig. 16.4 Spectrogram of the watermarked audio (jazz_tatoue.wav)

16.3.1.2 Evaluation of the Sound Quality by PEAQ

The PEAQ algorithm [11] allows for an objective evaluation of sound quality. It generates as output a note of objective difference grade (ODG). This algorithm compares the original signal and the watermarked signal and assigns a score between 0 and −4. The Table 16.1 presents the meaning of each note.

We note from Fig. 16.5 that the notes of ODG vary between 0 (Imperceptible) and −0.35 (Perceptible but not annoying). These values are very interesting and show that our watermarking system degrades very little the sound quality of extracts and proves that the proposed technique provides a good criterion for inaudibility of the brand during the insertion process.

16.3.1.3 Evaluation of the Sound Quality by Calculating the SNR

Another way to demonstrate the inaudibility of the mark is to calculate the Signal-to-Noise Ratio (SNR). It is a measure that calculates the similarity between the original audio and the watermarked audio.

The results for this technique are shown in the Fig. 16.6.

From the results displayed in Fig. 16.6 we can see that the values of SNR show more the inaudibility provided by our technique. These values vary between 74.1546 and 82.7722 db, they are very interesting and confirm the results previously obtained by PEAQ.

Table 16.1 Signification notes of ODG

Signification	ODG
Imperceptible	0.0
Perceptible but not annoying	−0.1 to −1
Slightly annoying	−1.1 to −2
Annoying	−2.1 to −3
Very annoying	−3.1 to −4

The results of this evaluation are shown in Fig. 16.5

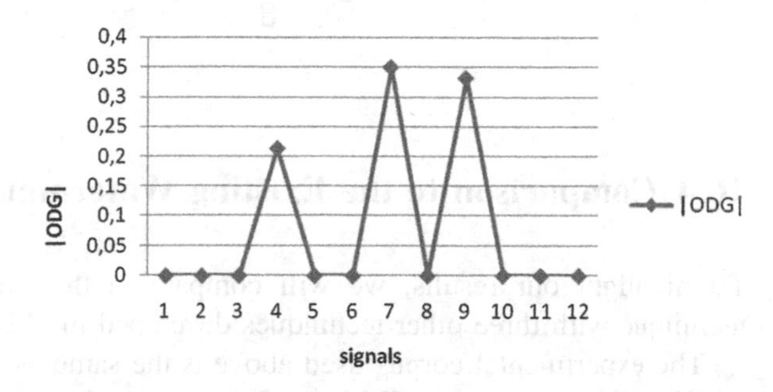

Fig. 16.5 Graphical representation of the absolute values of ODG notes

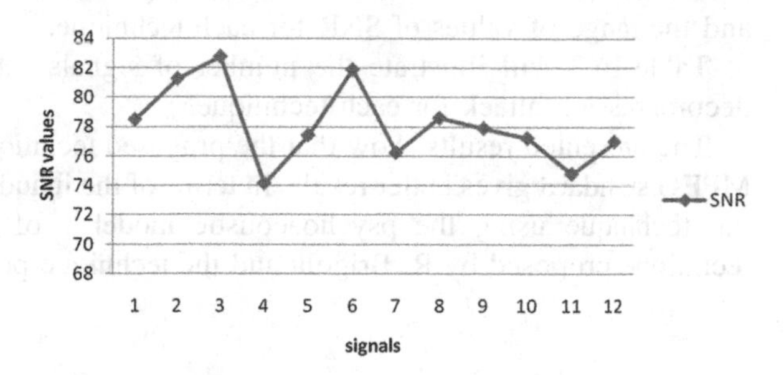

Fig. 16.6 Graphical representation of the SNR values

16.3.2 Robustness

16.3.2.1 Robustness Against Compression/Decompression MP3

The compression/decompression MP3 is performed by "lame.exe" at three different rates: 128, 96, and 64 Kbit/s. Test results are displayed in the Fig. 16.7.

From the results displayed in Fig. 16.7, we note that the technique is always robust against the attack of compression/decompression MP3 for the two compression rate 128 and 96 Kbit/s. The strength decreases for a rate of 64 Kbit/s but still very interesting (9 records/12 records are robust against attack).

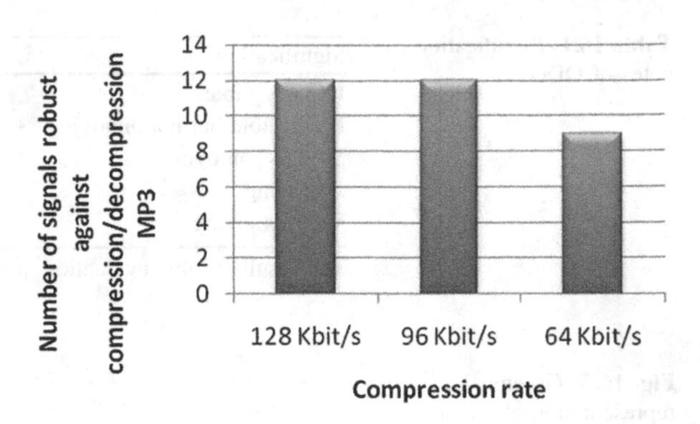

Fig. 16.7 Robustness of the technique against the attack of compression/ decompression MP3

16.4 Comparison to the Existing Watermark Techniques

To highlight our results, we will compare in this section the detailed above technique with three other techniques developed in [12].

The experimental corpus used above is the same as that used in [12].

We will present in the Table 16.2 the range of values of ODG given by PEAQ and the range of values of SNR for each technique.

Table 16.3 will illustrate the number of signals robust against compression/ decompression attack for each technique.

The presented results show that the proposed technique using the MPH2 of the MPEG standard gives better results in terms of the inaudibility and robustness than the technique using the psychoacoustic model 1 of the MPEG standard, the technique proposed by R. Brigola and the technique proposed by L. Rosa.

Table 16.2 Range of values of ODG and SNR for each technique

	ODG	SNR
Proposed watermarking technique + MPH2	0 to −0.35	74.1546 –82.7722 db
Watermarking technique + MPH1	0 to −0.819	23.1483–37.7138 db
Technique of R. Brigola	0 to −0.923	19.7856–35.8742 db
Technique of L. Rosa	−0.375 to −1.98	25.6629–32.2940 db

Table 16.3 Number of signals robust against compression/decompression MP3

	128 Kbit	96 Kbit	64 Kbit
Proposed watermarking technique + MPH2	12/12	12/12	9/12
Watermarking technique + MPH1	12/12	11/12	8/12
Technique of R. Brigola	8/12	0/12	0/12
Technique of L. Rosa	10/12	4/12	0/12

16.5 Conclusion and Perspectives

In this paper, we proposed a blind watermarking technique for audio (.wav) and which operates in the frequency domain. The time–frequency mapping is done by MDCT transformation applied to blocks of 1024 samples each. The inaudibility of the mark is favored by inserting bits in the LSB of components of the LF band which is under the final threshold of energy hearing calculated by the MPH2 of MPEG standard. The duplication of bits of the mark throughout the signal increases the robustness of the technique against attacks and allows having a high capacity of insertion. This important capability of insertion does not affect the sound quality of audio signals. In addition, the original brand is well identified in the detection phase. This detection is improved by using of Hamming coding. As perspective, we aim to test our technique against other types of attacks such as stirmark audio attacks.

References

1. Mu-Huo C, Yu-Hsin H (2003) Fast IMDCT and MDCT algorithms—a matrix approach. IEEE Trans Signal Process 51:221–229
2. Norme internationale, ISO/CEI 11172-3. Technologies de l'information codage de l'image animée et du son associé pour les supports de stockage numérique jusqu'à environ 1, 5 Mbit/s, partie 3: Audio
3. Hamming RW (1950) Error detecting and error correcting codes. Bell Syst Tech J 26(2):147–160
4. Bender W, Gruhl D, Morimoto N, Lu A (1996) Techniques for data hiding. IBM Syst J 35:313–336
5. Barnett R (1999) Digital watermarking: applications, techniques and challenges. Electron Commun Eng J 11(3):173–183
6. Baras C (2005) Tatouage informé de signaux audio numériques. Doctoral thesis, High National School of Telecommunications
7. Boney L, Tewfik AH, Hamdy KN (1996) Digital watermarks for audio signals. In: IEEE international conference on multimedia computing and systems, Hiroshima, Japan pp 473–480 June 17–23, 1996
8. Pinel J, Girin L, Baras C (2010) Une technique de tatouage haute capacité pour signaux musicaux au format CD-audio. In: 10 ème congrès français d'acoustique, 12–16, Lyon
9. Cvejic N, Seppanen, T (2003) Robust audio watermarking in wavelet domain using frequency hopping and patchwork method. In: Proceedings of the 3rd international symposium on image and signal processing and analysis, Rome, Italy
10. Charfeddine M, El Arbi M, Ben Amar C (2008) A blind audio watermarking scheme based on neural network and psychoacoustic model with error correcting code in wavelet domain. In: ISCCSP, Malta, pp 12–14
11. Union Internationale des Télécommunications (UIT): Recommandation B.S. 1387: Méthode de mesure objective de la qualité du son perçu (2001)
12. Bellaaj M, Ouni K (2012) Comparative analysis of audio watermarking technique in MDCT domain with other references in spectral domain. In: 9th international multi-conference on systems, Signals and Devices, Chemnitz

Chapter 17
A Wargame Data Visualization Algorithm Based on Regular Radius and Constrained Random Direction

Xiangli Xu, Xiaofeng Hu and Xiaoyuan He

Abstract Wargame data visualization is an important problem in war simulation field. In this paper, aiming at the big data generated by war gaming, a visualization algorithm based on regular radius and constrained random direction is proposed. First, we preprocess the original wargame data to extract the records to be visualized; second, the center vertices are determined for visualizing the data in a more regular way; third, we draw all the vertices and edges of the data based on regular radius and constrained random direction. Experiment results show that the proposed algorithm is effective and feasible.

Keywords Wargame · Data visualization · Regular radius · Constrained random direction

17.1 Introduction

Along with the development of tense war situation, many countries are speeding up the military construction [1, 2]. Many military exercises have been hold, such as joint naval exercise, peace drill, and so on. But actual exercises have the

X. Xu (✉) · X. Hu · X. He
The Department of Information Operation and Command Training,
NDU of PLA, Beijing 100091, China
e-mail: xuxiangli@aliyun.com

X. Hu
e-mail: xfhu@vip.sina.com

X. He
e-mail: bingling1922@sina.com

X. Xu
No. 65047 Troops of PLA, Shenyang 110000, China

A. A. Farag et al. (eds.), *Proceedings of the 3rd International Conference on Multimedia Technology (ICMT 2013)*, Lecture Notes in Electrical Engineering 278,
DOI: 10.1007/978-3-642-41407-7_17,

limitations on damage, scale, consumption, and many other aspects, and war gaming [3] supports an effective way. Although war gaming is a feasible way, it also brings many difficult problems at the same time, the wargame data processing is one of the most important problems. Wargame data belongs to the big data [4] field, and how to visualize the big data in a proper way is a basic and important problem. Because the applied field is so special, the research in this field is very few but must be solved urgently.

In this paper, aiming at the wargame data visualization problem, we propose a simple, effective, and directed visualization algorithm based on regular radius and constrained random direction. First, we preprocess the original wargame data to extract the records to be visualized; second, the center vertices are determined for visualizing the data in a more regular way; at last, we draw all the vertices and edges of the data based on regular radius and constrained random direction. Experiment result has shown the effect of the proposed algorithm.

The rest of this paper is organized as follows: We review the big data and wargame data in Sect. 17.2. Section 17.3 specifies current visualization algorithms and clarifies the difficulties of wargame data visualization. Section 17.4 details our proposed visualization algorithm based on regular radius and constrained random direction. In Sect. 17.5 we detail our experiments and discuss the results. Finally, we conclude our work in Sect. 17.6.

17.2 Big Data and Wargame Data

In recent years, big data has attracted high attention of industry, technology, and government. High fusion of triple world of human, machine, and object, causes bomb incensement of data size and high complexity of data mode, which indicates that our world has entered the big data period [4].

In general meaning, big data is the data set that cannot be perceived, achieved, managed, processed, and served in tolerant time, by traditional IT technology, soft and hard tools. The features of big data can be summarized by "4 V + 1C" [5] as follows:

(1) Variety. Big data include multi-types data, such as structured data, semi-structured data, and unstructured data. The process and analysis mode are very different.
(2) Volume. The data is generated by various kinds of intelligent equipments, which must be measured by PB and TB.
(3) Velocity. Big data requires fast processing, because the data is with timeliness.
(4) Vitality. In the network period, the process and analysis model for big data must adapt the requirements of new business.
(5) Complexity. Because of the above 4 V, the process and analysis for big data is more difficult and complex.

War system is a type of typical complex system, and wargame data generated by war simulation is typical big data. Compared with other large-scale warfare simulation systems, wargame as a traditional type of warfare simulations, has the advantages of low cost, convenience, practicability, etc. [3]. Unlike most games which have fixed rules, the rules for wargame can contain uncertainty. This uncertainty makes wargame generate huge amount of data, which is very difficult to process and analyze [6], and the data belongs to big data. For example, in a general simulation, the data size can achieve 10G per day on an average level, so how to process the wargame data becomes an urgent problem to be solved.

17.3 Wargame Data Visualization

Data visualization technology was proposed in 1980s, it can be defined as: technology using computer graphic and image processing to represent data in chart, map, tag cloud, animation, and any graphic mode, which can make the contents expressed by data more easily to be understood. 80 % information achieved from world by people is from the visual way, so it has important signification to represent large, complex, and multidimensional data in the visualization way [7]. In past years, many visualization methods have been proposed by scientists, which can be summarized as the following five classes [8]: (1) use multi-child windows to represent different combination of data dimensions, such as scatter plot matrices [9, 10] and pixel-oriented techniques; (2) rearrange all the data dimension in low-dimensional space, such as star coordinates [11] and parallel coordinates [12]; (3) do level division for low-dimensional space according with all the data dimensions, such as dimensional stacking [9] and treemap [9]; (4) adopt icon with multi-visible features, and each visible feature can represent a data dimensions, such as Chernoff and stick figures [9]; (5) map data to low-dimensional space, and maintain the relationships between data as much as possible, such as principal component analysis, and locally linear embedding [13].

Although there are many visualization technologies have been proposed, toward big data generated from war gaming, because of the special application, there are some special requirements and difficulties, which represents in the following aspects:

(1) Wargame data belongs to the big data field, so the visualization algorithm is more complex than general visualization algorithms, and much more efficient algorithms are required.
(2) Wargame data is dynamically and confrontly generated by the operational SoS, and most data belongs to the complex networks, so features of complex networks should be considered plenty to represent the visible features on the SoS level.

(3) Wargame data belongs to the military operational field, in order to represent military application significance, we must visualize the data combining the operational function of military requirements.

So aiming at the wargame data visualization problem, directed visualization algorithms are required.

17.4 Wargame Data Visualization Algorithm Based on Regular Radius and Constrained Random Direction

17.4.1 Preprocessing

The wargame data is big data with different formats and military meanings, so we must preprocess the data before visualization. First, extract the records to be visualized from the big data set,and then define the key elements in each item according with the visualized requirement. Toward wargame, the kernel content is the system of systems, which expressed as networks. Suppose the system of systems is expressed as G, we define the basic parameters as: vertices in active status $v_1, v_2, \cdots, v_N$, N is the number of vertices in active status; adjacency matrix $M_{N\times N} = \begin{pmatrix} m_{11} & m_{12} & \cdots & m_{1N} \\ m_{21} & \ddots & & \\ \vdots & & & \\ m_{N1} & & & m_{NN} \end{pmatrix}$, if v_i and v_j has the relationship R in G, $m_{ij} = 1$, else $m_{ij} = 0$; the in-degree of v_i is denoted as d^i_{in}, the out-degree of v_i is denoted as d^i_{out}.

Based on the basic mathematics definition above, we can do the visualization process further.

17.4.2 Center Vertices Determination

In order to visualize the wargame data in a regular way, the center vertices should be determined first. We define the center vertices of a network as the out-degree exceeding the threshold value δ_{out}, and it can be expressed as follows:

$$\begin{aligned} &\text{CenterVertices} = \left\{ v_i | d^i_{\text{out}} > \delta_{\text{out}}, i = 1, 2, \cdots, N \right\} \\ &d^i_{\text{out}} = \sum_{j=1}^{N} m_{ij} \end{aligned} \tag{17.1}$$

After determining the center vertices, we can start to design method to draw the other vertices and edges in a regular way.

17.4.3 Visualization Based on Regular Radius and Constrained Random Direction

Because of the huge amount of vertices, if drawing the vertices randomly the visualization effect will be very orderless. So we should define some rules to specify the positions of them. In our visualization algorithm, we define two main parameters: regular radius denoted by r, and constrained random angle denoted by θ. The core thinking of the algorithm is like this:

(1) The position of center vertices should be determined first. For example, use homogeneous random x-coordinates and y-coordinates denote positions of center vertices. The center vertices of system of systems G are denoted as $C_1, C_2, \cdots, C_M$, M is the number of center vertices, the position of C_i can be determined as follows:

$$\begin{cases} x_{c_i} = x_{\min} + (x_{\max} - x_{\min}) * \text{rand} \\ y_{c_i} = y_{\min} + (y_{\max} - y_{\min}) * \text{rand} \end{cases} \tag{17.2}$$

where $x_{\min}$ and $x_{\max}$ are the minimal and maximum x-coordinate of the visualized area, $y_{\min}$ and $y_{\max}$ are the same.

(2) Because the amount of vertices connected to center vertices is large, in order to avoid intersecting of vertices connected to different center vertices, we define a regular radius r to limit the distances between linked vertices.

(3) For vertices connected to a certain center vertex, the amount is also large, so in order to avoid overlap of them, we should let the positions have certain random characteristics. So we define a constrained random angle θ to distinct the position of them. So the position of a vertex V_k connected to a center vertex C_t is defined as follows:

$$\begin{cases} \theta = -180^\circ + \text{rand} \\ x_{V_k} = x_{C_t} + r \cdot \cos\theta \\ y_{V_k} = y_0 - \sqrt{r^2 - (x_{V_k} - x_{C_t})^2} \end{cases} \tag{17.3}$$

We call θ as constrained random angle because we constrain the angle θ in $[-180, 0]$ for a more regular effect.

Until now, we can visualize the wargame data in a regular way.

17.5 Experiment and Result Analysis

In order to test our proposed visualized algorithm, we select a command and control system of systems as the typical wargame data to do experiments, which is selected from a certain air sea battle war gaming. There are more than 1G data in the initial file, and through preprocessing and further analysis, about 7000 records are left, and the experiments are done in matlab environment.

We first visualize the records in an usual way, which is full random, and the visualization result is shown in Fig. 17.1.

From Fig. 17.1 we can see that the visualization effect is very terrible, the vertices and edges are crossing and overlapping with each other, almost no relationships can be seen clearly in the visualization result.

Then we visualize the records by the algorithm proposed in this paper, and the visualization result is shown in Fig. 17.2.

From Fig. 17.2 we can see that the visualization effect is very clear, the vertices and edges are regular, and the directions are also regular but with certain random. The center vertices are also can be seen clearly, and we can determine the important vertices more easily, which facilitate us to analyze the system of systems further.

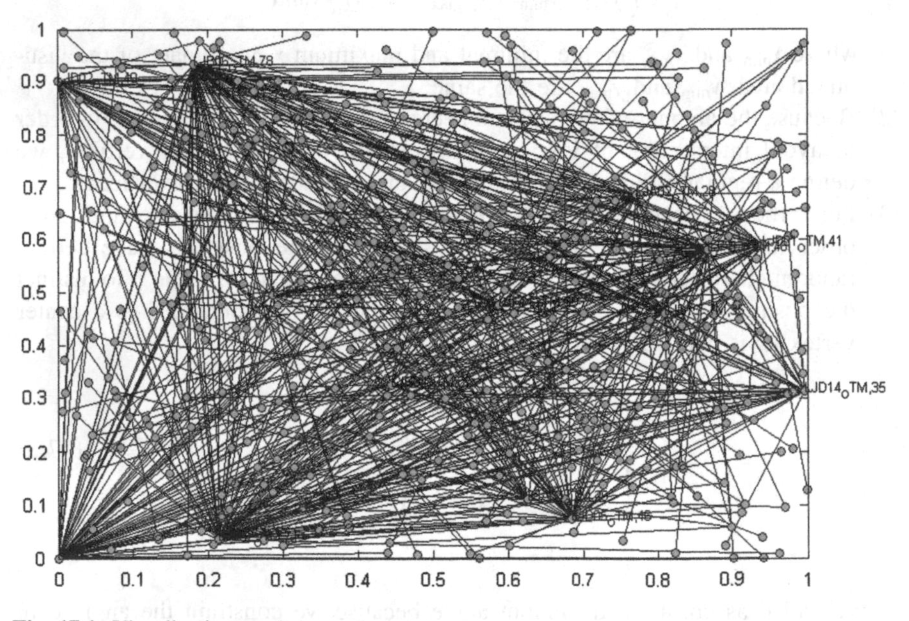

Fig. 17.1 Visualization result in random way

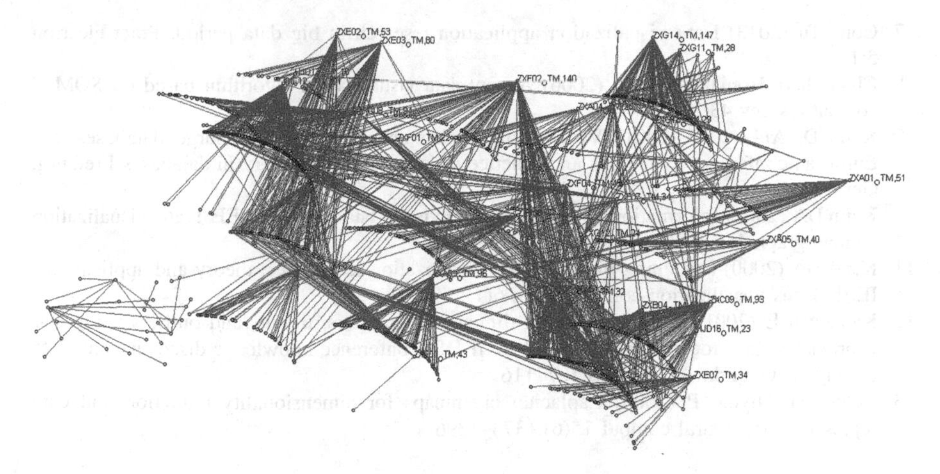

Fig. 17.2 Visualization result by the algorithm proposed in this paper

17.6 Conclusion

We propose a visualization algorithm based on regular radius and constrained random direction in this paper. In the progress, we first preprocess the original wargame data to extract the records to be visualized, then determine the center vertices and draw all the vertices and edges of the data based on regular radius and constrained random direction, and at last verify the reasonableness through experiment. Experiment results show that the proposed algorithm is effective and feasible, and it can bring facilitation of analyzing the system of systems further.

Acknowledgments This work is supported by the National Natural Science Foundation of China (No.71073172, 61174156, 61174035, 61273189, 61203140).

References

1. van Tol Jan, Gunzinger Mark, Krepinevich Andrew, Thomas Jim (2010) Airsea battle: a point-of-departure operational concept. Center for Strategic and Budgetary Assessments, Washing
2. Clarke Richard A, Knake Robert (2010) Cyber war: the next threat to national security and what to do about it. HarperCollins publisher, NewYork
3. Perla Peter P (1990) The art of wargaming. US naval institute press, Annapolis
4. Steve L (2012) The age of big data. New York Times
5. Editorial Office of Journal of China Academy of Electronics and Information Technology (2013) The big data age. J china acad electron Inf Technol 8(1):27–31
6. Zhijun Han, Shaojun Liu, Yubo Tang, Min Jing (2011) Study on computer wargaming system. Comput simul 28(4):10–13

7. Cong Tu (2013) Data visualization application research in big data period. Pract electron 5:118
8. Chao Shao, Houkuan Huang (2006) A new data visualization algorithm based on SOM. J comput res dev 43(3):429–435
9. Keim D, Ankerst D (2001) Visual data mining and exploration of large databases. 5th European conference principles and practice of knowledge discovery in databases Freiburg, Germany
10. Keim DA (2002) Information visualization and visual data mining. IEEE Trans visualization comput graph 1(1):1–8
11. Keim DA (2000) Designing pixel-orented visualization techniques: theory and applications. IEEE Trans visualization comput graph 6(1): 59–78
12. Kandogan E (2001) Visualizing multi-dimensional clusters, trends, and outliers using star coordinates. In: Proceedings 7th ACM SIGKDD Conference knowledge discovery and data mining. New York: ACM press,107–116
13. Belkin M, Niyogi P (2003) Laplacian eigenmaps for dimensionality reduction and data representation. Neural comput 15(6):1373–1396

Chapter 18
Analysis of Different Information Dissemination Ways for Disaster Prewarning: A Case Study of Beijing

Nan Zhang, Hong Huang, Boni Su and Bo Zhang

Abstract Knowing the mechanisms of different information dissemination ways and having an efficient information dissemination of disaster prewarning plays a very important role in reducing the loss and ensuring the safety of disaster carrier. In this paper, first, the transmission characteristics of different information media are analyzed. Second, the mathematical models of information dissemination are established for three typical information media including television, telephone, and email. Finally, the information dissemination capability of each medium considering different ages, genders, and residential areas is simulated and the characteristic of each medium is studied. The models and the results are essential for improving the efficiency of information dissemination and making the emergency plan, which should considerably reduce the possibility injury, deaths, and other losses in a disaster.

Keywords Information dissemination · Disaster prewarning · Television · Telephone · Email

18.1 Introduction

In recent years, the frequent natural and manmade disasters have seriously threatened the security of human life and properties. The government could make the efficient emergency plan and the people could evacuate and take protective measures before or during the disaster if the prewarning information can be disseminated to all the people in a fast and reliable way [1]. All of information formats like words, voice, and images could transmit the disaster information and

N. Zhang · H. Huang (✉) · B. Su · B. Zhang
Department of Engineering Physics, Institute of Public Safety Research,
Tsinghua University, Beijing, China
e-mail: hhong@mails.tsinghua.edu.cn

A. A. Farag et al. (eds.), *Proceedings of the 3rd International Conference on Multimedia Technology (ICMT 2013)*, Lecture Notes in Electrical Engineering 278,
DOI: 10.1007/978-3-642-41407-7_18, © Springer-Verlag Berlin Heidelberg 2014

share the data [3]. Therefore, the study on the information dissemination for disaster prewarning among the people is of great theoretical and practical significance. Now, with the development of technology, the network information dissemination including microblog, email, and news portal is generally popularized because of fine timeliness [8]. Two major ways including media networks and straightforward approaches, occupy the almost routes of information dissemination. Among the media network which spreads the information very fast but with low stability, telephones, messages, televisions, emails, microblogs, radios, and news portals are still the dominance. As for straightforward approaches including sound trucks, stationary loudspeakers, newspapers, and oral communications have the characteristics of stability, but with high cost or long delay time.

Nowadays, there are few researches on information dissemination of disaster prewarning and majority of them are focused on the single method for information dissemination. With the rapid development of technology, information dissemination media tend to diversity. Considering the different ways of information dissemination could improve the effectiveness of information acquisition and meet the realistic demands. In this paper, taking Beijing as an example, the characteristics of different information dissemination ways are studied. Television, telephone, and email are chosen as typical cases for study. Different information media have different characteristics. As for telephone, the users can be provided a rapid feedback at distance [5], however, it is easy to misunderstand the information and generally create the rumor in the process [7]. TV was thought capable of effectively communicating news images and information and presented instantaneously [6] but the number of audience is associated with the time periods in a large margin [9]. Email is faster, cheaper, and more effective in the information dissemination [4] but some users send the email irresponsible because of the convenience of it [2]. In this study, the capabilities and mechanisms of the three information dissemination media are studied considering the different ages, genders, and residential areas. Through the simulation and effectiveness analysis of the three information dissemination media in Beijing, the optimized plan and suggestions are put forward to improve the effectiveness of information dissemination in emergencies. The results of this research are helpful to achieve a comprehensive information dissemination system to transmit the information effectively.

18.2 Model Establishment

In this study, three status including ignorance, receiver, and believer are set to the people who are facing the disaster. Ignorance expresses the people who haven't received the disaster information. Receiver indicates that the people get the information but don't believe. Believer is the condition that the people received the information and believed. The information dissemination models for the three media are developed as follows:

18.2.1 Telephone

From the 150 times telephone-calling experiments, the probability *P(A)* and the average waiting time (t_{delay1}) of people answering telephone, the probability *P(B)* of the busy line, and the average delay time (t_{delay2}) for answering telephone after busy line, the probability *P(C)* of no answering and the average delay time (t_{delay3}) for re-answering telephone, the probability *P(D)* of powering off and the delay time (t_{delay4}) for re-answering time were obtained. The probability $P_{pt}(i)$ and the number $n_p(i)$ of target people forwarding the information with telephone are got through the questionnaires. Also, credibility of telephone $B_p(i)$ is a very important value. With all the data above, the information dissemination model for telephone is established and the some parameters are calculated. The main program flow is listed below:

1. Set all the parameters to the target people and create 5 information believers treated as information sources.
2. Search for the target people who have qualification to convey the information through the telephone. In these target people, 4 conditions should be satisfied that the line is not busy, the person is the information believer, wants to convey the information using telephone and hasn't reached the expected number target of information transmission.
3. The information transmitters call his target people who can answer the telephone. For other people who couldn't answer in time, the delay time is used in the different conditions.
4. The information transmitters continue to call until they reach the number target.
5. Record the number of new additional information receivers and information believers $n_{pb}(t)$ in this step and summary of them.

Using this information dissemination model of telephone, the average delay time T_{call}(min) considering different conditions and the probability of information acquisition $P_p(i)$ are calculated respectively by Eqs. 18.1 and 18.2.

$$T_{\text{phone-int}} = P(A) * t_{\text{delay1}} + P(B) * t_{\text{delay2}} + P(C) * t_{\text{delay3}} + P(D) * t_{\text{delay4}} \quad (18.1)$$

$$P_p(i) = B_p * P(A) * P(\text{receive}) = B_p(i) * P(A) * \frac{\overline{P_{pt}} * \sum_{t=t_1}^{t_1+1} n_{pb}(t)}{N_{\text{phone}}} \quad (18.2)$$

where *P*(receive) is the probability of target people receiving the information in the step and N_{phone} is the total number of people using the telephone.

18.2.2 Television

The distribution function $T_i(t_1-t_2)$ of television-watching periods of target people and credibility of television $B_t(i)$ were obtained from the questionnaires. In each time periods (per 3 h), the starting time of people watching television is treated as an uniform distribution in $(180-t_{t1-t2}(i))$. We assumed that the information is broadcasted each 1 h through the TV station. The information dissemination model of television is established and the program flow is listed below:

1. Set all the parameters which are from questionnaires to target people.
2. Create the starting time of each person in each time period.
3. The target people will receive the information when they watched the information from television. With probability of believe $B_t(i)$, they can turn to information believers.
4. Record the number of information receivers and information believers in this step and summary of them.

Through this model, the average delay time of television-watching T_{tv-int} (min) is calculated by Eq. 18.3. Also, in the time periods t to t + 3, the probability of ignorance turn to believer $P_t(i)$ is calculated by Eq. 18.4.

$$T_{\mathrm{TV-int}} = \frac{\left(180*(8+n_{k1}-n_{kp})-\frac{t_{k1}}{2}-\frac{t_{kp}}{2}\right)^2+\sum_{i=2}^{p}\left(180*\left(n_{ki}-n_{k(i-1)}\right)-\frac{t_{ki}}{2}-\frac{t_{k(i-1)}}{2}\right)^2}{2*60*24} \tag{18.3}$$

$$P_t(i)=\begin{cases} T_{3t\sim 3t+3}60*B_t(i) & T_{3t\sim 3t+3}<60 \\ 1-(1-B_e(i))\left\{\frac{(T_{3t\sim 3t+3}\mathrm{mod}60)}{60}*\left[\frac{T_{3t\sim 3t+3}}{60}\right]+\left(1-\frac{(T_{3t\sim 3t+3}\mathrm{mod}60)}{60}\right)*\left[\frac{T_{3t\sim 3t+3}}{60}+1\right]\right\} & T_{3t\sim 3t+3}\geq 60 \end{cases} \tag{18.4}$$

where n_{ki} is the sequence number of time periods (1–8) for the people who watched TV at that time periods and t_{ki} is the watching time in the n_{ki} time periods. p is the sequence number of last time periods for the people who watched TV in 1 day.

18.2.3 Email

Four parameters are considered to establish the information dissemination model of email. (a) The credibility of email (B_{i-e}). (b) The frequency of users using emails per day $n_{\mathrm{eu}}(i)\,\mathrm{times/day})$. (c) Forwarding people number $(\mathrm{n_{ef}(i)}))$ (d) Forwarding probability of people who get the disaster information. All of the parameters can be obtained through questionnaires. The model is listed below:

1. Set all the parameters which are from questionnaires to target people and create 5 initiative believers which are regarded as the information sources forwarding the information through email.
2. Search for the target people who have qualification to forward emails in this step. In these target people, 4 conditions should be satisfied including the

person is using email at this step, the person is the information believer, the person wants to forward the email and the person hasn't forward the email yet.
3. Update all the online conditions of email usage.
4. The email-online target people i check the mailbox. The probability of believe $b_e^{'}(i)$ is related with the number of received information $n_{er}(i)$ (Eq. 18.5).

$$B_e^{'}(i) = 1 - (1 - B_e(i))^{n_{er}(i)} \tag{18.5}$$

5. Record the total number of information receiver n_{er} and information believer $\mathrm{n_{eb}}$ in this step and calculate the summary of information receiver n_{er} and information believer n_{eb}.

Through the procedure of information dissemination of email, the average time interval of email usage $T_{\text{emain-int}}(\text{min})$ is calculated by Eq. 18.6 and the probability of ignorance turn to believer in each step $P_e(i)$ is calculated by Eq. 18.7. These two parameters reflect the capability for information acquisition.

$$T_{\text{email-int}} = \begin{cases} 60 * \frac{12 * n_{eu}(i) + 3}{(n_{eu}(i)+1)^2} & n \geq 1 \\ 60 * \frac{12}{n_{eu}(i)} & n < 1 \end{cases} \tag{18.6}$$

$$P_e(i) = \frac{n_{eu}(i)}{16 * 60} * \left(1 - (1 - Bi)^{n_{er}(i)}\right) \tag{18.7}$$

where n_{ki} is the sequence number of time periods (1–8) for people who watched TV at that time periods and t_{ki} is the watching time in n_{ki} time periods. P is the sequence number of last time periods for people who watched TV in 1 day.

A survey was conducted for credibility, usage frequency, forwarding times and number, and some relevant data of each medium. 370 questionnaires were filled out, and 350 of them were available.

18.3 Results

The simulation is performed based on the models developed above and different curves are obtained to judge the capability of information dissemination. Three influencing factors including ages, genders, and the residential areas are regarded as typical factors to study. Considering the children and elders are very limited to acquire the information in emails and telephone, the age of main research objects is from 16 to 55 years old. Considering the actual number of inhabitants in Beijing, 25 million people are set in the simulation. With the usage conditions of different ways, the detail results of information dissemination are analyzed below:

18.3.1 Telephone

The curves of information dissemination by telephone are similar with the emails. From the Figs. 18.1 and 18.2, comparing the curves of email, the delay time at the beginning of the information dissemination is reduced significantly. The increasing number of people getting the information accords with Logistic curve, because of the popularity and the convenience of the telephone.

Through the analysis to Fig. 18.1, the telephone usage of young people (16–35) is obviously more active than it of middle-aged people (36–55). From Fig. 18.2, it can be seen that the female has higher capability of information dissemination by telephone than it of male. Because the inconvenient telephone communication, lower economy, and some special occupations of rural areas, urban areas have a better information dissemination capability when an emergency is occurred. To analyze the synthesis curve of telephone and consider the time of communication and busy line, more than 10 million people could be informed through the telephone communication within 6 h and the curves turned to plain after a short time. As of 1,000 min, the final value tended to 16.8 million. 7.2 million people haven't

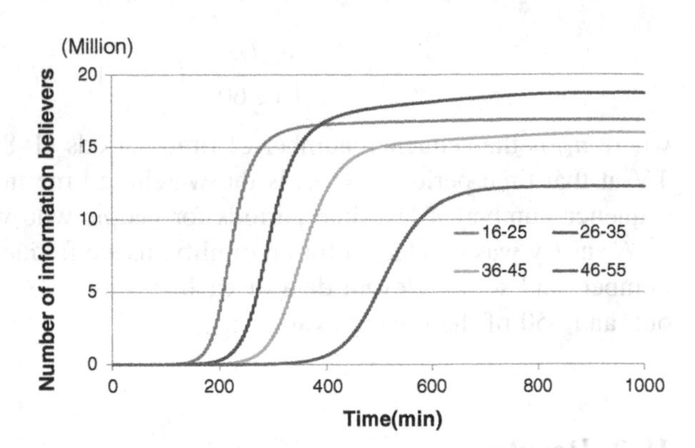

Fig. 18.1 Information dissemination by telephone with different ages

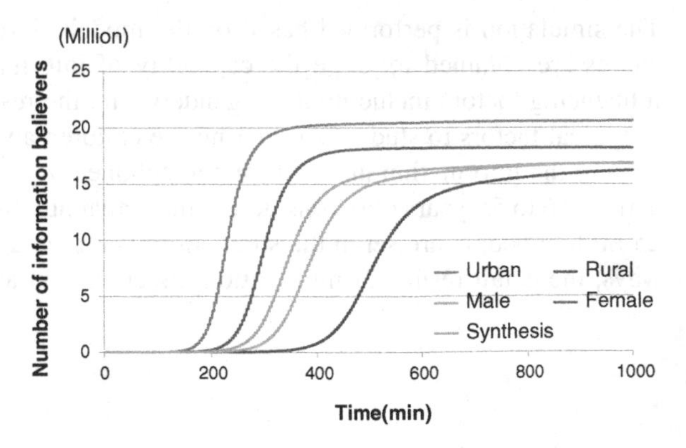

Fig. 18.2 Information dissemination by telephone with synthesis, different genders, and residential areas

turned into the information believers because the credibility of telephone is lower and usage ratio of rural areas is not very high. And, the increasing number of people getting the information accords with Logistic curve.

Through the data analysis to synthesis curve, the speed of information dissemination by telephone can be calculated with the approximate exponential Eq. 18.8 and $R^2 = 0.997$.

$$N_{\text{phone}}(t) = -1.291 \times 10^6 \times e^{-0.0003432t} + 1.41 \times 10^6 \tag{18.8}$$

18.3.2 Television

From the Figs. 18.3 and 18.4, it could conclude that the speed of information dissemination by TV is strong related to different time periods. A large number of people accustomed to watch TV between 6 p.m. and 12 p.m., however, at other

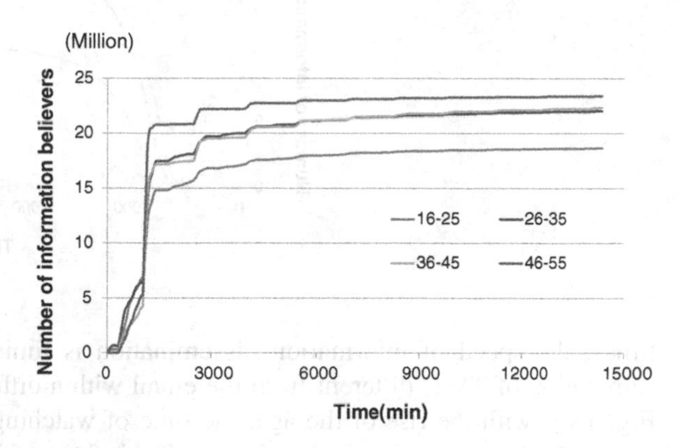

Fig. 18.3 Information dissemination by TV with different ages

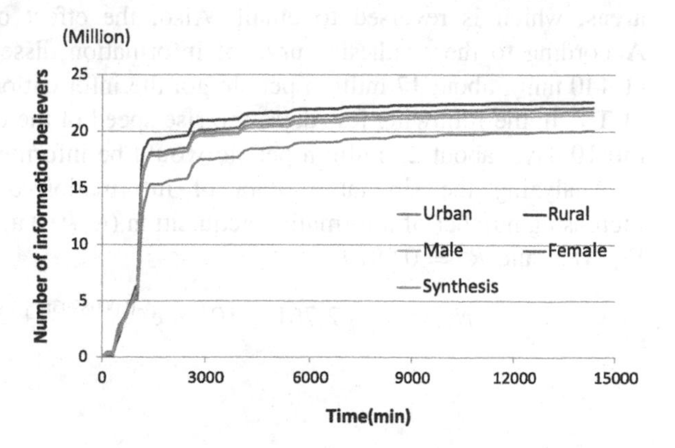

Fig. 18.4 Information dissemination by TV with synthesis, different genders, and residential areas

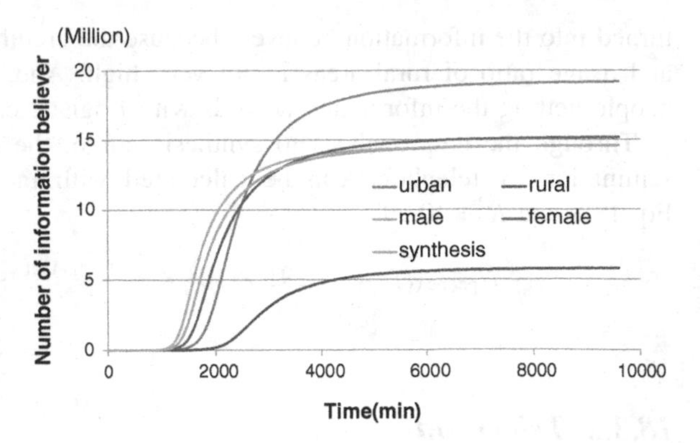

Fig. 18.5 Information dissemination by email with different ages

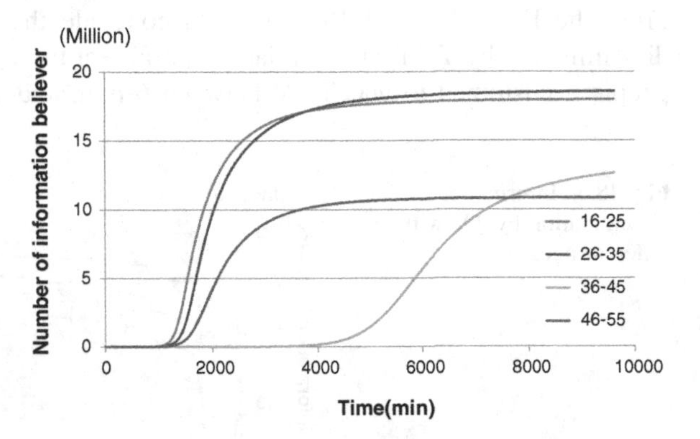

Fig. 18.6 Information dissemination by email with synthesis, different genders, and residential areas

times, the speed of information dissemination is limited. The information dissemination of TV is different from the email within different ages. As it shown in Fig. 18.3, with the rise of the age, the time of watching TV is increased and the speed of information acquisition through TV is faster. Observing the Fig. 18.4, the ability of information dissemination in rural areas of TV is better than it at urban areas, which is reversed to email. Also, the effect of genders is very small. According to the synthesis curve of information dissemination, in the first day (1,440 min), about 17 million people got the information with the high credibility of TV. In the following few days, the rise speed of the curve is slower. Finally, in the 10 days, about 22 million people would be informed through TV.

Analyzing the simulation data of information dissemination by TV, the increasing number of information acquisition ($N_{TV}(t)$) agrees with the approximate Eq. 18.9 and $R^2 = 0.9979$.

$$N_{TV}(t) = -2.761 \times 10^7 \times \mathrm{e}^{-0.0008993t} + 2.142 \times 10^7 \tag{18.9}$$

18.3.3 Email

Figure 18.5 shows the results of information dissemination through emails with different ages. It can be seen that the age of 36–45 is a vulnerable group while the groups of 16–35 has stronger capability of information dissemination with email. From Fig. 18.6, comparing the influence of genders and residential areas, it is easy to see that the capability of information dissemination in urban areas much larger than it in rural areas, however, the influence of gender can be ignored. Observing the synthesis information dissemination curve of emails, the starting time is longer which is close to 1,200 min. But with the fast speed of information dissemination after that, the curve becomes stationary at 4,000 min. At last, the number of information believers by emails tends to 14.2 million based on the usage ratio and credibility of mailbox.

The number of information-receiving in mailbox is according with the Logistic curve. However, considering the frequency of mailbox usage and credibility of emails, the increasing number of information believers by emails ($N_{\mathrm{email}}(t)$) agrees with the approximate Eq. 18.10 ($R^2 = 0.9979$).

$$N_{\mathrm{email}}(t) = \begin{cases} 36.79 \times e^{0.007377t} & 0 \le t < 1500 \\ -8.213 \times 10^7 \times e^{-0.001274t} + 1.444 \times 10^7 & t \ge 1500 \end{cases} \tag{18.10}$$

18.4 Conclusions

In this study, taking Beijing as an example, the three models of information dissemination including email, TV, and telephone were developed. The capability of information dissemination of each medium was calculated based on the transmission models. The regression models of information dissemination were also obtained to quantify the detailed values of received and believed information approximately. According to the simulation results, the capability of information dissemination of each medium is expressed through the transmission curves. The mechanisms of different influencing factors including ages, genders, and residential areas were also studied based on the actual data from the 370 questionnaires. These results associated with an actual situation, will contribute to reducing losses in a disaster and improving the safety of disaster carriers. These models and simulation methods could be used in many other regions. More ways of information dissemination and influencing factors would be considered likes the maximum information carrying capacity and the vulnerability of each medium in the future work.

Acknowledgments This work was supported by National Natural Science Foundation of China under (Grant No. 71173128, 91224008) and Ministry of Science and Technology of the People's Republic of China under Grant No. 2011BAK07B02.

References

1. Basher R (2006) Global early warning systems for natural hazards: systematic andpeople-centred. Philos Trans Roy Soc 364:2167–2180
2. Berghel H (1997) Email- the good, the bad, the ugly. Commun ACM 50:11–15
3. David C, Tia G, David W (2006) Information collection and dissemination: toward a portable, real-time information sharing platform for emergency response. In: Proceedings of the annual symposium of American medical information association, 898
4. Farnham S, Pedersen E, Kirkpatrick R (2006) Observation of Katrina/Rita groove deployment: addressing social and communication challenges of ephemeral groups. In: Proceedings f the 3rd International ISCRAM conference, 39–49
5. Lengel RH, Daft RL (1989) The selection of communication media as an executive skill. Acad Manag 2:225–232
6. Robinson JP, Davis DK (1990) Television news and the informed public: an information-processing approach. J Commun 40:106–119
7. Si GY, Zhang F, Luo P (2012) Modeling and simulation on information communication based on networks of handsets. Complex Syst Complex Sci 9:10–19
8. Tong YL, Yang XY, ChenY (2010) The research on control strategy of information spreadingin mass emergency based on theory of complex network. In: 3rd International conference of computer intelligence industrial applications, 257–260
9. Wei JC, Zhao DT, Yang F et al (2010) Timing crisis information release via television. Disasters 34:1013–1030

Chapter 19
Human Detection with EOH-OLBP-Based Multi-Level Features

Yingdong Ma and Liang Deng

Abstract In this work we aim to develop an onboard monocular pedestrian detection system by employing a classifier based on multi-level features. Different from most state-of-the-art detectors, our system employs a set of multi-level features to ensure high detection rate. More specifically, a set of EOH and OLBP based multi-level features is used to describe cell-level and block-level structure information. Multi-level features capture larger scale structure information which is more informative for pedestrian localization. Experiments on the INRIA dataset and the Caltech pedestrian detection benchmark demonstrate that the new pedestrian detection system is not only comparable to the existing pedestrian detectors but it also performs at a faster speed.

Keywords Human detection · Structure information · Multiple–level features

19.1 Introduction

In many applications of pedestrian detection, such as online human detection for robotics and automotive safety, both efficiency and accuracy are important. The goal of this paper is to find a tradeoff between accuracy and efficiency. For this purpose, solutions of two important questions need to be found:

This work was supported in part by Program of Higher-level talents of Inner Mongolia University.

Y. Ma (✉)
College of Computer Science, Inner Mongolia University, Hohhot, China
e-mail: csmyd@imu.edu.cn

L. Deng
Shenzhen Institutes of Advanced Technology, Shenzhen, China
e-mail: liangdeng.2008@gmail.com

A. A. Farag et al. (eds.), *Proceedings of the 3rd International Conference on Multimedia Technology (ICMT 2013)*, Lecture Notes in Electrical Engineering 278,
DOI: 10.1007/978-3-642-41407-7_19,

- A highly discriminative feature (or a feature set), which captures salient features of humans and it can be computed quickly.
- The feature has the ability to capture larger scale structure information beyond small 3×3 (e.g., LBP) or 8×8 (e.g., HOG) local range.

A multi-level coding scheme is discussed to tackle these problems. Instead of using the cell-level histogram features directly in a Boosting classifier, we present a set of multi-level descriptors to extract both intra-block information and inter-block information. The new descriptors are designed to capture an object's larger scale structure information for efficient object localization.

19.2 Previous Work

In the case of onboard pedestrian detection, many human detection systems use sliding window based classification methods. Recent search on the sliding window based method focus on developing more discriminative features to improve system detection rate. Some well-known features include Histograms of oriented gradients (HOG) [1], shapelet [2], edge orientation histogram (EOH) [3], edgelet [4], region covariance [5], LBP [6], and poselet [7].

Resent literature indicates some new trends in human detection. First, the occurrence of part-based deformable model substantially improves detection performance [8–10]. In a deformable model, different parts can move spatially to allow both local and global shape deformations. The second trend is the use of Multiple Kernel Learning (MKL) method for better human and object detection. A typical MKL based object detection work is introduced in [11], in which multiple features are combined in a three stage MKL classifier. Third, various context models are widely used in object detection to gain higher accuracy [12]. Existing studies have shown that context information is critical for reducing the uncertainty in object detection, especially in images with cluttered background.

However, higher detection accuracy usually means increased computational costs. As an example, the MKL based method proposed in [11] requires roughly 67 s per image, even with the carefully designed three stage cascade. Moreover, both part-based deformable models and context models introduce additional computational cost. It is infeasible to apply these complicated human detectors directly in onboard human detection applications.

In our previous work [13], the Oriented LBP feature was introduced. It was used directly in a Boosting algorithm on INRIA dataset. In this paper, we derive a set of intra-block and inter-block level Oriented LBP features to enhance its ability of capturing larger scale structure information.

19.3 EOH-OLBP Based Multi-Level Features

19.3.1 Gradient-Based Feature

To speed up feature extraction, we choose the EOH feature proposed in [3]. Calculation of the EOH feature begins by performing edge detection in the image, which computes the gradient magnitude and gradient orientation of each pixel. The gradient orientation is evenly divided into K bins. The gradient orientation histograms in each orientation bin k of cell C_i, $E_k(C_i)$, are obtained by summing all the gradient magnitudes whose orientations belong to bin k.

We define a set of k EOH features of a single cell as the ratio of the bin value of a single orientation to the sum of all bin values:

$$EOH_{C_{i,k}} = \frac{E_k(C_i) + \varepsilon}{\sum_{j=1}^{K} E_j(C_i) + \varepsilon} \tag{19.1}$$

where ε is a small value that avoids the denominator being zero. The EOH features are computed with the cell size of 8 × 8 pixels. The integral-image method [14] can be used to speed up EOH feature computation.

19.3.2 Oriented Local Binary Patterns

In the Oriented LBP feature, the magnitude The magnitude $m(x,y)$ and the orientation $\theta(x,y)$ of a pixel are defined as the number of continuous "1" bits of its neighbors and the principle direction of these "1" bits, respectively, (see our previous work [13] for more detail). The pixel orientation is evenly divided into n bins over 0° to 360° The orientation histograms $F_k(C_i)$ in each orientation bin k of cell C_i are obtained by summing all the pixel magnitudes whose orientations belong to bin k in C_i.

$$F_k(C_i) = \sum_{\substack{(x,y) \in C_i \\ \theta(x,y) \in bin_k}} m(x,y) \tag{19.2}$$

In the case of 8 bins, we calculate a set of k Oriented LBP features for each cell as:

$$OLBP_{C_i,n} = \frac{F_n(C_i) + \varepsilon}{\sum_{j=1}^{N} F_j(C_i) + \varepsilon} \tag{19.3}$$

Since the computation of OLBP is similar to that of uniform LBP, it can be computed faster than HOG. Moreover, construction of uniform LBP feature needs 59 bins to model the variant of binary bits distribution. However, the OLBP histogram feature only has 8 bins, which further accelerates the computation of OLBP feature.

19.3.3 Intra-block and Inter-block Features

The EOH feature and the LBP feature are cell-level histogram features. Most up-to-date human detection systems either use them in a boosted cascade [6, 15] or concatenate them to form a high-dimensional feature vector for a SVM-based classifier [16]. Instead of using these cell-level histogram features directly, we introduce a method which operates on these features to extract intra-block and inter-block information (For convenience, we refer them as IB–IB information for short). These block-level features capture larger scale structure information which is not only more informative for object localization, but also robust in the case of partial occlusion. In this work, a block is defined as a group of 2×2 cells and each cell has the size of 8×8 pixels.

19.3.3.1 Intra-block Features

Let C^i_{EOH} be the sum of EOH feature of cell C_i and C^i_{OLBP} be the sum of OLBP feature of cell C_i:

$$C^i_{EOH} = \sum\nolimits_k E_k(C_i) \tag{19.4}$$

$$C^i_{OLBP} = \sum\nolimits_n F_k(C_i) \tag{19.5}$$

where k is the bin number of EOH feature and n is the bin number of OLBP feature. Let $cell_{i,j}(i = [1,2]; j = [1,2])$ be four cells of a block (see Fig. 19.1).

We define the intra-block EOH feature set as:

$$S_1 = \delta \left| C^{cell_{i,j}}_{EOH} + C^{cell_{i,j+1}}_{EOH} + C^{cell_{i+1,j}}_{EOH} + C^{cell_{i+1,j+1}}_{EOH} \right| \tag{19.6}$$

$$S_2 = \delta \left| C^{cell_{i,j}}_{EOH} + C^{cell_{i,j+1}}_{EOH} - C^{cell_{i+1,j}}_{EOH} - C^{cell_{i+1,j+1}}_{EOH} \right| \tag{19.7}$$

$$S_3 = \delta \left| C^{cell_{i,j}}_{EOH} + C^{cell_{i+1,j}}_{EOH} - C^{cell_{i,j+1}}_{EOH} - C^{cell_{i+1,j+1}}_{EOH} \right| \tag{19.8}$$

$$S_4 = \delta \left| C^{cell_{i,j}}_{EOH} - C^{cell_{i+1,j+1}}_{EOH} \right| \tag{19.9}$$

$$S_5 = \delta \left| C^{cell_{i+1,j}}_{EOH} - C^{cell_{i,j+1}}_{EOH} \right| \tag{19.10}$$

where δ is a normalizing factor. The set of intra-block OLBP features $(S_6, S_7, S_8, S_9, S_{10})$ are defined in the same way as the intra-block EOH features. The calculation of S_3 and S_4 is illustrated in Fig. 19.1 (second row).

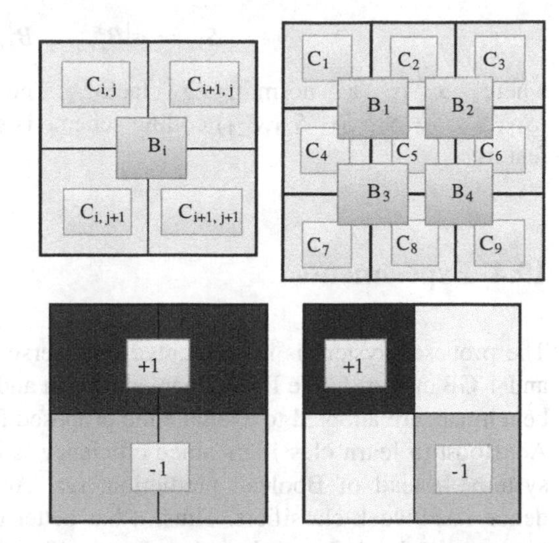

Fig. 19.1 Computation of Intra-block and Inter-block features. First row: $C_{i,j}$ are four cells of block B_i (*left*), C_1 to C_9 are nine cells of four adjacent blocks: B_1 and (*right*); Second row: computation of the intra-block features S_3 (*left*) and S_4 (*right*)

19.3.3.2 Inter-block Features

Let the four cells of block B_i are: C_1, C_2, C_3 and C_4. We define the EOH feature of block B_i $\left(B^i_{EOH}\right)$ and the OLBP feature of block B_i $\left(B^i_{OLBP}\right)$ as:

$$B^i_{EOH} = \sqrt{{C^1_{EOH}}^2 + {C^2_{EOH}}^2 + {C^3_{EOH}}^2 + {C^4_{EOH}}^2} \tag{19.11}$$

$$B^i_{OLBP} = \sqrt{{C^1_{OLBP}}^2 + {C^2_{OLBP}}^2 + {C^3_{OLBP}}^2 + {C^4_{OLBP}}^2} \tag{19.12}$$

The coding scheme of inter-block feature is different from that of intra-block feature because a block contains much more pixels than a cell. This may lead to a very large block-level feature value. Suppose that four adjacent blocks $\{B_1, B_2, B_3, B_4\}$ contains nine cells $\{C_i, i = [1, \ldots, 9]\}$ as shown in Fig. 19.1. Then we can define the inter-block EOH features as:

$$S_{11} = \sigma\left|B^1_{EOH} + B^2_{EOH} + B^3_{EOH} + B^4_{EOH}\right| \tag{19.13}$$

$$S_{12} = \sigma\left|B^1_{EOH} - B^2_{EOH}\right| \tag{19.14}$$

$$S_{13} = \sigma\left|B^3_{EOH} - B^4_{EOH}\right| \tag{19.15}$$

$$S_{14} = \sigma\left|B^1_{EOH} - B^3_{EOH}\right| \tag{19.16}$$

$$S_{15} = \sigma\left|B^2_{EOH} - B^4_{EOH}\right| \tag{19.17}$$

$$S_{16} = \sigma\left|B^1_{EOH} - B^4_{EOH}\right| \tag{19.18}$$

$$S_{17} = \sigma \left| B_{EOH}^{3} - B_{EOH}^{2} \right| \tag{19.19}$$

where α is a normalizing factor. The inter-block OLBP features $(S_{18}, S_{19}, S_{20}, S_{21}, S_{22}, S_{23}, S_{24})$ coding scheme is similar with the inter-block EOH features.

19.4 Experiments

The proposed system is implemented on a personal computer with 3.0 GHz CPU and 4 GB memory. The INRIA person dataset and the Caltech pedestrian detection benchmark are adopted to evaluate the proposed features. In this work, we use real AdaBoost to learn classifiers since efficiency is an important requirement of our system. Instead of Boolean prediction, real AdaBoost algorithm returns confidence-rated weak classifiers. Thus, it has better discriminative character than the binary valued AdaBoost algorithm. For a 640 × 480 video sequence, the proposed approach runs at about 5–6 frames per second. Performance evaluation on the INRIA dataset is based on the false positive per window (FPPW) metric, whereas the false positive per image (FPPI) metric is mainly used in the Caltech pedestrian detection benchmark. which is important for onboard pedestrian detection systems.

19.4.1 Performance Comparison on INRIA Dataset

System performance comparison between the proposed approach and some state-of-the-art methods is illustrated in Fig. 19.2. These methods include HOG-SVM [1], RecEOHED-Cascade [15], Semantic-LBP (SLBP)-SVM [6], HOG-LBP-SVM (without occlusion handling) [16], CS-LBP [17], Fast IKSVMs [18], the partial least squares based multiple-feature method (PLS) [19], and the Hybrid feature cascade [20].

Our method achieves 91.8 % detection rate at 10^{-5} FPPW on INRIA dataset, higher than SLBP-SVM (67.3 %), HOG-SVM (80.7 %), RecEOHED-Cascade (83.0 %), and CS-LBP (86.2 %). The HOG-LBP-SVM, Fast IKSVMs, PLS, and the Hybrid feature cascade have better performance than our method. For example, the HOG-LBP-SVM method has detection rate of 93.0 % at 10^{-5} FPPW, 1.3 % higher than the proposed system. However, please note that the purpose of our method is to find an optimal tradeoff between accuracy and efficiency, whereas these top performance methods focus on obtaining higher detection rate by using multiple features. For instance, the Fast IKSVMs method uses multi-level histograms of oriented edge energy feature and a histogram intersection kernel SVM classifier to achieve high detection rate. Its processing time is about five to six times slower than the linear SVM [18].

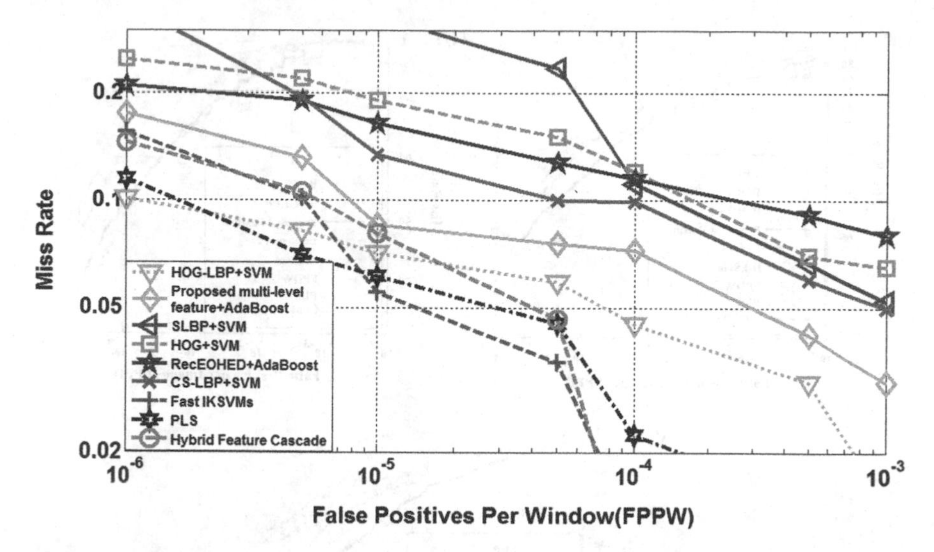

Fig. 19.2 Performance comparison between the proposed approach and state-of-the-art methods

19.4.2 Performance Comparison on Caltech Pedestrian Dataset

On Caltech pedestrian dataset, detections are considered as true positives if the overlap area of their bounding boxes and corresponding ground truth exceed 50 %. We evaluate performance on the "Near," "Reasonable," and "Partial occlusion" subsets. All classifiers are trained using the INRIA training dataset.

19.4.2.1 Performance Comparison on "Near" Subset

Figure 19.3a shows the results on the "Near" subset, in which pedestrians are 80 pixels or taller and without occlusion. The proposed approach performs the best with miss rates of 0.37 at 0.1 fppi and 0.15 at 1 fppi, respectively. The HOG-LBP-SVM method achieves the best FPPW score in INRIA dataset but it only slightly outperforms HOG on Caltech Pedestrian dataset.

This is caused by the lower resolution video frames of Caltech Pedestrian dataset. Moreover, video frames are blurring than still pictures. The blurring frames make pedestrian detection more difficult, particularly for classifiers based on contour information (e.g., LBP). It means that contribution from LBP feature is limited in this case. This is also demonstrated by the LBP curve. LBP alone has the highest miss rate, e.g., 61 % at 1 fppi, about 46 % higher than our detector. Comparing to HOG-LBP-SVM, our detector gains lower miss rate in Caltech Pedestrians: at 0.1 and 1 fppi the improvements are 9 and 11 %, respectively. The higher performance is obtained due to the combination of orientation information

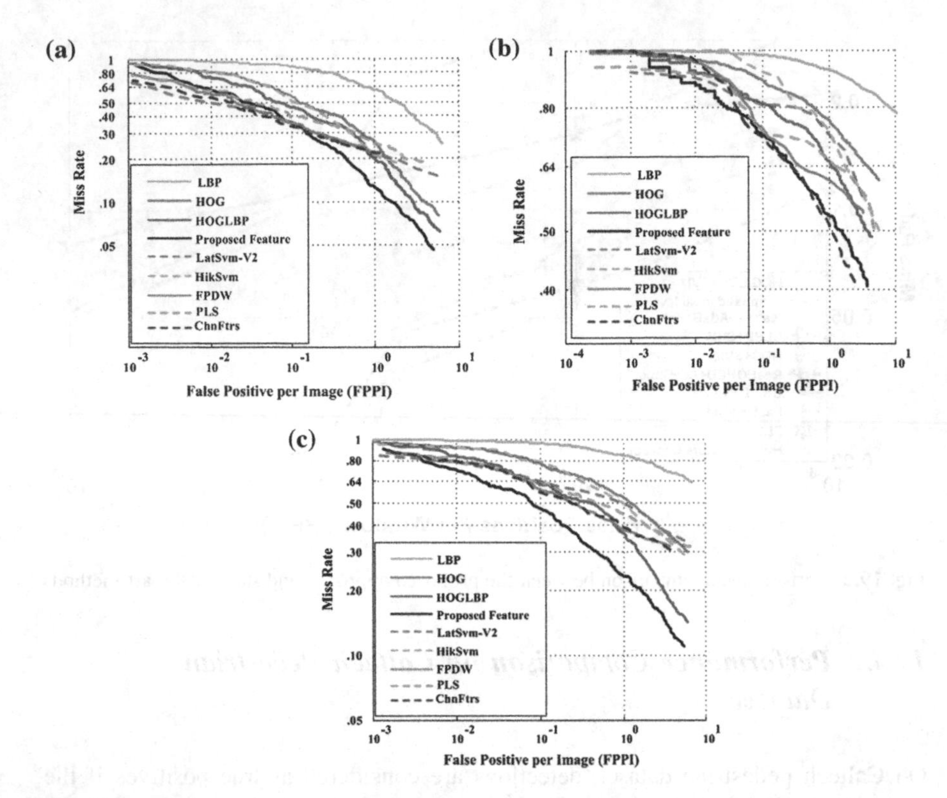

Fig. 19.3 Evaluation results under three different conditions on the Caltech Pedestrian Dataset. "Near" subset (**a**), "Partial Occlusion" subset (**b**), "Reasonable" subset (**c**)

with pixel intensity difference. It makes our detector not very sensitive to lower quality images.

We compare our method with several representative state-of-the-art pedestrian detectors, including the histogram intersection kernel (HikSVM) [18], the Latent SVM (LatSVM-V2) [8], the Fastest Pedestrian Detector in the West (FPDW) [21], the PLS method [19], and the integral channel features method (ChnFtrs) proposed in [22]. HikSVM has similar performance with HOG but lower than the proposed method by 16 % at 0.1 fppi. Latent SVM and ChnFtrs have higher accuracy than our method in the range of lower false positive rates (e.g., less than 0.1 fppi), but worse than our method when false positive rate is higher than 0.1 fppi.

19.4.2.2 Performance Comparison on "Partial Occlusion" Subset

In Fig. 19.3b, the performance on the "Partial Occlusion" subset (50 pixels or taller, 1–35 % occluded) is shown.

We observe obvious performance degradation of all evaluated algorithms. Although the performance degradation of our detector is also considerably (e.g., about 35 %), it still outperforms others on "Partial Occlusion" subset. Multiple-level features, especially the inter-block large area structure information, helps in the presence of partial occlusion. Significant performance degradation of HOG feature demonstrates that occlusion has more influence on local features: occlusion of a local region is likely to affect its similarity seriously to all other local regions.

The relative improvement with our method is noticeable: at 0.1 and 1 fppi, the miss rates of our detector are 71 and 58 %, lower than HOG-LBP by about 11 and 8 %, respectively. Our method has similar performance with the ChnFtrs method when false positive rates are higher than 0.1 fppi but outperforms other methods in the rage of lower false positive rates (e.g., less than 0.1 fppi).

19.4.2.3 Performance Comparison on "Reasonable" Subset

Figure 19.3c compares to the results of all detectors on the "Reasonable" subset. It consists of pedestrians at least 50 pixels tall and they are fully visible or less than 35 % occluded. Performances of all evaluated algorithms on the "Reasonable" subset are substantially better than that of the "Partial Occlusion" subset.

Similar to the "Near" subset, the HikSVM shows almost the same performance to HOG. LatSVM-V2 and the PLS method are slightly better than our method when false positives are close to 0.001 fppi. Our detector has higher accuracy, in particular in the range of higher false positive rate (e.g., more than 0.01 fppi). Comparing to the HOG-LBP method, we observe improvement provided by our detector clearly at 0.1 and 1 fppi: about 9 and 11 %, respectively.

19.5 Conclusion

In this work a multi-level features based pedestrian detection system is introduced, which aims at extracting human objectives from video streams with high efficiency and accuracy. We demonstrate that the proposed human detection algorithm outperforms some state-of-the-art approaches and, more importantly, it achieves efficient human detection which is critical in many pedestrian detection applications.

This is achieved because of the coding scheme of multi-level features. Instead of using local histogram features directly, the coding scheme of Intra-block and Inter-block features is discussed. Extensive experiments show that the block-level structure features are complementary to cell-level local features as they describe an object's larger scale structure information. The proposed detection system using these multi-level features achieves high detection rate. Moreover, comparing to gradient histogram based features, e.g., HOG, both EOH and OLBP features can be computed at a faster speed, which leads to an efficient pedestrian detection system.

References

1. Dalal N, Triggs B (2005) Histograms of oriented gradients for human detection. In: Proceedings of the IEEE international conference on computer vision and pattern recognition
2. Sabzmeydani P, Mori G (2007) Detecting pedestrians by learning shapelet features. In: Proceedings of the IEEE international conference on computer vision and pattern recognition, pp 1–8
3. Levi K, Weiss Y Learning (2004) Object detection from a small number of examples: the importance of good features. In: Proceedings of the IEEE international conference on computer vision and pattern recognition, pp 53–60
4. Wu B, Nevatia R (2005) Detection of multiple, partially occluded humans in a single image by Bayesian combination of edgelet part detectors. In: Proceedings of the IEEE international conference on computer vision, pp.90–97
5. Paisitkriangkrai S, Shen C, Zhang J (2008) Fast pedestrian detection using a cascade of boosted covariance features. IEEE Trans Circuits and Syst Video Technol 18(8):1140–1151
6. Mu Y, Yan S, Liu Y, Huang T, Zhou B (2008) Discriminative local binary patterns for human detection in personal album. In: Proceedings of the IEEE international conference on computer vision and pattern recognition, pp 1–8
7. Bourdev L, Malik J (2009) Poselets: body part detectors trained using 3D human pose annotations. In: Proceedings of the IEEE international conference on computer vision, pp 1365–1372
8. Felzenszwalb PF, Girshick RB, McAllester D, Ramanan D (2010) Object detection with discriminatively trained part based models. IEEE Trans Pattern Recogn and Mach Intell 32(9):1627–1645
9. Felzenszwalb P, McAllester D, Ramanan D (2008) A discriminatively trained, multiscale, deformable part model. In: Proceedings of the IEEE international conference on computer vision and pattern recognition
10. Zhu L, Chen Y, Yuille A, Freeman W (2010) Latent hierarchical structural learning for object detection. In: Proceedings of the IEEE international conference on computer vision and pattern recognition, pp 1062–1069
11. Vedaldi A, Gulshan V, Varma M, Zisserman A (2009) Multiple kernels for object detection. In: Proceedings of the IEEE 12th international conference on computer vision, pp 606–613
12. Zheng W, Gong S, Xiang T (2012) Quantifying and transferring contextual information in object detection. IEEE Trans Pattern Anal Mach Intell 34(4):762–777
13. Ma Y, Chen X, Jin L, Chen G (2011) A monocular human detection system based on eoh and oriented lbp features. In: Proceedings of the 7th international conference on Advances in visual computing, vol I, pp 551–562
14. Porikli F (2005) Integral histogram: a fast way to extract histograms in Cartesian spaces. In: Proceedings of the IEEE international conference on computer vision and pattern recognition
15. Chen Y, Chen C (2008) Fast human detection using a novel boosted cascading structure with meta stages. IEEE Trans Image Process 17(8):1452–1464
16. Wang X, Han T, Yan S (2009) An HOG-LBP human detector with partial occlusion handling. In: Proceedings of the IEEE international conference on computer vision
17. Heikkilä M, Pietikäinen M, Schmid C (2009) Description of interest regions with local binary patterns. Pattern Recogn 42(3):425–436
18. Maji S, Berg A, Malik J (2008) Classification using intersection kernel SVMs is efficient. In: Proceedings of the IEEE international conference on computer vision and pattern recognition
19. Schwartz W, Kembhavi A, Harwood D, Davis L (2009) Human detection using partial least squares analysis. In: Proceedings of the IEEE international conference on computer vision

20. Wu B, Nevatia R (2008) Optimizing discrimination-efficiency tradeoff in integrating heterogeneous local features for object detection. In: Proceedings the IEEE international conference on computer vision and pattern recognition, pp 1–8
21. Doll'ar P, Belongie S, Perona P (2010) The fastest pedestrian detector in the west. In: Proceedings of the British machine vision conference
22. Doll'ar P, Tu Z, Perona P, Belongie S (2009) Integral channel features. In: Proceedings of the British machine vision conference

Chapter 20
A Gray Model Application Method for H.264/AVC Intra Prediction Coding

Ruifang Hu, Ju Liu, Hui Yuan, Chuan Ge and Wei Liu

Abstract Gray System Theory (GST) is employed to solve problems with less data, little sample, and insufficient experience, it has been successfully applied to medicine, industry technology, and so on. In this paper, a Gray Model (GM) based intra prediction method for H.264/AVC is proposed. For a certain block, when all the predicted values (obtained from the existing intra prediction methods) are similar, they are sent to a Gray system so as to obtain the final prediction results. Accordingly, the rate distortion decision procedure could be saved; meanwhile the mode information could also be saved. Since the encoder and decoder can determine whether different prediction results are similar or not, no additional flag information should be included into the bit-stream. Experimental results demonstrate that by integrating the GM-based prediction method into the H.264/AVC Joint Model version 15.1 (JM15.1), an average 2.994 % bit rate and 39.28 % coding time can be saved while maintaining the same quality of reconstructed videos.

Keywords Gray model application · Intra prediction · H.264/AVC

20.1 Introduction

Gray System Theory (GST) was proposed in order to solve the uncertain problems lack of data and information, which was widely used in social life. It was established based on the cognitive information principle, difference information principle, and minimum information theory [1]. Gray Model (GM) (1, 1) is the definition type of GM.

R. Hu · J. Liu (✉) · H. Yuan · C. Ge
School of Information Science and Engineering, Shandong University, Jinan 250100, China
e-mail: juliu@sdu.edu.cn

W. Liu
Hisense State Key Laboratory of Digital Multi-Media Technology, Qingdao 266061, China

A. A. Farag et al. (eds.), *Proceedings of the 3rd International Conference on Multimedia Technology (ICMT 2013)*, Lecture Notes in Electrical Engineering 278,
DOI: 10.1007/978-3-642-41407-7_20,

H.264/AVC gives higher coding efficiency than any other existing video coding standards. One important improvement of H.264/AVC [2] standard is that it contains many complex modes both in intra and inter-frames. As might be expected, the increase in coding efficiency and flexibility comes at the expanse of an increase in complexity with respect to earlier standards. In order to further increase the coding gain, a new standard, i.e., High Efficiency Video Coding (HEVC) [3] is carrying out now. The main difference of intra prediction among those standards is the various forms of prediction directions. In this work, an intra prediction method based on GM is proposed. Motivated by the observation that different prediction methods can give analogous predicted values, and the predicted pixels have the characteristics (less data, little sample, and incomplete) which correspond with the GST [4], GM is utilized to refine the prediction results so as to increase the coding efficiency.

The remainder of this paper is organized as follows. A brief introduction of H.264/AVC and HEVC intra prediction method is given in Sect. 20.2. In Sect. 20.3, the proposed GM-based intra prediction method is presented in detail. Experimental results and conclusions are given in Sects. 20.4 and 20.5, respectively.

20.2 Intra Prediction in H.264/AVC and HEVC

The choice of intra prediction is critical to the efficiency of video coding. In H.264/AVC intra prediction, a predicted block (with the size of 4 × 4, 8 × 8, 16 × 16, etc.) is formed based on previously encoded and reconstructed pixels (neighbouring pixels) [5]. Since the improvement of H.264/AVC, a HEVC standard has been carried out.

The main features being investigated are the size of the blocks and the types of intra prediction. The number of prediction directions used in H.264/AVC is 9 (including DC and directional predictions), Fig. 20.1a. The higher number 33 of prediction directions used in HEVC, Fig. 20.1b [6, 7]. HEVC increased angular intra prediction modes over H.264/AVC by specifying 17 modes for 4 × 4 blocks, 34 modes for 8 × 8, 16 × 16, and 32 × 32 blocks, as well as three modes for 64 × 64 blocks. The increasing number of directions contributes to the coding gain with 5 % BD-rate on average [8].

Based on the description of H.264/AVC and HEVC intra prediction procedure, the predicted pixel $p(i, j)$ could be represented as:

$$p(i,j) = f_r(s_x, s_y) = \frac{\sum_x (\alpha_{r,x} \times s_x)}{\sum_x \alpha_{r,x}} or \frac{\sum_y (\beta_{r,y} \times s_y)}{\sum_y \beta_{r,y}}, \tag{20.1}$$

where S_x, S_y represent the xth and yth adjacent predicted pixels, respectively, is a function that predicts the pixels by utilizing the predicted ones such as S_x, S_y. $\alpha_{r,x}$ and $\beta_{r,x}$ are the corresponding weighted coefficients of xth and yth prediction pixel

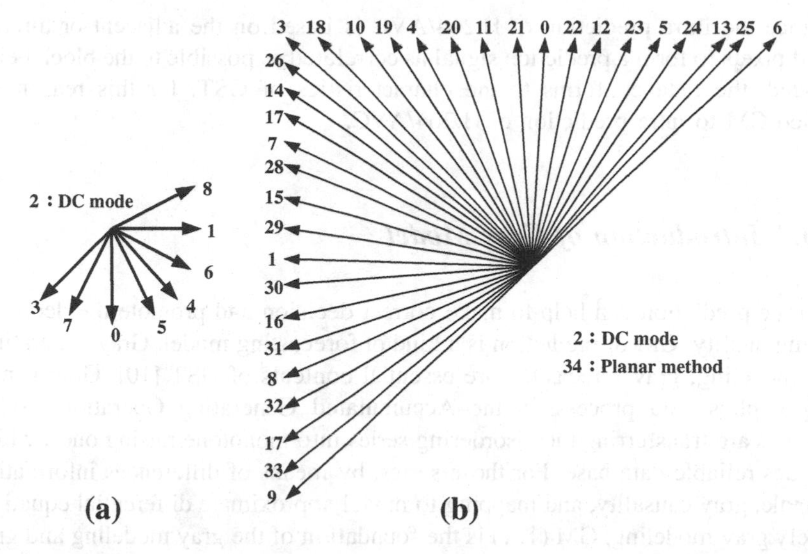

Fig. 20.1 Mapping between intra predictions and modes. **a** Intra prediction modes in H.264/AVC. **b** Intra prediction directions in HEVC

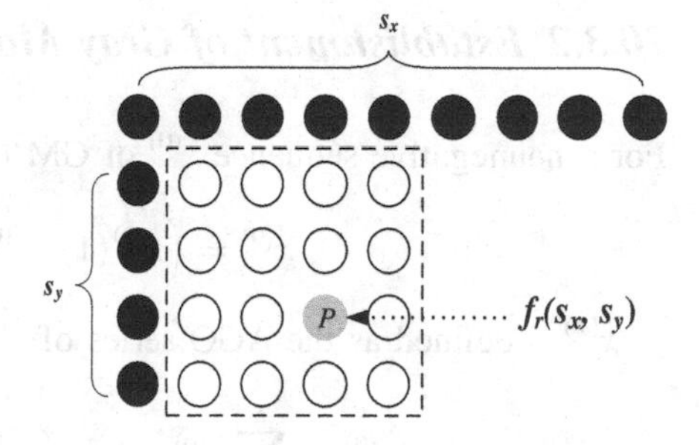

Fig. 20.2 The current predicted in intra prediction of H.264/AVC

under the *r*th prediction mode. The function f_r can be described in Fig. 20.2. The final coding mode of each block is determined by rate distortion optimization (RDO) [9]. Then the residuals of current block are transformed, quantized, and entropy coded to generate the bit-stream.

20.3 Gray Model Method Application for Intra Prediction

The uncertainty problems existing universal in the field of nature and thinking, the probability theory can solve the problems with myriad sample, otherwise the fuzzy mathematics can deal with the kenning uncertainty. However, there also exists another set of issues on uncertainty in less data, incomplete information, which is just suitable to be solved by GST.

Since the intra prediction of H.264/AVC is based on the adjacent or already coded pixels to form a prediction signal as correlated as possible to the block being encoded, the data conforms to the characteristics of GST, for this reason we applied GM to intra prediction of H.264/AVC.

20.3.1 Introduction of Gray Model

Accurate prediction can help to make correct decision and promote the decision-making quality. GM of prediction is a kind of forecasting model. Gray generating, gray modeling, gray forecasting are essential contents of GST [10]. Gray generating implies data processing, the Accumulated Generating Operation (AGO) functions are transferring the disordering series into monotone raising ones, which provides reliable data base. For those series, by means of differences information rationale, gray causality, and mapping to model approximate differential equation, namely gray modeling. GM (1, 1) is the foundation of the gray modeling and gray forecasting, whose distinguishing features are modeling by less four data [11].

20.3.2 Establishment of Gray Model

For a nonnegative sequence $x^{(0)}$ of GM (1,1) is defined as:

$$x^{(0)} = \left(x^{(0)}(1), x^{(0)}(2), \cdots, x^{(0)}(n)\right) \tag{20.2}$$

$x^{(1)}$ is defined as the AGO series of $x^{(0)}$, which is computed as:

$$x^{(1)} = \text{AGO}x^{(0)} = \sum_{m=1}^{k} x^{(0)}(m) = \left(x^{(1)}(1), x^{(1)}(2), \cdots, x^{(1)}(n)\right), \forall k. \tag{20.3}$$

$z^{(1)}$ is defined as the MEAN sequence of $x^{(1)}$, for each $z^{(1)}(k) \in z^{(1)}$ is computed as:

$$z^{(1)}(k) = \text{MEAN}x^{(1)} = 0.5x^{(1)}(k) + 0.5x^{(1)}(k-1), \tag{20.4}$$

The GM can be abstracted the differential equation shown as:

$$\frac{dx^{(1)}}{dt} + ax^{(1)} = b, \tag{20.5}$$

Equation (20.5) is a transition equation, b represents the gray input, obtaining from identifying, a represents developing coefficient, reflects the growing tendency of $x^{(0)}$ and $x^{(1)}$. Via (20.6) and (20.7), the final predicted value is obtained as:

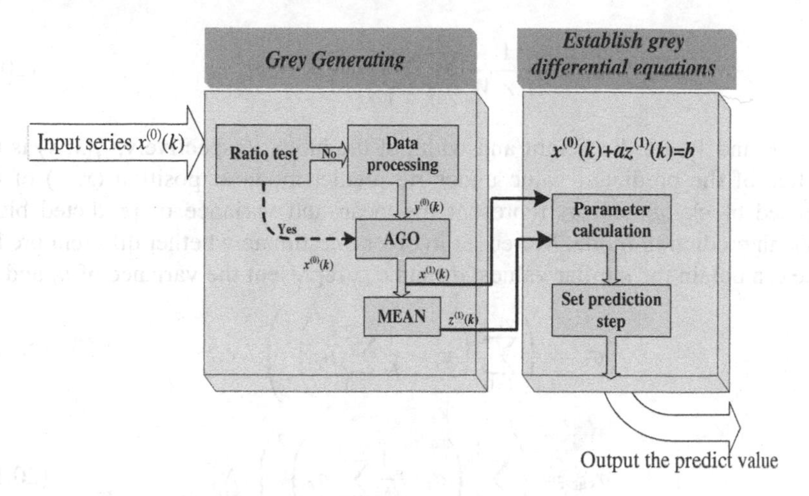

Fig. 20.3 The diagram of gray modeling arithmetic

$$\hat{x}^{(1)}(k+1) = \left(x^{(0)}(1) - \frac{b}{a}\right)e^{ak} + \frac{b}{a}, \tag{20.6}$$

$$\hat{x}^{(0)}(k+1) = \hat{x}^{(1)}(k+1) - \hat{x}^{(1)}(k), \tag{20.7}$$

where $\hat{x}^{(1)}(k+1)$ and $\hat{x}^{(0)}(k+1)$ represent predict values, (20.6) and (20.7) are defined as white model of GM (1, 1), which stands for real differential equation, the high accuracy predict value can be got [11]. The diagram of GM (1, 1) establishment is shown in Fig. 20.3 GM (1, 1).

20.3.3 Intra Prediction Method Based on Gray Model

The establishment of GM was described in [8]. Figure 20.3 shows the logic diagram of GM (1, 1). In (20.1), when S_x and S_y are the same, the predicted value $p(i, j)$ are the same, and even if S_x and S_y are different, the $p(i, j)$ may still be similar because the function f is different under varying prediction modes [12]. Accordingly the complexity process of RDO can be omitted; fewer bits are required to label different prediction modes.

For ordinary applications, means and variances are used to judge whether different predicted blocks similar or not. The mean and variance of each predicted block are computed as [12],

$$\mu_r = \frac{1}{H \times W}\sum_{x=1}^{H}\sum_{y=1}^{W} f_r(x, y), \tag{20.8}$$

$$\sigma_r = \frac{1}{H \times W} \sum_{x=1}^{H} \sum_{y=1}^{W} [f_r(x,y) - \mu_r]^2, \tag{20.9}$$

where H and W are the height and width of the block, respectively, $f_r(x, y)$ is the function of the predicted value under rth predict mode at position (x, y) of the predicted block, μ_r and σ_r represent the mean and variance of predicted block under rth prediction mode. Moreover in order to estimate whether different predict mode can obtain the similar values, σ_u and σ_{var} represent the variance of μ_r and σ_r,

$$\sigma_\mu = \left(\sum_{i=0}^{N-1} \left(\mu_i - \frac{1}{N} \sum_{r=0}^{N-1} \mu_r \right)^2 \right) / N, \tag{20.10}$$

$$\sigma_{\text{var}} = \left(\sum_{i=0}^{N-1} \left(\sigma_i - \frac{1}{N} \sum_{r=0}^{N-1} \sigma_r \right)^2 \right) / N, \tag{20.11}$$

where N is the number of prediction modes of the block. When σ_u and σ_{var} are less than threshold T there is no need to do the RDO process for each predict mode. T is preset according to the quantization step (QP), because even if the predicted values are a little farther from each other, the quantized values are still similar. To correspond to QP, T is set to be $QP/2$ empirically [12]. However, which predict mode can be the best one is not sure, therefore GM (1, 1) is applied to forecasting an accurate value substitute for the best mode, a GM function is modeled as,

$$P_{\text{final}}(i,j) = f_{\text{GM}(1,1)}(p_1, p_2, \cdots, p_r), \tag{20.12}$$

where P_{final} (i, j) represents the final predicted value obtained by GM (1, 1) at position (i, j) of the current block, and the modeling data $(p_1, p_2, \ldots, p_r)$ can be obtained from rth predict mode.

In a bit-stream of H.264/AVC, mode information of all the blocks of an MB are followed by a coded block pattern (CBP) and coefficient bits, as shown in Fig. 20.4a, which means more bits should be used without considering whether the block is suitable for GM. To reduce the information bits, the bit-stream of the MB is changed as in Fig. 20.4b [12], where CBP of the MB is decoded first, and then whether the different predicted mode predicts the similar value is determined, that is whether the block is suitable for GM. The P_{final}, shown in (20.12), is the final prediction result, and the following bits are decided as coefficient.

20.4 Experimental Results

In order to verify the performance of the proposed method, extensive experiments were implemented. The proposed method is implemented in H.264/AVC reference software Joint Model (JM) version 15.1. Ten video sequences are utilized in this

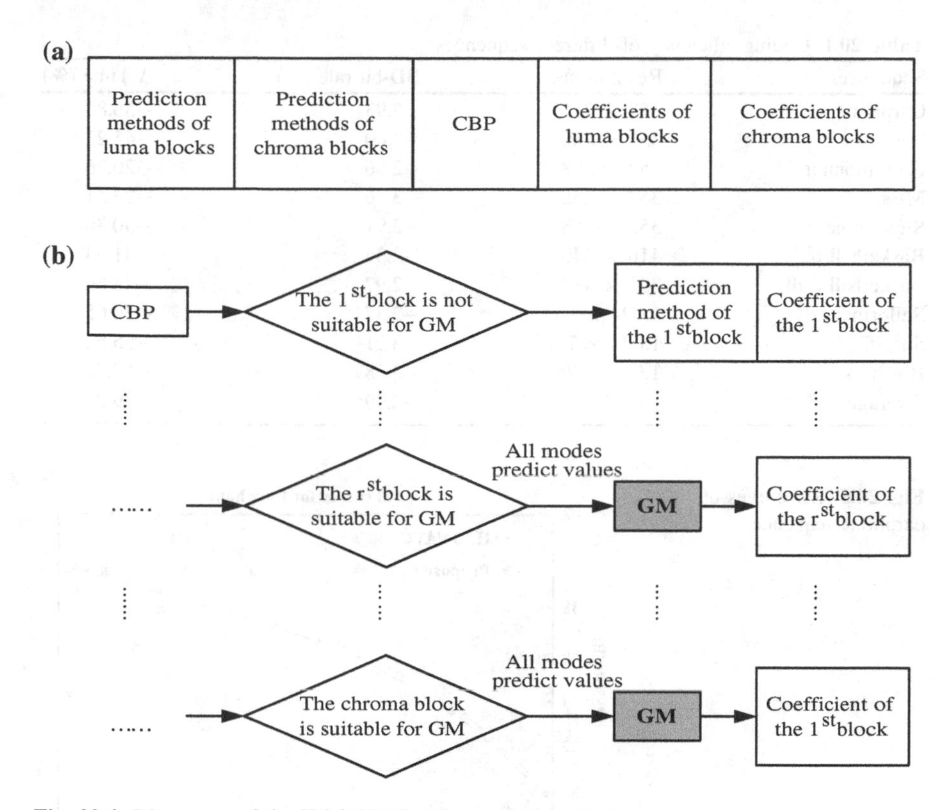

Fig. 20.4 Bit-stream of the H.264/AVC and proposed method

experiment. Each sequence is encoded at four different quantization parameters (QP: 27, 32, 37, 42) [13]. All the frames are intra coded, and CABAC or CAVLC entropy coding is used. The coding efficiency is evaluated by the percentage of videos (BD bit rate [14]). Experimental results are shown in Table 20.1, where negative BD-rate values indicate higher gain. From Table 20.1, it can be seen that by integrating GM (1, 1) into JM15.1, an average of −2.994 % BD bit rate and the 39.28 % coding time could be achieved.

In order to show the results perceptually, rate distortion curves for the *Carphone* sequence is presented in Fig. 20.5, from which we can see the proposed method outperforms JM15.1.

Table 20.1 Coding efficiency of different sequences

Sequences	Resolution	BD-bit rate (%)	Δ Time (%)
Carphone	352 × 288	−3.93	−36.85
Foreman	352 × 288	−2.60	−45.21
Hall_monitor	352 × 288	−2.56	−20.76
Missa	352 × 288	−3.46	−27.51
Sign_Irene	352 × 288	−2.91	−30.36
Basketball pass	416 × 240	−2.35	−41.54
Basketball drill	832 × 480	−2.97	−44.56
Sailormen	1,024 × 768	−2.37	−30.65
Sheriff	1,024 × 768	−3.21	−55.62
Bigships	1,024 × 768	−3.58	−59.77
Average		−2.99	−39.28

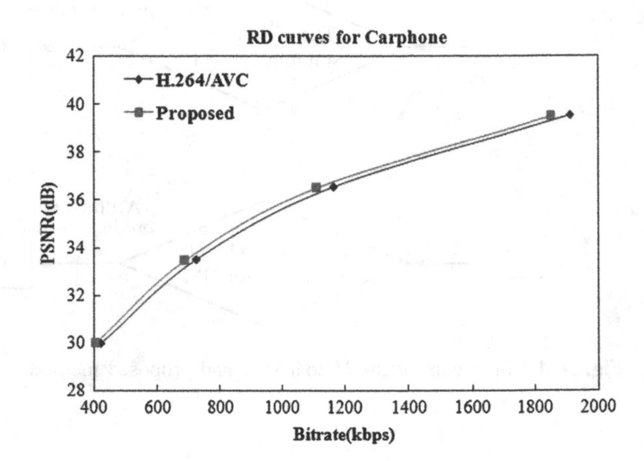

Fig. 20.5 RD curves of carphone sequence

20.5 Conclusions

A GM application method for H.264/AVC intra prediction coding is proposed. In the proposed method, whether different prediction modes of a block predicted values are similar is determined first, and then all the predicted values are input GM so as to obtain the predicted pixels. By combining GM (1, 1) with intra prediction, more accurate prediction result could be obtained. Experimental results show that coding complexity is decreased, and the coding efficiency is improved as well.

Acknowledgments This work was supported partially by the National Basic Research Program of China (973Program.2009CB320905, 2010CB735906), the National Natural Science Foundation of China (61201211, 61001180); by a grant from the Ph.D. Programs Foundation of Ministry of Education of China (No. 20120131120032), the Excellent Youth Scientist Award Foundation of Shandong Province (No. BS2012DX021); China Postdoctoral Science Foundation funded project (2013M530320); China Postdoctoral Special Foundation funded project (2012T50629);

Shandong Postdoctoral Innovation Project Fund (201203053); the Science and Technology Basic Research Project of Qingdao (10-3-4-12-1-jch), thc Cultivation Fund of the Key Scientific and Technical Innovation Project (708059), Independent Innovation Foundation of Shandong University (2011GN061).

References

1. Deng J (2008) Introduction to gray system theory. Wuhan, China
2. Richardson IE (2003) H.264 and MPEG-4 video compression: video coding for next-generation multimedia. John Wiley and Sons, New York, pp 160–184
3. Gabriellini A, Flynn D, Mrak M (2011). Combined intra-prediction for high-efficiency video coding. IEEE J Sig Process 5(7)
4. Deng J (2005) The primary methods of gray system theory. Wuhan, China
5. Joint Video Team of ITU-T and ISO/IEC JTC 1 (2003) Final draft international standard of joint video specification. ISO/IEC 14496-10, ITU-T Rec. H.264
6. Ugur K, Andersson K, Fuldseth A (2010) Video coding technology proposal by tandberg, nokia, and ericsson. document JCTVC-A119 of joint collaborative team on video coding (JCT-VC)
7. Davies T (2010) BBC's response to the call for proposals on video compression technology. Document JCTVC-A 124 of joint collaborative team on video coding (JCT-VC)
8. Sugimoto K (2011) CE10: summary of CE10 on number of intra prediction directions. Document JCTVC-D100 of joint collaborative team on video coding (JCT-VC)
9. Wiegand T, Schwarz H, Joch A, Kossentini F, Sullivan G (2003) Rate-constrained coder control and comparison of video coding standards. IEEE Trans Circ Syst Sig Process 688–703
10. Deng J (2005) Less data principle in gray theory. Wuhan, China
11. Deng J (2002) Gray prediction and gray decision, Revised edn. Wuhan, China
12. Yuan H, Chang Y, Lu Z, Li M (2010) Fast and efficient intraprediction method for H.264/AVC. Opt Eng 49(4):1–3
13. Wallendael G, Leuven S, Cock J, Bruls F, Walle R (2012) 3D video compression based on high efficiency video coding. IEEE Trans Consum Electron 58(1):137–145
14. Bjontegaard G (2001) Calculation of average PSNR diffcrences between RD-curves. Document VCEG-M33 of ITU-T VCEG

Chapter 21
The Research of SVR Algorithms Based on Several Loss Functions and the Application in Exchange Rate Prediction

Fuyong Wan and Lijing Shen

Abstract Support vector machine (SVM) with the introduction of loss function is used to solve the regression problems, which is called support vector regression (SVR). ε-insensitive loss function, Gauss loss function, Huber loss function, and Laplacian loss function are the popular loss functions. The theory and practical application of SVR algorithm based on ε-insensitive loss function are relatively more mature than the other SVR algorithms, so the theory of SVR algorithms based on the other three kinds of loss function still needs further improvement. This paper firstly describes the theory of SVR algorithm based on ε-insensitive loss function. Then the other three kinds of loss function SVR algorithm are deduced from their original optimization problems according to Lagrange dual theory. At last, through data analysis and prediction of exchange rate between China and the USA, we compare SVR algorithms based on the four different loss functions with the advantages of SVR algorithm and Auto Regression (AR) model.

Keywords Support vector regression · ε-insensitive loss function · Gauss loss function · Huber loss function · Laplacian loss function

21.1 Introduction

Support vector machine is a new learning machine proposed by Vapnik et al. [1, 2] in the early 1990s, which is based on statistics learning theory and structural risk minimization principle. It is an important research result in the field of machine

F. Wan (✉) · L. Shen
Department of Mathematics, East China Normal University,
Shanghai 200241, China
e-mail: fywan@math.ecnu.edu.cn

L. Shen
e-mail: 1285870235@qq.com

A. A. Farag et al. (eds.), *Proceedings of the 3rd International Conference on Multimedia Technology (ICMT 2013)*, Lecture Notes in Electrical Engineering 278,
DOI: 10.1007/978-3-642-41407-7_21,

learning and is a new method in data mining. Because of its excellent learning ability to solving the small sample, nonlinear, high dimension problems, support vector machine (SVM) has received wide attention and has been used more and more in pattern recognition, function estimation [3–5], predictive modeling [6–8].

According to the SVM is used to solve the classification and regression problems are respectively called support vector classification (SVC) and support vector regression (SVR). This paper mainly studies the SVR algorithm based on the different loss functions. The common loss functions mentioned here are the ε-insensitive loss function, Gauss loss function, Huber loss function, and Laplacian loss function. Through many references, we find the SVR algorithm theory based on ε-insensitive loss function is mature and widely used in many real data experiments. Relatively speaking, several other loss functions and their applications in data processing need further study. This paper does some improving work in this area, and puts the theory into the practical operation, and realizes the forecast of exchange rate data by Matlab programs. Then the experimental results are compared and analyzed, which give a more comprehensive understanding of all kinds of loss function. The methods of choosing some parameters involved there will be some inspiration for the need to carry out data processing by using this theory.

The data used in this paper is from Wind Information, which is the exchange rate of RMB against the dollar during the period from July 1, 2005 to September 20, 2010 including 1,280 data. Using the method of general data processing, we divide all data into two parts: the first part as a training set, and the latter as a test set. Then we use the AR model to determine the autoregressive order and construct the SVR model for them.

This paper mainly includes the following contents: the first section is an introduction, the second section is the SVR theory, the third section is the research on SVR algorithm with different loss functions, the fourth section is the analysis and results of numerical experiments, and the final section gives the summary and prospect.

21.2 SVR Machine Theory

Given a training set $T = \{(x_1, y_1), \cdots, (x_l, y_l)\} \in (X \times Y)^l$, where $x_i \in X = R^n$, $y_i \in Y = \{-1, 1\}$, $i = 1, \cdots, l$, we try to seek a real-valued function $g(x)$ on $X = R^n$. Through the decision function $y = f(x) = \mathrm{sgn}(g(x))$, for any pattern x, we can determine its category by the value y. In fact the classification problem is to find the rules which can divide points on R^n into two parts. Classification problems we mentioned is commonly the problems dealing with two classes. Intuitively, there are three types of two-class classification problems, namely linear separable problems, approximately linear separable problems and essentially linear nonseparable problems.

Given a training set $T = \{(x_1, y_1), \cdots, (x_l, y_l)\} \in (X \times Y)^l$, where $x_i \in X = R^n$, $y_i \in Y = \mathbf{R}$, $i= 1, \cdots, l$, we try to seek a real-valued function $f(x)$ for the decision function $y = f(x)$ valued on $X = R^n$ to infer the corresponding values y of arbitrary pattern x. With regard to the two-classification problems, there are only two output values, while in the regression problems the output can be arbitrary real numbers. By referring to the method for solving classification problems, regression problem will be smoothly and easily solved. The main idea is converting the regression problem into a classification problem. The specific process is as follows: first we construct the positive class point set D^+: $D^+ = \{(x_i^T, y_i + \varepsilon')^T, i = 1, \cdots, l\}$, and the negative class point set D^-: $D^- = \{(x_i^T, y_i - \varepsilon')^T, i = 1, \cdots, l\}$, which come from the training set T by increasing ε' and decreasing ε' from each training point values y. Now we get a new training set T':$\{((x_1^T, y_1 + \varepsilon')^T; 1), \cdots ((x_l^T, y_l + \varepsilon')^T; 1), ((x_1^T, y_1 - \varepsilon')^T; -1), \cdots, ((x_l^T, y_l - \varepsilon')^T; -1)\}$, where the original outputs y_i are regarded as input indicators, and we add output index $z_i = 1$ or $z_i = -1$, namely, $z_i \in Z = \{-1, 1\}$. So we get two kinds of positive and negative points of the two-class classification problem. There are three types of the classification problems, while there are only two types of regression problems, namely, linear regression and nonlinear regression problem.

21.3 Research on SVR Algorithm with Different Loss Functions

SVM based on the introduction of the loss function is applied to solve the regression problems, which is named the SVR machine. At present, either in theory or in practical applications studies are more focused on relatively simple form of the ε-insensitive loss function. Analyses of the results based on the same data sample in the practical applications using different loss functions in SVR machine are different. The theory we will refer to SVR algorithm based on the ε-insensitive loss function, use Lagrange duality theory to deduce the other three SVR algorithms corresponding to the Gauss loss function, Robust loss function, and Laplacian loss function.

21.3.1 SVR Algorithm Based on ε-Insensitive Loss Function

In SVR algorithm, the most commonly used loss function is the ε-insensitive loss function, i.e., $L_\varepsilon(\xi) = \begin{cases} 0, & |\xi| < \varepsilon; \\ |\xi| - \varepsilon, & |\xi| \geq \varepsilon. \end{cases}$

(1) For the linear regression problem, we suppose the regression hyperplane is $z = (w \cdot x) + b$, the corresponding original optimization problem [9, 10] is:

$$\text{(P1)} \min_{w,b} \frac{1}{2}\|w\|^2 + \frac{C}{l}\sum_{i=1}^{l}(\xi_i + \xi_i^*)$$

$$\begin{aligned} s.t. \quad (w \cdot x_i) + b - z_i \leq \varepsilon + \xi_i, \quad & i = 1, \cdots, l; \\ z_i - (w \cdot x_i) - b \leq \varepsilon + \xi_i^*, \quad & i = 1, \cdots, l; \\ \xi_i, \xi_i^* \geq 0, \quad & i = 1, \cdots, l; \end{aligned}$$

By using the Lagrange duality theory to derive its dual problem, we get:

$$\text{(P1}'\text{)} \min_{\alpha^{(*)} \in R^{2l}} \frac{1}{2}\sum_{i,j=1}^{l}(\alpha_i^* - \alpha_i)(\alpha_j^* - \alpha_j)(x_i \cdot x_j) + \varepsilon\sum_{i=1}^{l}(\alpha_i^* + \alpha_i) - \sum_{i=1}^{l} z_i(\alpha_i^* - \alpha_i)$$

$$s.t. \quad \sum_{i=1}^{l}(\alpha_i^* - \alpha_i) = 0,$$

$$0 \leq \alpha_i, \ \alpha_i^* \leq \frac{C}{l}, \ i = 1, \ 2, \cdots, l;$$

Suppose that the optimal solution of (P1′) is $\hat{\alpha}^{(*)} = (\hat{\alpha}_1, \ \hat{\alpha}_1^*, \ \cdots, \ \hat{\alpha}_l, \hat{\alpha}_l^*)^T$, then $w = \sum_{i=1}^{l}(\hat{\alpha}_i^* - \hat{\alpha}_i)x_i$. If we note the index set as $\Lambda = \{i|\hat{\alpha}_i \in (0, \ C)\}$, $\Lambda^* = \{j|\hat{\alpha}_j \in (0, \ C)\}$, then we can prove that $\Lambda \cap \Lambda^* = \Phi$. Now we denote $|\Lambda|$ as the number of indexes contained in the set Λ, then we get the threshold:

$$b = \frac{1}{|\Lambda| + |\Lambda^*|}\left[\sum_{k \in \Lambda \cup \Lambda^*} z_k - \varepsilon(|\Lambda^*| - |\Lambda|) - \sum_{k \in \Lambda \cup \Lambda^*}\sum_{i=1}^{l}(\alpha_i^* - \alpha_i)(x_i \cdot x_k)\right]$$

(2) For nonlinear regression problems, we can set up a mapping $\Phi(x) : X \rightarrow H$, which H is a high dimensional feature space and the regression problem dealing with $\{\Phi(x_i), \ i = 1, \cdots, l\}$ is linear in H. Now we can replace the regression hyperplane with a regression hypersurface: $z = (w \cdot \Phi(x)) + b$. Therefore, the above algorithm can be generalized by replacing the inner product $(x_i \cdot x_j)$ with a kernel function $K(x_i \cdot x_j)$ so that it can handle the nonlinear regression problems. In addition, because all of the algorithms based on different loss functions can be generalized in the same way, the description (2) are no longer described in the following algorithm.

21.3.2 SVR Algorithm Based on the Gauss Loss Function

For the linear regression problem, the original problem based on Gauss loss function $L_{\text{quad}}(\xi) = \frac{1}{2}\xi^2$ is:

$$(\text{P2}) \min_{w,b} \quad \frac{1}{2}\|w\|^2 + C\sum_{i=1}^{l}(\xi_i^2 + \xi_i^{*2})$$

$$s.t. \quad (w \cdot x_i) + b - y_i \leq \xi_i,\ i = 1, \cdots, l;$$
$$y_i - (w \cdot x_i) - b \leq \xi_i^*,\ i = 1, \cdots, l;$$

By using the Lagrange duality theory to derive its dual problem, we get:

$$(\text{P2}') \max_{\alpha^{(*)} \in R^{2l}} \quad -\frac{1}{2}\sum_{i,j=1}^{l}(\alpha_i^* - \alpha_i)(\alpha_j^* - \alpha_j)(x_i \cdot x_j) + \sum_{i=1}^{l} z_i(\alpha_i^* - \alpha_i)$$
$$-\frac{1}{2C}\sum_{i=1}^{l}(\alpha_i^2 + \alpha_i^{*2})$$

$$s.t. \quad \sum_{i=1}^{l}(\alpha_i^* - \alpha_i) = 0$$
$$\alpha_i^{(*)} \geq 0, i = 1, \cdots, l.$$

Suppose that the optimal solution of (P2′) is $\hat{\alpha}^{(*)} = (\hat{\alpha}_1, \hat{\alpha}_1^*, \cdots, \hat{\alpha}_l, \hat{\alpha}_l^*)^{\mathrm{T}}$, then $w = \sum_{i=1}^{l}(\hat{\alpha}_i^* - \hat{\alpha}_i)x_i,\ b = \frac{1}{|\Lambda|+|\Lambda^*|}\left[\sum_{k\in\Lambda\cup\Lambda^*} z_k - \sum_{k\in\Lambda\cup\Lambda^*}\sum_{i=1}^{l}(\alpha_i^* - \alpha_i)(x_i \cdot x_k)\right]$, where $\Lambda = \{i|\alpha_i > 0\},\ \Lambda^* = \{j|\alpha_j^* > 0\}$.

21.3.3 SVR Algorithm Based on the Huber Loss Function

For the linear regression problem, subject to constraints $\xi \geq 0$, we can get the Huber loss function: $L_{\text{huber}}(\xi) = \begin{cases} \frac{1}{2u}\xi^2, & 0 \leq \xi < u; \\ \xi - \frac{u}{2}, & \xi \geq u. \end{cases}$

The original problem based on the Huber loss function is:

$$(\text{P3}) \min_{w,b} \quad \frac{1}{2}\|w\|^2 + \frac{C}{u}\sum_{i\in I_1}\left(\frac{1}{2}\xi_i^2 + \frac{1}{2}\xi_i^{*2}\right) + \frac{C}{u}\sum_{i\in I_2}(u\xi_i + u\xi_i^* - u^2)$$

$$s.t. \quad (w \cdot x_i) + b - z_i \leq \xi_i,\ i = 1, \cdots, l;$$
$$z_i - (w \cdot x_i) - b \leq \xi_i^*,\ i = 1, \cdots, l;$$
$$\xi_i, \xi_i^* \geq 0,\ i = 1, \cdots, l;$$

where $I_1 = \{i \in \{1,2,\cdots,l\} | \xi_i^{(*)} < u\}$, $I_2 = \{i \in \{1, 2,\cdots,l\} | \xi_i^{(*)} \geq u\}$.

By using the Lagrange duality theory to derive its dual problem, we get:

$$\text{(P3}') \max_{\alpha^{(*)} \in R^{2l}} \quad -\frac{1}{2}\sum_{i,j=1}^{l} (\alpha_i^* - \alpha_i)\left(\alpha_j^* - \alpha_j\right)(x_i \cdot x_j) + \sum_{i=1}^{l} z_i(\alpha_i^* - \alpha_i)$$

$$-\frac{u}{2C}\sum_{i=1}^{l} (\alpha_i^2 + \alpha_i^{*2})$$

$$s.t. \quad \sum_{i=1}^{l} (\alpha_i^* - \alpha_i) = 0,$$

$$0 \leq \alpha_i^{(*)} \leq C, \; i = 1,2,\cdots,l;$$

Suppose the optimal solution of (P3′) is $\hat{\alpha}^{(*)} = (\hat{\alpha}_1, \hat{\alpha}_1^*, \cdots, \hat{\alpha}_l, \hat{\alpha}_l^*)^{\mathrm{T}}$, then $w = \sum_{i=1}^{l} (\hat{\alpha}_i^* - \hat{\alpha}_i)x_i$, $b = \frac{1}{|\Lambda| + |\Lambda^*|}[\frac{u}{C}(\sum_{i \in \Lambda} \alpha_i - \sum_{j \in \Lambda^*} \alpha_j^*) + \sum_{k \in \Lambda \cup \Lambda^*} z_k - \sum_{k \in \Lambda \cup \Lambda^*} \sum_{i=1}^{l} (\alpha_i^* - \alpha_i)(x_i \cdot x_k)]$,
where $\Lambda = \{i | \hat{\alpha}_i \in (0, C)\}$, $\Lambda^* = \{j | \hat{\alpha}_j \in (0, C)\}$.

21.3.4 SVR Algorithm Based on the Laplacian Loss Function

For the linear regression problem, the original problem based on the Laplacian loss function $L_{\text{laplace}}(f(x) - y) = |f(x) - y|$ is:

$$\text{(P4)} \min_{w,b} \quad \frac{1}{2}\|w\|^2 + C\sum_{i=1}^{l} (\xi_i + \xi_i^*)$$

$$s.t. \quad (w \cdot x_i) + b - y_i \leq \xi_i, \; i = 1,\cdots,l;$$

$$y_i - (w \cdot x_i) - b \leq \xi_i^*, \; i = 1,\cdots,l;$$

$$\xi_i^{(*)} \geq 0, \; i = 1,\cdots,l;$$

By using the Lagrange duality theory to derive its dual problem, we get:

$$\text{(P4}') \max_{\alpha^{(*)} \in R^{2l}} \quad -\frac{1}{2}\sum_{i,j=1}^{l} (\alpha_i^* - \alpha_i)(\alpha_j^* - \alpha_j)(x_i \cdot x_j) + \sum_{i=1}^{l} z_i(\alpha_i^* - \alpha_i)$$

$$s.t.\ \sum_{i=1}^{l}(\alpha_i^* - \alpha_i) = 0,$$

$$0 \le \alpha_i^{(*)} \le C,\ i = 1,\ 2, \cdots, l;$$

Suppose that the optimal solution of (P4′) is $\hat{\alpha}^{(*)} = (\hat{\alpha}_1,\ \hat{\alpha}_1^*, \cdots, \hat{\alpha}_l,\ \hat{\alpha}_l^*)^{\mathrm{T}}$, then $w = \sum_{i=1}^{l}(\hat{\alpha}_i^* - \hat{\alpha}_i)x_i$, $b = \frac{1}{|\Lambda|+|\Lambda^*|}[\sum_{k\in\Lambda\cup\Lambda^*} z_k - \sum_{k\in\Lambda\cup\Lambda^*}\sum_{i=1}^{l}(\alpha_i^* - \alpha_i)(x_i \cdot x_k)]$, where Λ and $\Lambda^*$is defined in (P3′).

21.4 Numerical Experiments and Results Analysis

For the experimental data, we can get the optimal solution by selecting the appropriate penalty parameter C, kernel functions, different loss functions in the SVR algorithm by Matlab programs. In order to evaluate the prediction of the SVR model results and parameters (the kernel functions, penalty parameter C, epsilon, loss functions) involved in the algorithm, we use some common statistics as evaluation indexes: (a) Mean Absolute Deviation: MAD = $\frac{1}{n}\sum_{i=1}^{n}|y_i - \widehat{y}_i|$; (b) Mean Square Error: MSE = $\frac{1}{n}\sum_{i=1}^{n}(y_i - \widehat{y}_i)^2$; (c) The regression model coefficient of determination: $R^2 = 1 - \sum_{i=1}^{n}(y_i - \widehat{y}_i)^2 / \sum_{i=1}^{n}(y_i - \overline{y})^2$. Among them, $\{y_i\}$ are the actual data, $\{\widehat{y}_i\}$ are the prediction data, $\overline{y}$ is the mean of $\{y_i\}$, n is the data capacity of $\{y_i\}$. We believe that the smaller the value of MAD, MSE (close to 0) and the bigger the value of R^2 (close to 1) is, the more accurate the model prediction is.

21.4.1 Data Preprocessing

In numerical experiments, we are using a total of 1,280 data of the US dollar versus RMB exchange rates dating from July 1, 2005 to September 20, 2010 (Fig. 21.1). The 1,280 data are divided into two parts. The first 640 data (2005.7.1–2008.2.13) are used as the training set and denoted as a dataset (A); the other 640 data (2008.2.14–2010.9.20) are used as a test set and denoted as a dataset (B). After the logarithmic difference, the nonstationary time series will be transformed into stationary time series. Then we use the AR model and the regression model using SVR algorithm based on different loss functions to get the test set results, finally by the data processing inversion we will get the prediction results. According to mean absolute deviation (MAD), the mean square error (MSE) and

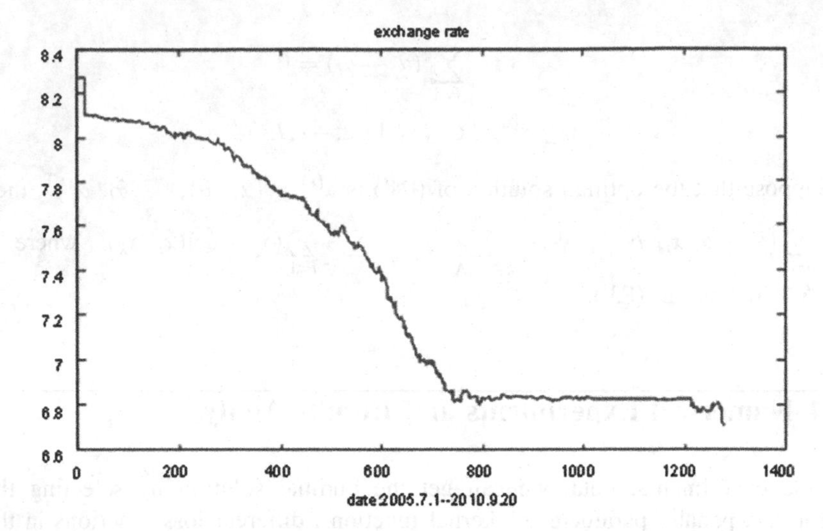

Fig. 21.1 US dollar vesus RMB exchange rates from 2005.7.1 to 2010.9.20

the regression model coefficient of determination (R^2) and other evaluation index, we select parameters suitable to the test of dataset (B).

From the dataset (A), we choose the first 300 data as the training set and calculate the autoregressive order $n = 16$. However, the pole figure shows that not all the poles are inside the unit circle, so the autoregressive model corresponding to the number is unstable. The data after the logarithmic difference will result to the autoregressive order $n = 16$. The pole diagram is as shown in Fig. 21.2. All the

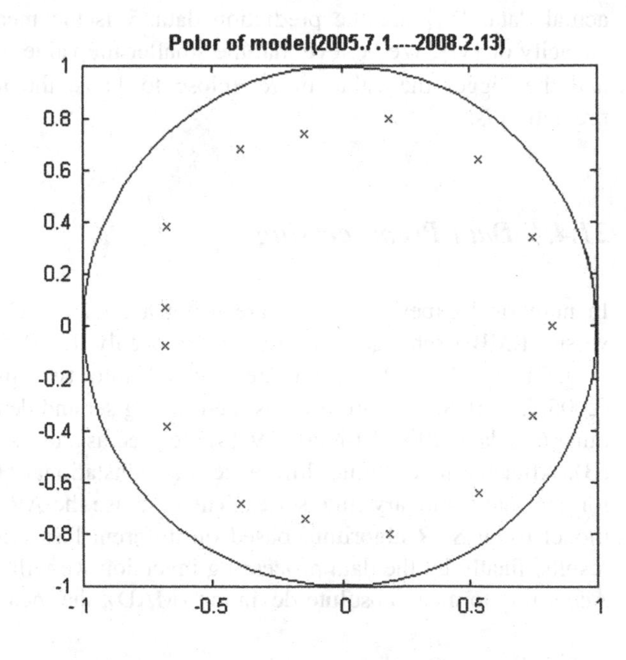

Fig. 21.2 Polor of log differential of exchange rate dataset (A)

poles are inside the unit circle, showing that the corresponding autoregressive model is stable.

For the dataset (B) 2008.2.14–2010.9.20 do the similar processing: logarithmic differencing the data for training and calculating autoregressive order $n = 13$.

21.4.2 The Model Parameters Training for Exchange Rate

First we explain several parameters of the numerical experiments used in Matlab programs: (1) Kernelfun refers to the kernel function, including linear (linear kernel), poly (polynomial kernel function), RBF (Gauss radial basis kernel function); (2) Cz refers to the penalty parameter C; (3) the eps_width has different meanings in different loss functions: in ε-insensitive loss function of SVR algorithm the eps_width is ε width, in the Gauss loss function of SVR algorithm is without reference, in Huber loss function of SVR algorithm it refers to the parameters u, in Laplacian loss function of SVR algorithm its value is zero; (4) lossfun refers to the loss function, i.e., einsensitive (the ε-insensitive loss function), quadratic (the Gauss loss function), Huber (the Huber loss function), and Laplace (the Laplacian loss function).

Through massive progressive experiments and results analysis, ultimately we determine the most appropriate parameters: (1) using a linear kernel function, (2) penalty parameter $C = 50$, (3) another small parameter ε (or u) = 0.0250, (4) the ε-insensitive loss function.

21.4.3 Application of Exchange Rate Prediction Model

Now the above parameters are used in the actual exchange rate data (i.e., dataset (B)) prediction. We list all the four loss functions and the test results in Table 21.1.

From this set of data, the algorithm of SVR based on the ε-insensitive loss function (see the third line of Table 21.1) has more advantages than other three loss function SVR algorithms: (1) the MAD, MSE is very close to others', but the accuracy R^2 is largest; (2) the number of support vectors (NSV) is relatively small.

Table 21.1 Application of exchange rate prediction model

Kernelfun: linear, $C = 50.0000$, $\varepsilon = 0.0250$

Loss function	MAD	MSE	R^2	NSV
Einsensitive	0.00169274	0.00001582	0.96610153	189
Quadratic	0.00231463	0.00001701	0.96355001	299
Laplace	0.00231514	0.00001702	0.96353973	299
Huber	0.00231366	0.00001701	0.96354718	299

Therefore, SVR algorithm based on the ε-insensitive loss function can be the best algorithm when we use SVR machine to solve exchange rate forecasting problems.

21.5 Summary and Prospect

This paper briefly describes the SVC and SVR theory. We finally get four SVR algorithms based on four kinds of different loss functions, deriving the dual problems from the original optimization problems, and constructing the optimal solutions of the dual problems by the original optimization solutions.

We use the AR model to determine the autoregressive order and construct a combination algorithm of SVR model. Through training data and progressive experiments in SVR algorithms based on the different parameters (kernel functions, penalty parameter C, epsilon and loss functions), we get the optimal choice: using the linear kernel function, penalty parameter $C = 50$, another small parameter ε (or take u) = 0.0250, choosing the ε-insensitive loss function. Through the experiments, we find the NSV (the number corresponding of the support vectors), the fourth index of SVR machine based on the ε-insensitive loss function of is relatively small, thus greatly reducing the calculation. Perhaps this is an important reason of choosing the ε-insensitive loss function as the best loss function in the SVR algorithm. Through the actual exchange rate data (i.e., the dataset (B)) prediction analysis, choosing these parameters turns out a more accurate result.

The selection of experimental parameters is not only very difficult, with reference to some experience, but also needs a lot of numerical experiments. If we select a better theory or method to guide the various parameters, we can improve the efficiency of data processing to achieve a better effect. The problem remains to be further discussed and solved.

References

1. Vapnik VN (2000) The nature of statistical learning theory, 2nd edn. Springer, New York
2. Vapnik VN (1998) Statistical learning theory. Wiley, New York
3. Min JH, Lee Y-C (2005) Bankruptcy prediction using support vector machine with optimal choice of kernel function parameters. Expert Syst Appl 28:603–614
4. Mukherjee S, Osuna E, Girosi F (1997) Nonlinear prediction of chaotic time series using support vector machines. In: Proceedings of the IEEE NNSP'97, Amelia Island, pp 24–26
5. Vert J-P (2002) Support vector machine prediction of signal peptide cleavage site using a new class of kernels for strings. Pacific Symp Biocomput 7:649–660
6. Wang D, Wu W (2003) Application of support vector machines regression in predicting Shanghai stock composite index. Wuhan Univ J Nat Sci 8(4):1126–1130
7. Tay FEH, Cao LJ (2001) Application of support vector machines in financial time series forecasting. Omega 29:309–317

8. Kim K-J (2003) Financial time series forecasting using support vector machines. Neurocomputing 55:307–319
9. Deng N, Tian Y (2009) Support vector machine: theory, algorithm and expanding. Science Press, Beijing (in Chinese)
10. Gunn SR (1998) Support vector machines for classification and regression

Chapter 22
Design of 2 MHz CMOS G_m-C Complex BPF with Automatic Tuning for WSN Node

Lingjie Meng and Zhiqun Li

Abstract A fourth-order complex band-pass continuous-time filter with automatic tuning for a WSN transceiver is presented. The active transconductance-capacitor (G_m-C) filter was chosen for low power consumption and designed by series of two two-order filter units. The two-order filter is achieved with the source follower structure, which could get the image rejection as large as possible under the lowest power consumption. The center frequency and bandwidth deviation due to the process corner, aging and temperature deviation are adjusted by an automatic frequency tuning circuit. The core of the filter in a 0.18 μm RF CMOS technology consumes about 0.27 mA current from a 1.8 V power supply. The simulation results show that the bandwidth is about 2.2 MHz and the center frequency is about 2 MHz with automatic frequency tuning. The voltage gain of filter is about −0.8 dB, the ripple in the pass band is lower than 0.1 dB, and the image rejection ratio is larger than 30 dB at −2 MHz.

Keywords Complex band-pass continuous-time Filter · Low power · Automatic frequency tuning · WSN

22.1 Introduction

Wireless Sensor Network (WSN) is one of the three high-tech industries in the future, which is composed of a large number of cheap micro sensor nodes. It is a self-organization network system formed through a wireless communication

L. Meng · Z. Li
Institute of RF and OEICs, Southeast University, Nanjing 210096, China

L. Meng · Z. Li (✉)
School of Integrated Circuit, Southeast University, Nanjing 210096, China
e-mail: zhiqunli@seu.edu.cn

L. Meng · Z. Li
Key Laboratory of Jiangsu Province Sensor Network Technology, Wuxi 214135, China

A. A. Farag et al. (eds.), *Proceedings of the 3rd International Conference on Multimedia Technology (ICMT 2013)*, Lecture Notes in Electrical Engineering 278,
DOI: 10.1007/978-3-642-41407-7_22,

network with the aim of sensing, collecting, and processing the information about objects in the network coverage area and sending them to the viewer. WSN has brought a revolution in information perception with its low-power consumption, low cost, distributed, and self-organization.

RF transceiver module is an important part of the WSN chip, whose performance and power consumption directly affect the whole system. Complex band pass filter (BPF) is a module of RF receiver with the function of allowing the IF signal through, attenuating mirror signal, and adjacent channel signals. Because of the special applications, the WSN RF transceiver chip should be low power consumption and low cost. As a major module of RF receiver, low-power design of complex BPF can help to low power consumption of whole WSN chip. The block diagram of the transmitter is shown as Fig. 22.1.

It is an effective means of reducing power consumption by reducing circuit complexity, which is used in this design. As the operating frequency of the filter is relatively low, the active G_m-C structure is adopted in this complex BPF design. The G_m-C structure is simply made up by several transconductance and capacitors. Without inductances, the area of the complex BPF can be reduced. The complex BPF operates at a supply voltage of 1.8 V, the main circuit DC power consumption is approximately 0.49 mW. This paper first introduces the circuit structure and the basic working principle of the complex BPF, and then an automatic tuning circuit is presented to ensure that the frequency response of the complex BPF is insensitive to fabrication process tolerances, operating temperature tuning and aging. The simulation results of the complex BPF and the automatic tuning circuit are provided in the end.

22.2 Complex BPF Design

The design of the complex BPF is carried out in two steps. First, the architecture of complex BPF is introduced. Then, the circuit of the BPF is provided.

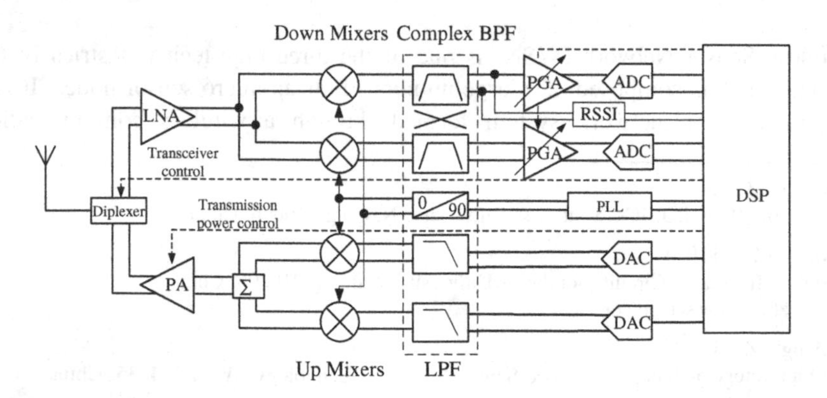

Fig. 22.1 Block diagram of the RF transmitter

22.2.1 Complex BPF Architecture

Figure 22.2 shows how to make a linear frequency transformation of a LPF to the complex BPF and the change of poles and zeros [1–3]. The frequency transfer function Eq. (22.1) derives the structure of a complex BPF based on replacing each pair of the capacitors in I and Q branches. With the transformation, the LPF frequency response is shifted around ω_{IF} and the image signal at $-\omega_{IF}$ outside the pass band of the filter is attenuated.

$$H(s) \rightarrow H(s - j\omega_{IF}) \tag{22.1}$$

As small group delay variation (0.6 μs) and no pass-band ripple, the fourth Butterworth LPF is chosen to realize the complex BPF [3]. Because fully differential circuit realizes the same function with less sub-circuit and has a better inhibition on common mode signal interference, the complete G_m-C complex BPF which is synthesized as shown in Fig. 22.3 is fully differential.

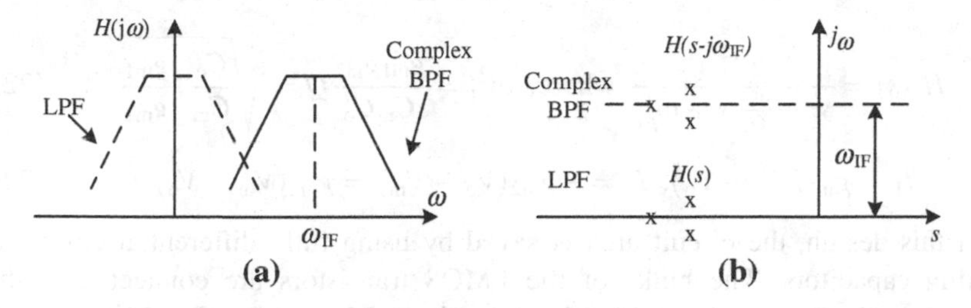

Fig. 22.2 The LPF shifted to BPF, **a** Frequency transformation, **b** The changes of poles and zeros

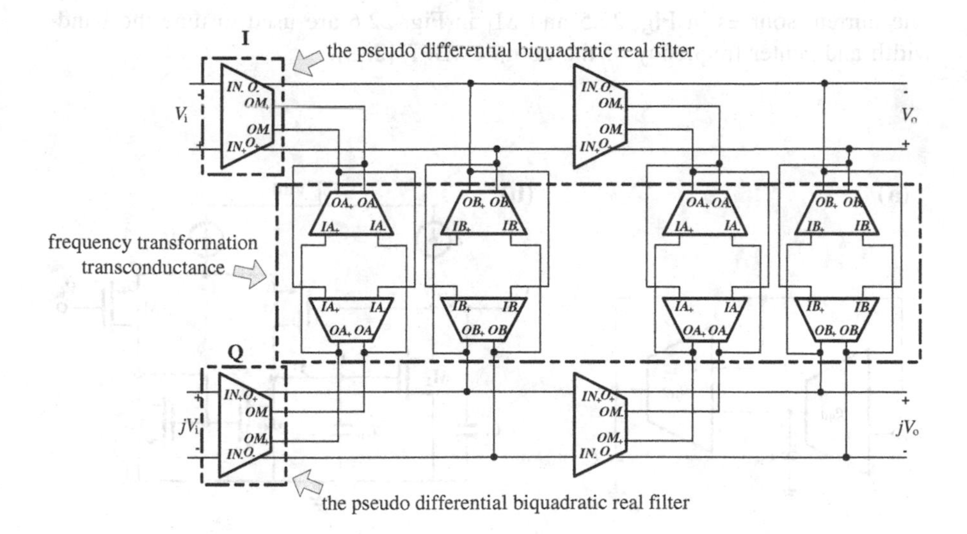

Fig. 22.3 Complete G_m-C complex BPF

22.2.2 Circuit of the BPF

The transfer function of the fourth-order Butterworth LPF is described by Eq. (22.2). For example, the transfer function of the two-order LPF is written as Eq. (22.3), which is realized by the block diagram shown in Fig. 22.4a. The circuits operating at high frequency are simpler, the frequency response is better and the noise is lower. Accordingly, we should choose simple circuit structure which is shown in Fig. 22.4b to realize the two-order LPF. M_1, M_2 in Fig. 22.4b represents g_{m1}, g_{m2} in Fig. 22.4a. The effect of M_3 is buffering the input signal V_{in} to the source of M_1 and M_3, M_1 converts the difference between V_o and V_{in} to the drain current I_1. M_2 subtracts the voltage V_o and V_m and converts it into a current I_2, I_1 and I_2 are calculated by Eq. (22.4). The pseudo differential G_m-C biquadratic real filter used in this paper is shown in Fig. 22.5 [2, 3].

$$H(s) = \frac{a_0}{s^4 + a_3 s^3 + a_2 s^2 + a_1 s + a_0} = \frac{\omega_{p1}^2}{s^2 + \frac{\omega_{p1}}{Q_1} s + \omega_{p1}^2} \cdot \frac{\omega_{p2}^2}{s^2 + \frac{\omega_{p2}}{Q_2} s + \omega_{p2}^2} \quad (22.2)$$

$$H_1(s) = \frac{\omega_{p1}^2}{s^2 + \frac{\omega_{p1}}{Q_1} s + \omega_{p1}^2} \text{ where, } \omega_{p1}^2 = \frac{g_{m1} g_{m2}}{C_m C_o} Q = \sqrt{\frac{C_o}{C_m} \cdot \frac{g_{m1}}{g_{m2}}} \quad (22.3)$$

$$I_1 = g_{m1}(V_{in} - V_o),\ I_2 = -g_{m2}(V_o - V_m) = g_{m2}(V_m - V_o) \quad (22.4)$$

In this design, the circuit area is saved by using fully differential circuit and floating capacitors. The bulks of the PMOS transistors are connected to their sources, so there is good matching between them. Meanwhile, Fig. 22.6 shows the linear frequency transformation realized by single transistor. The output voltage of the transconductance circuit is stabilized by the common mode feedback (CMFB). The current sources in Fig. 22.5 and M_1 in Fig. 22.6 are used to tune the bandwidth and center frequency of the complex BPF [2, 3].

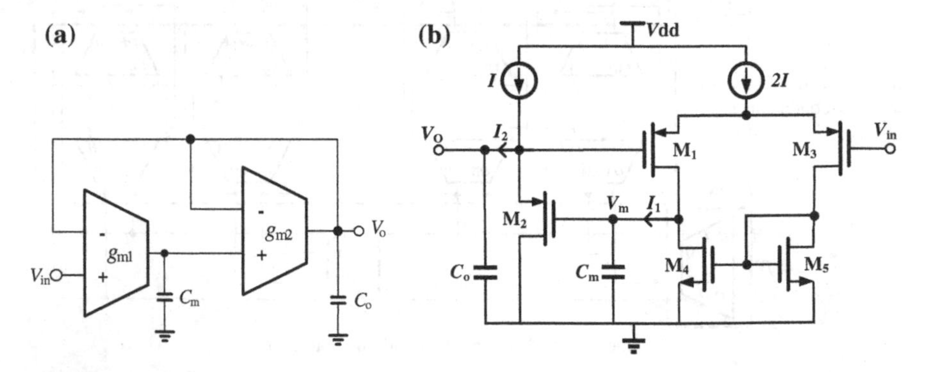

Fig. 22.4 Single-ended G_m-C biquadratic real LPF, **a** The block diagram, **b** The circuit

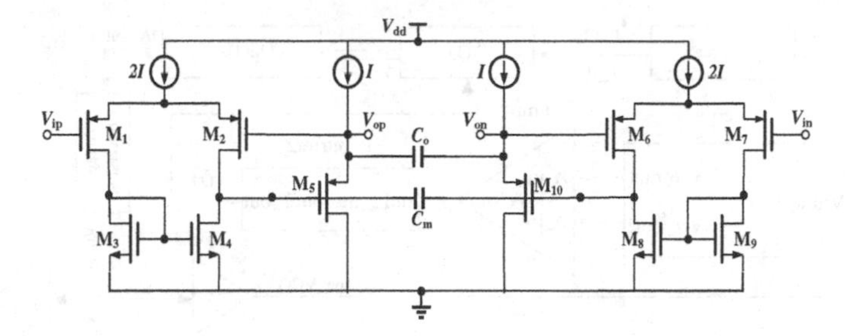

Fig. 22.5 The pseudo differential G_m-C biquadratic real filter

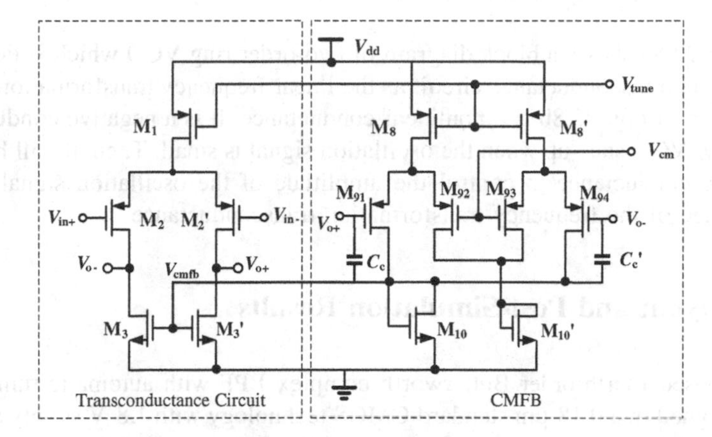

Fig. 22.6 Transconductance for linear frequency transformation

22.3 Automatic Frequency Tuning Circuit Design

Automatic tuning is very important for an active filter to achieve accurate frequency response. The block diagram of the automatic frequency tuning circuit which is consisted of VCO, amp, DFD, 7-bit DAC, voltage conversion, and buffers is shown in Fig. 22.7 [1, 4]. It is used for both center frequency tuning and bandwidth tuning. The external reference frequency is 16 MHz.

The automatic frequency tuning circuit works as follows. First, the digital frequency discriminator (DFD) produces digital signals which are used to control the 7-bit digital-to-analog convert (DAC) by detecting the frequency difference between the signal generated by voltage-controlled oscillator (VCO) and the external reference signal. Then, the analog signal which is generated by the DAC under the control of the digital signals converts to the voltage V_{tune}. Under this control voltage, the frequency of the signal generated by the VCO and the frequency response are changed. This process doesn't stop until the frequency of the oscillation signal and the center frequency of the complex BPF are equal.

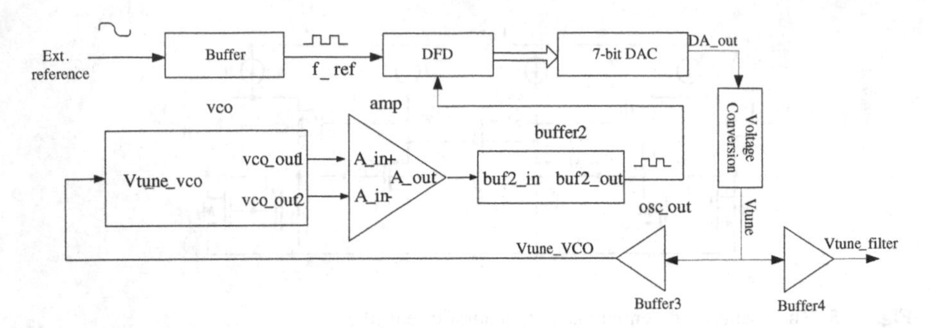

Fig. 22.7 The block diagram of automatic frequency tuning

Figure 22.8a shows a block diagram of two-order ring VCO which is designed by the same transconductance circuit as the linear frequency transformation [5, 6]. G_{NL} shown in Fig. 22.8b is a nonlinear conductance. It is a negative conductance to help the VCO start-up when the oscillation signal is small. Then, it will become a positive conductance to control the amplitude of the oscillation signal in the linear range of the frequency transformation transconductance.

22.4 Layout and Post-Simulation Results

The proposed fourth-order Butterworth complex BPF with automatic tuning has been designed in a 0.18 μm standard CMOS technology with 1.8 V supply voltage and simulated with Cadence. Figure 22.9 shows the layout of the whole circuit

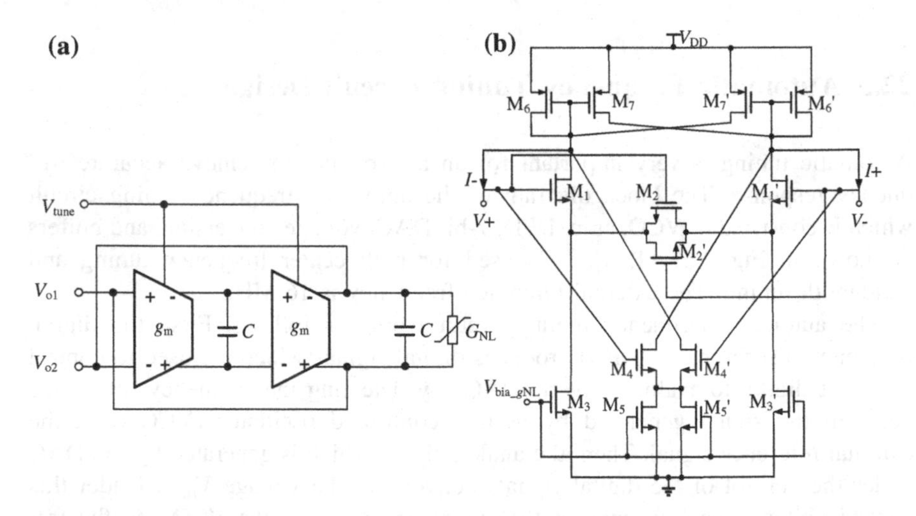

Fig. 22.8 VCO and GNL, **a** The block diagram of two-order ring VCO, **b** Nonlinear conductance

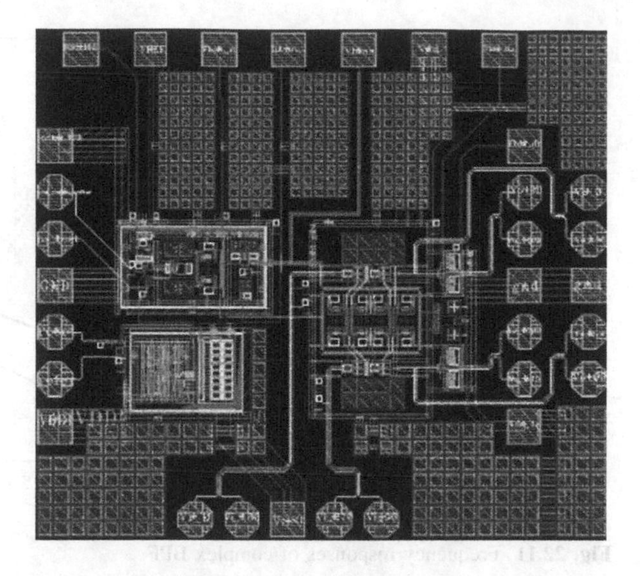

Fig. 22.9 Layout of the whole circuit

which size is about 1.22 mm × 1.08 mm, including ESD and pads. We must note the match between I and Q branches, and the match between g_m cells in VCO and in the complex BPF.

Under the temperature of 27° and TT corner, the transient waveforms of the DAC out and the control signal V_{tune} produced by the automatic tuning circuit are shown in Fig. 22.10. The value of V_{tune} finally reaches 1.188 V.

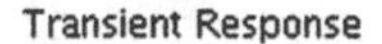

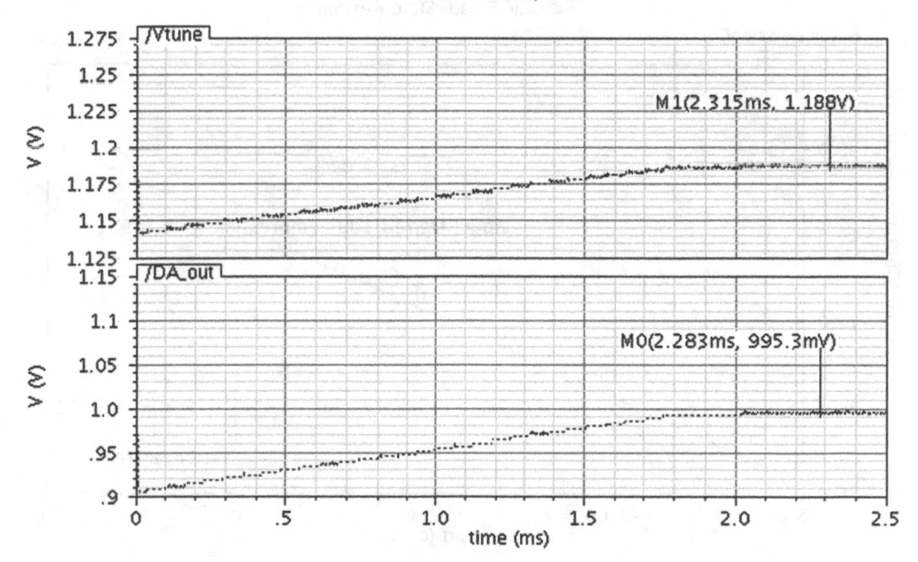

Fig. 22.10 Transient waveforms of the DAC_out and V_{tune}

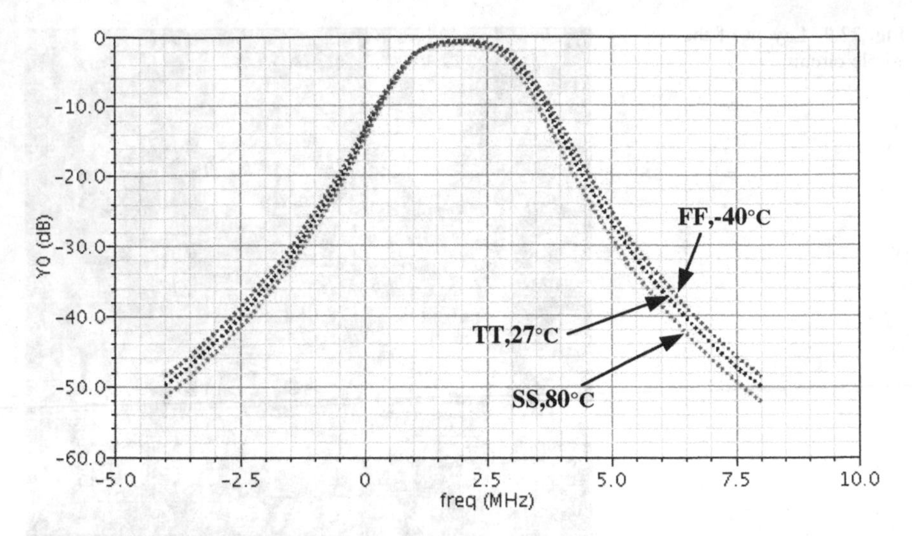

Fig. 22.11 Frequency responses of complex BPF

Figure 22.11 shows the frequency responses of the complex BPF under various simulation conditions. From this figure, we can see that the frequency response of the filter is relatively stable with the tuning error which is less than 5 %.

Under the temperature of 27° and TT corner, the $IP_{1\mathrm{dB}}$ of the complex BPF is –4.77 dBm, as shown in Fig. 22.12. The bandwidth is about 2.2 MHz and the center frequency is about 2 MHz with the automatic frequency tuning circuit working.

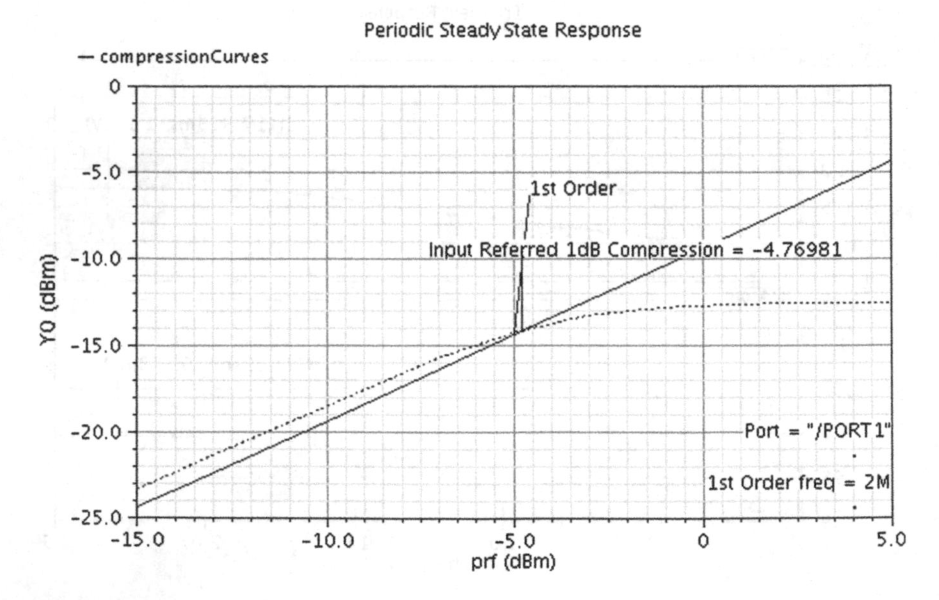

Fig. 22.12 Simulation $IP_{1\mathrm{dB}}$

Table 22.1 Simulation results summary of complex BPF (Temperature: 27°, TT corner)

Parameters	Value
Filter response	Butterworth
Order	4
Center frequency	2 MHz
Bandwidth	2.2 MHz
Pass-band gain	−0.8 dB
Input 1 dB compression point	−4.77 dBm
Image rejection	36.58 dB
Pass-band ripple	<0.1 dB
Supply voltage	1.8 V
Consumed current (with automatic tuning)	0.54 mA
Frequency tuning error	<5 %

The voltage gain of filter is about −0.8 dB, the ripple in the pass band is lower than 0.1 dB, and the image rejection ratio is about 36 dB at −2 MHz. The complex BPF consumes 0.27 mA and the automatic frequency tuning consumes 0.27 mA from a 1.8 V power supply, so the total power dissipation of the complex BPF with automatic frequency tuning is about 0.97 mW for a single supply voltage of 1.8 V. Table 22.1 summarizes the simulation results of the complex BPF with automatic frequency tuning.

22.5 Conclusions

This paper has designed a fully differential Butterworth complex BPF with automatic frequency tuning which consumes 0.54 mA with 1.8 V single supply voltage. According to the simulated results, this design satisfies the requirements of WSN nodes.

References

1. Emira AA, Sanchez-Sinencio E (2003) A pseudo differential complex filter for bluetooth with frequency tuning. IEEE Trans Circuits Syst II Analog and Digit Signal Process 50(10):742–754
2. Liu H, Fu Z, Lin F (2012) A low power G_m-C complex filter for ZigBee receiver. In: ICMMT 2012 international conference on microwave and millimeter wave technology, pp 1–4. doi: 10.1109/ICMMT.2012.6229961
3. Zanbaghi R, Atarodi M (2006) An ultra low power G_m-C complex filter for low-IF wireless PAN applications. In: TENCON 2006-IEEE region 10th conference, 14–17 Nov 2006, pp 1–4. doi:10.1109/TENCON.2006.344131
4. Sun J, Li Z, Li Z, Li W (2011) Design of tuning circuit for a G_m-C complex band-pass filter. J Southeast Univ (Natural Sci Ed) 41(4):682–686. doi:10.3969/j.issn.1001-0505.2011.04.005

5. Krummenacher F, Joehl N (1988) A 4 MHz CMOS continuous-time filter with on-chip automatic tuning. IEEE J Solid-State Circuits 23(3):750. doi:10.1109/4.315
6. Wan C, Li Z (2009) A 2 MHz CMOS G_m-C complex filter with automatic tuning for wireless sensor networks application. In: 15th Asia-Pacific conference on communications (APCC 2009), pp 199–202. doi:10.1109/APCC.2009.5375657

Chapter 23
Real-Time Quality Monitoring for Networked H.264/AVC Video Streaming

Jiarun Song and Fuzheng Yang

Abstract In this paper a packet-layer model for video quality assessment is proposed to monitor the service quality of networked H.264/AVC video streaming in real-time. Using the information extracted from packet header, the frame type is first predicted, based on which the structure of the group of picture (GOP) is estimated since quality degradation caused by packet loss is significantly affected by the loss position. Then the number of impaired frames due to packet loss is calculated as an input of the proposed model. Moreover, since the video quality significantly relies on the motion characteristic of the video content, an algorithm to measure the temporal complexity is designed and incorporated into the proposed model. Eventually, the quality of each GOP can be calculated and the correspondingly video quality is derived using the temporal pooling strategy. Experimental results show that the proposed model remarkably outperforms the model recommended as ITU-T G.1070.

Keywords Video quality · Packet-ayer model · Networked video streaming · Packet loss · Temporal complexity

23.1 Introduction

With the explosive development of technologies in networks and H.264/AVC video compression, the networked video application has become popular in our daily life. However, the quality of these applications cannot be guaranteed in an IP network due to its best-effort delivery strategy. The networked videos are often subject to packet loss, delay, and jitter when transmitted over the network,

J. Song (✉) · F. Yang
State Key Laboratory of Integrated Service Networks, Xidian University, Xi'an, China
e-mail: jrsong@stu.xidian.edu.cn

A. A. Farag et al. (eds.), *Proceedings of the 3rd International Conference on Multimedia Technology (ICMT 2013)*, Lecture Notes in Electrical Engineering 278,
DOI: 10.1007/978-3-642-41407-7_23,

therefore, it is essential to employ quality evaluation of the network video for quality of service (QoS) planning or control in the involved applications [1].

Packet loss is one of the major factors which influence the perceived quality for video streaming over the IP network. A variety of objective quality assessment methods have been reported to evaluate quality degradation caused by packet loss. In [2] and [3] for example, the packet loss rate is employed to estimate the video quality. Considering burst packet loss, more accurate estimations of the related distortion, resorting to both the packet loss rate and the average burst length, have been introduced in [4] and [5]. However, due to the fact that the motion compensation scheme has been commonly used in the video codec, packet loss impacts not only the current video frame but also its successive frames by error propagation. Therefore, since only statistical parameters are considered, the detailed influence of lost packets on a specific video cannot be well captured by these models. To find a more accurate mapping between packet loss and its impact on quality degradation, the temporal complexity, representing the moving character of a video, is taken into account in [6] and [7] by using quantization parameter (QP) and motion vector (MV), respectively. However, these parameters cannot be accessed if the video streaming is encrypted and correspondingly it is difficult to obtain the temporal complexity in this situation. Therefore, how to efficiently and accurately evaluate the video quality affected by the packet loss is still left open.

Since the packet-layer model only utilizes information from packet headers, it is very efficient in quality monitoring due to its low complexity, especially suitable for real-time quality monitoring at network internodes [8]. In this paper, a novel packet-layer assessment model is proposed to evaluate the quality of networked video considering packet loss. The main contributions of this paper can be summarized as follows: (1) though limited information is obtained from the packet header, the motion characteristic of the video content is considered and an efficient estimated temporal complexity is introduced to evaluate the video quality affected by packet loss, (2) knowing the error position caused by packet loss, the quality of each GOP is accurately evaluated and the video quality is assessed by temporal pooling strategy. Specially, in this work it is assumed that direct packet loss always leads to frame loss, which holds true when a corrupted frame is to be discarded by the error control strategy, and is always valid for low bit-rate transmissions where a packet typically contains an entire frame.

The remainder of this paper is organized as follows: Sect. 23.2 presents a detailed description of the frame type detection. The quality assessment model is addressed in Sect. 23.3. Performance evaluation and conclusions are given in Sects. 23.4 and 23.5, respectively.

23.2 Frame Type Detection

For a packet-layer model, the coding type of each frame is not directly available since the payload information cannot be accessed. However, due to the fact that different frames have different sensitivity to packet loss, the knowledge of the frame type information will consolidate the quality assessment model [9]. Accordingly, the analysis of the video frame distortion caused by packet loss will be more accurate than that using the packet loss rate only.

As a general principle, video coding exploits the spatial redundancy through intra-frame coding and resorts to inter-frame coding to remove temporal redundancy, where inter-frame coding modes are more efficient in redundancy removal. Accordingly, the I-frame size is always much larger than P-frame size. Therefore, a threshold is set to distinguish I-frames from the other frame types. For example, Fig. 23.1 shows the frame size of the sequence "Grandma" and "Soccer" in a Group of Picture (GOP) with I and P frame at the bit-rate of 128 kbps, where we can see that for each sequence the size of an I-frame is larger than that of a P-frame.

To identify the threshold value of the I-frame, experiments have been carried out using 10 standard QCIF sequences, i.e., "Soccer", "Carphone", "Foreman", "Paris", "Grandma", "Hall", "News", "City", "Mother-Daughter", and "Highway". All the sequences were encoded using x264 codec with the bit-rate ranged from 48 to 512 kbps. The GOP structure was "IPPP" with the length of 60. The length of the sliding window was set at 200 frames. By analyzing the frame size of each coding sequence, the threshold of the I-frame T_I can be expressed as follows:

$$T_I = r_1 \cdot \frac{1}{\lceil r_2 \cdot M \rceil} \cdot \sum_{j=1}^{\lceil r2 \cdot M \rceil} F_{\text{size}}(n_j) \tag{23.1}$$

where $\lceil \ \rceil$ is the top integral function. M is the length of the sliding window which is set at 200 in the experiment. $F_{\text{size}}(n_j)$ is the size of frame n in the sliding window which is the jth frame in descending order, r_1 and r_2 are constants obtained by the amounts of experiments through regression.

By using the threshold of T_I, the I-frames can be distinguished from other frame types. If the size of a frame is larger than T_I, the frame is predicated as an I-frame; otherwise, the frame is predicated as a P-frame. The corresponding accuracy of frame detection is 99.25 % in average, which demonstrates the effectiveness of the proposed frame type detection method.

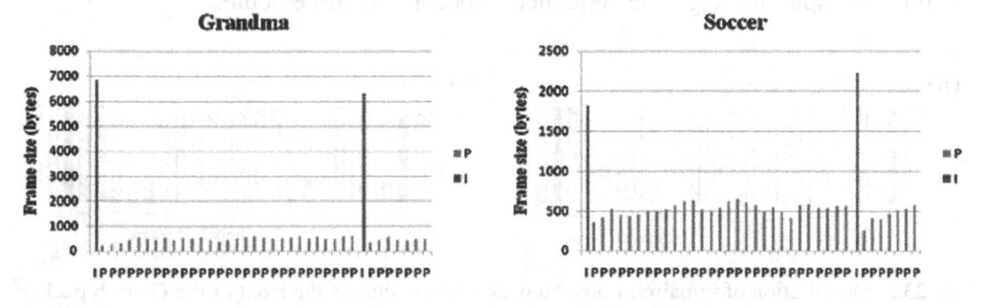

Fig. 23.1 Frame size of different sequences

23.3 Estimation of Distortion Caused by Packet Loss

When a packet gets lost, not only the current video frame is subject to errors but also the subsequent frames will be affected due to error propagation. The severity of video quality degradation caused by packet loss also depends on the GOP structure and different manners of errors propagate. In this section, key factors that greatly affect the quality of the video will be discussed, based on which the quality of each GOP is estimated and then further integrated to obtain the overall video quality.

23.3.1 Effect of the Number of Impaired Frames

The impaired frame is defined as the frame which is contaminated by the packet loss or by error propagation. When a packet gets lost, the number of impaired frames is not specific since it may occur at different positions. Figure 23.2 shows the distribution of the impaired frames with different positions of packet loss. It is clear that when packet loss occurs in the front of a GOP as shown in Fig. 23.2a, the impaired frames are more than those in Fig. 23.2b which the packet loss occurs closely to the end of the GOP. Consequently, the distortion in these two situations differs.

To evaluate the impact of impaired frames number, the sequences "Soccer", "Carphone", "City", and "Grandma" with a lost packet in the second GOP were examined and the loss position is chosen to lead to different numbers of impaired frames. Then the second GOP is extracted for subjective test, which is described in detail in the next section. The subjective quality of the corresponding video without packet loss was also evaluated, denoted as coding quality. Figure 23.3 shows the relationship between the quality degradation and the number of impaired frames in the second GOP of each sequence at the bit-rate of 150 kbps with the GOP size of 2 s. It can be found that the quality degradation increases approximately linearly with the growth in the number of impaired frames at first, and tends to steady when the number reaching a certain value. In addition, the slope of the linear piece is different for different video sequences. Compared with a low temporal complexity sequence, e.g., "Grandma", the quality degradation of the sequence with a higher temporal complexity, e.g., the sequence "Soccer", is more acute.

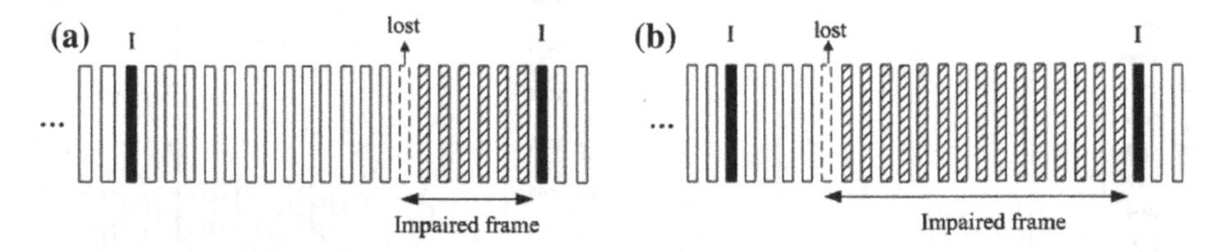

Fig. 23.2 Distribution of impaired frames **a** packet loss occurs in the front of the GOP; **b** packet loss occurs close to the end of the GOP

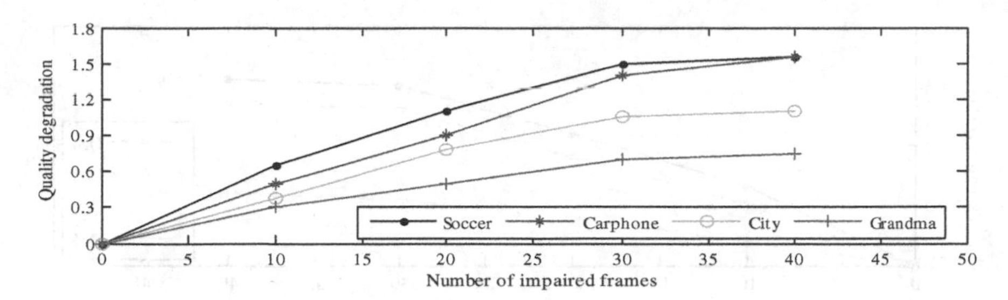

Fig. 23.3 Relationship between quality degradation and impaired frames number for different sequences

From the Fig. 23.3, it is clear that for different movement characteristics of the sequence, the degree of quality degradation due to the packet loss is not identical. The temporal complexity, which generally means the acuteness of changes in the temporal domain, is of great significance to evaluate the quality of a specific video. In [10], the magnitude of motion vectors is used to interpret the temporal complexity. It can also be evaluated based upon the difference between the pixel values (of the luminance plane) at the same spatial location but at successive times or frames as recommended by ITU-T P.910 [11]. However, for a packet layer model, neither the motion vectors nor the residuals are available for quality assessment. Therefore, in our earlier work [12], the temporal complexity δ_T is estimated by the ratio of the frame size for coding I-frames and P-frames as follows:

$$\delta_T = \frac{a_1 \cdot \ln(BR) + a_2}{\ln(B_I) - \ln(B_P)} \tag{23.2}$$

where BR is the average bit-rate which can be calculated by the information extracted from packet header, B_I is the frame size of the I-frame, B_P is the average size of P-frames, and a_1, a_2 are constants empirically obtained by experiments.

23.3.2 Quality Degradation Caused by Coding

Apart from the number of impaired frames and temporal complexity, the degree of quality degradation caused by packet loss is also varying with the video quality related to coding. To examine the impact of the video coding quality, the sequence "Soccer" was encoded at the different target bit-rates with a fixed number of impaired frames. Then the second GOP is extracted for subjective tests to measure its quality degradation. It should be noticed that the video coding quality Q_c can be obtained when the number of the impaired frame is set to zero.

Figure 23.4 shows the relationship between the quality degradation and the number of the impaired frames at different coding bit-rates 48, 80, 128, and 150 kbps, corresponding to different values of coding quality Q_{c1}, Q_{c2}, Q_{c3} and

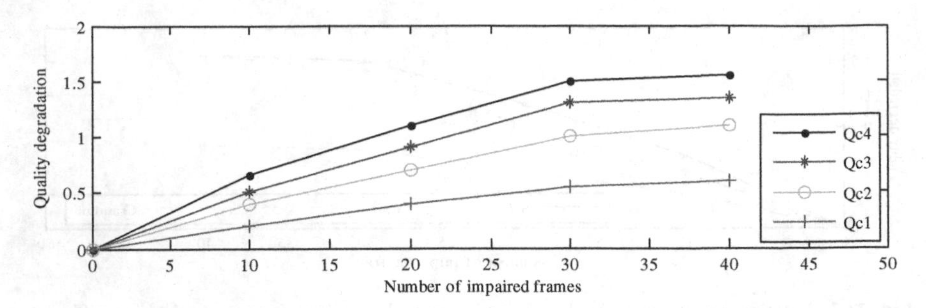

Fig. 23.4 Relationship between the quality degradation and the number of impaired frames at different bit-rates, e.g., "Soccer"

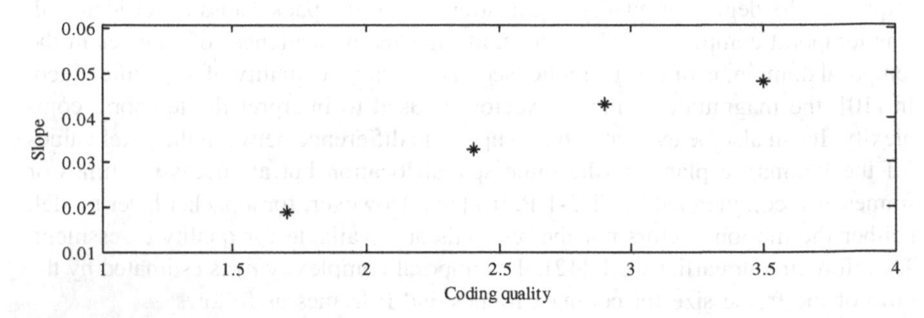

Fig. 23.5 Relationship between the slope of the linear piece in Fig. 23.4 and the coding quality

Q_{c4}. When a packet loss occurs at a high bit-rate, even a slight error due to packet loss will be rather visible, while it may not be very noticeable in a sequence of low quality due to quantization [6]. It can be seen that there is a linear piece in each curve in Fig. 23.4 when the number of impaired frames is less than 30. Figure 23.5 gives the relationship between the slope of the linear piece and the video coding quality.

Based on the analysis above, the quality degradation of the GOP introduced by packet loss, i.e., D_g, can be predicted by the number of impaired frames, the video coding quality, and the temporal complexity, as follows:

$$D_g = \begin{cases} a_3 \cdot \delta_T \cdot \frac{N_{err}}{30} \cdot Q_c N_{err} 30 \\ a_3 \cdot \delta_T \cdot Q_c N_{err} \geq 30 \end{cases} \tag{23.3}$$

where N_{err} is the number of the impaired frames. Here a_3 is the parameter determined through training. Moreover, compared with the GOP which is not contaminated, the GOP with severe distortions affect the overall video quality much more. Thus, the sequence distortion D_s can be estimated by the following equation:

$$D_s = \left(\frac{1}{N} \cdot \sum_{k=1}^{N} (D_g(k))^P\right)^{\frac{1}{P}} \tag{23.4}$$

where N is the number of the GOP, D_g (k) is the distortion of the kth GOP, and P equals to 2. Therefore, the video quality Q_s can be calculated as follows:

$$Q_s = Q_c - D_s \tag{23.5}$$

where Q_c is the coding quality which can be obtained as described in [12].

23.4 Experimental Results

All the experiments in this paper used the x264 encoder and the sequences are all QCIF format at 30 frames per second (fps). In simulations, the RTP/UDP/IP protocol stack is employed to packet the encoded data. The parameters defined in the metric were empirically obtained through the least square error fitting, where $r_1 = 0.81$, $r_2 = 0.02$, $a_1 = -0.38$, $a_2 = 1.43$, $a_3 = 0.32$.

In order to verify the efficiency and accuracy of the proposed model for video quality assessment, standard test sequences were employed including "Hall", "Paris", "Mother-Daughter", "Highway", and "Football". The bit-rates were set at 64, 96, 128, and 150 kbps, respectively. In addition, the packet loss rates used in the experiments were set to 0.5, 1, 2, 3, 5, and 7 %, respectively, where a random packet loss model was employed to simulate packet loss distribution in IP networks.

Then a subject test was carried out using the Single Stimulate Method (SSM) specified by the Video Quality Experts Group (VQEG). The Mean Opinion Scores (MOS) of reconstructed sequences is obtained by the Absolute Category Rating (ACR) with a 5-point scale from bad to excellent. There were 24 nonexpert observers that rated the quality. The G.1070 model was used for comparison purposes, and the proposed model outperforms G.1070 by getting an increment about 0.015 in PCC and a decrement about 0.049 and 0.007 in RMSE and OR, respectively, as shown in Table 23.1. The scatter plots of the objective scores versus the subjective scores are shown in Fig. 23.6, where the same conclusion can be drawn that the proposed model is very accurate in quality evaluation.

Table 23.1 Performance comparison of proposed model and G.1070

Assessment model	PCC	RMSE	OR
Proposed model	0.913	0.353	0.028
G.1070 model	0.898	0.402	0.035

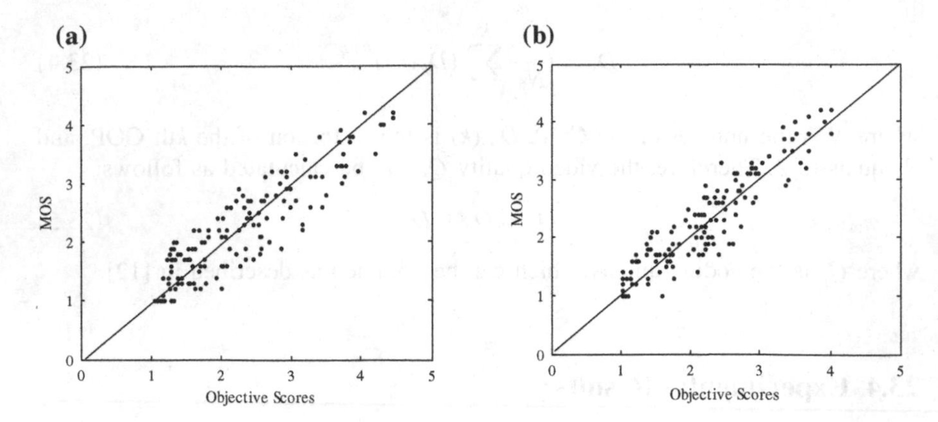

Fig. 23.6 Scatter plot of MOSs vs objective scores, **a** G.1070 Model; **b** Proposed Model

23.5 Conclusions

An accurate packet-layer model is proposed to evaluate the video quality. The impairment of packet loss has been evaluated, based on which the quality degradation of a specific video due to packet loss can be estimated using the number of impaired frames in each GOP. The proposed packet-layer model enables real-time and nonintrusive quality monitoring for networked video streaming.

References

1. Wu HR, Rao KR (2007) Digital video image quality and perceptual coding. CRC Press, London, pp 181–201
2. Verscheure O, Frossard P, Hamdi M (1999) User-oriented QoS analysis in MPEG-2 video delivery. Real-Time Imaging 5(5):305–314
3. Boyce JM, Gaglianello RD (1998) Packet loss effects on MPEG video sent over the public internet. In: Proceedings of 6th ACM multimedia, pp 181–190
4. You F, Zhang W, Xiao J (2009) Packet loss pattern and parametric video quality model for IPTV. In: Proceedings of the 2009 8th IEEE/ACIS international conference on computer and information science, pp 824–828
5. Frossard P, Verscheure O (2001) Joint source/FEC rate selection for quality-optimal MPEG-2 video delivery. IEEE Trans Image Process 10(12):1815–1825
6. Yang FZ, Wan S, Xie QP, Wu HR (2010) No-reference quality assessment for networked video via primary analysis of bit-stream. IEEE Trans Circuits Syst Video Technol 20(11):1544–1554
7. Wan S, Yang FZ, Xie ZQ (2010) Evaluation of video quality degradation due to packet loss. In: international symposium on intelligent signal processing and communication systems
8. Yamagishi K, Hayashi T (2008) Parametric packet-layer model for monitoring video quality of IPTV services. In: IEEE international conference on communications ICC

9. Yamada T, Yachida S, Senda Y (2010) Accurate video-quality estimation without video decoding. In: Proceedings of 2010 IEEE international conference on acoustics, speech and signal processing, pp 2426–2429
10. Feghali R, Speranza F, Wang D (2007) Video quality metric for bit rate control via joint adjustment of quantization and frame rate. IEEE Trans Broadcast 53(1):441–446
11. ITU-T Recommendation P.910 (2008) Subjective video quality assessment methods for multimedia applications
12. Su HL, Yang FZ, Song JR (2012) Packet-layer quality assessment for networked video. Int J Comput Commun Control 7(3):541–549

Chapter 24
A Hierarchical Registration Method of the Chang'E-1 Stereo Images

Mengjie Ye and Jing Huang

Abstract Aim at the characteristics of Chang'E-1 stereo images, a geometrically constrained based hierarchical registration method has been proposed. The method uses a coarse-to-fine hierarchical matching strategy with a combination of several image registration algorithms, which essentially consists of four steps: matching area determination, interest points extraction, geometrically constrained registration, and affine transformation matrix estimation. The experimental results show that the proposed algorithm can provide dense, well-distributed, and accurate corresponding points, and reduce the search space to achieve computationally efficient.

Keywords Chang'E-1 · DEM · Image registration feature detector

24.1 Introduction

Chang'E-1 lunar orbiter was launched by Long March 3A carrier rocket at 10:05 GMT on October 24, 2007 from the No. 3 launching tower at the Xichang Satellite Launch Center in Southwest China. The stereo camera onboard on the Chang'E-1 orbiter has successfully obtained 1098 tracks of CCD images covering the whole lunar surface [9, 10, 13]. A global digital elevation model (DEM) can be generated from the stereo images. But a key step is the registration of dense and well-distributed corresponding points with an appropriate registration method [23].

M. Ye
Space Science Institute, Macau University of Science and Technology, Macau, China
e-mail: yemengjie@gmail.com

J. Huang (✉)
Information Technology College, Beijing Normal University Zhuhai, Zhuhai, China
e-mail: huangjingzh@gmail.com

A. A. Farag et al. (eds.), *Proceedings of the 3rd International Conference on Multimedia Technology (ICMT 2013)*, Lecture Notes in Electrical Engineering 278,
DOI: 10.1007/978-3-642-41407-7_24,

Registration is a fundamental task in image processing used to match two or more pictures taken, for example, at different times, from different sensors, or from different viewpoints [3]. The transformation in stereo images can be solved by methods that perform image registration. A transformation must be found so that the points in one image can be related to their corresponding points in the other. Thus, the 3D information of objects can be computed. In photogrammetry and remote sensing, image registration techniques have been employed for automatic DEM generation.

On the contrast, establishing point correspondences is a crucial step for image registration. A typical image registration scheme includes feature detection, feature matching, transform model estimation, image resampling, and transformation. Here we will focus on feature detection and matching.

Feature detection is a basic step for image registration and image analysis. Feature which is also called feature points, interest points or corners, is usually extracted by some kinds of local "interest" detectors. Interest detectors extract salient image features, which are distinctive in their neighborhood and are reproduced in corresponding images in a similar way. At the same time, image windows around features usually contain sufficient image intensity variations, which can be used during the later image registration.

24.2 Related Works

In recent years, research concerning in the topic of features detection is particularly rich and an amount of methods have been proposed in the literature. The feature or interest point detectors here have been roughly classified into two types: image-space-based and scale-space-based.

The image-space-based methods mainly detect interest point in image space. The interest point detector can be tracked back to the work of Moravec [18]. Several interest point detectors [5, 6] (Förstner 1994; [20] are based on a matrix related to the autocorrelation matrix (also named second moment matrix). Forstner detector consists of two steps namely optical window selection and interest point localization. First, the optimal windows are selected using the autocorrelation matrix. And then the location of interest point is calculated by a differential edge intersection approach. Harris and Stephen improve the Moravec detector by using the autocorrelation matrix instead of the use of discrete shifts.

The image-space-based interest point detectors commonly are not scale-invariant. Although some detectors like Harris are invariant to rotation. So the image-space-based methods cannot provide a good basis for registering images of different sizes.

The scale-space-based approaches detect both location and scale of interest points in scale-space. In 1983, Witkin introduced the concept that the scale-space representation of an image can be constructed by convolution with a one-parameter family of Gaussian kernels of increasing width [21]. Later, it has been shown by [1, 8, 11] that Gaussian kernel is the optical and unique choice.

Lindeberg introduced the concept of automatic scale selection that allows detecting interest points in an image, each with their own characteristic scale [12]. He experimented with both the determinant of the Hessian matrix as well as the Laplacian (which corresponds to the trace of the Hessian matrix) to detect blob-like structures.

Scale Invariance Feature Transform (SIFT) introduced by Low [14] extended the local feature approach (Schmid et al. 1997) to achieve scale invariance. SIFT approximated the Laplacian of Gaussian (LoG) by Difference of Gaussian (DoG) for computationally efficient. This method also described a local descriptor that provided more distinctive features while being less sensitive to local image distortion such as 3D viewpoint change. And (Low [15]) provided a more in-depth development and analysis of his earlier work, while also representing a number of improvements in stability and feature invariance. Various refinements on SIFT have been proposed. Focusing on speed, a PCA-SIFT method which applied PCA on the gradient image has been represented by [7]. PCA-SIFT yields a 36-dimensional descriptor instead of 128-dimensional of SIFT. However, a study by Mikolajczyk [16] has proven that PCA-SIFT are less distinctive than SIFT. Inspired by the SIFT, a robust local feature detector has been proposed namely Speeded Up Robust Features (SURF) [2]. The standard version of SURF is several times faster than SIFT and claimed by its authors to be more robust against different image transformations than SIFT.

One should be noted that unlike conventional photographs, where all pixels in an image are exposed simultaneously and therefore can be measured by only one transformation matrix, each line of Chang'E-1 stereo images was acquired in a pushbroom manner at a different instant of time. This means that there is a different set of the exterior orientation parameters for each scan line. Since the stereo images were collected from different viewing angle, the geometric deformation also exists. Meanwhile, due to a number of craters on lunar surface, Chang'E-1 stereo images are consisting of very similar, neighboring texture patterns. Generally, these patterns are images of repetitive structures that have similar shape, size, and physical properties. The fact means that there are a lot of similar feature points which tend to confuse registration.

The image-space-based methods can acquire dense correspondences in stereo images, but cannot obtain accurate results. The SIFT descriptor can provide robust and accurate correspondences against errors caused by small geometric distortions. However, the SIFT detector can only extract blob-like interest points and provide sparse and uniformly registration results.

Aim at the characteristics of Chang'E-1 stereo images, a geometrically constrained based hierarchical registration method has been proposed. The method uses a coarse-to-fine hierarchical matching strategy with a combination of several image registration algorithms, which can provide dense, well-distributed, and accurate corresponding points, and reduce the search space to achieve computationally efficient.

24.3 Methods Overview

The method essentially consists of four steps shown as follows [22].

- *Matching area determination.* Split the original images up into small image patches as reference image, and determine the rough matching area from backward or forward image using area-based method.
- *Interest points extraction.* For each image obtained by the mentioned above steps, two sets of interest points need be extracted: one is robust and accurate, but sparse interest points with the descriptors obtained by the SIFT detector. This set of interest points can be directly used to determine the corresponding points by geometrically constrained registration method. The other one is dense and well-distributed interest points obtained by the Forstner detector. This set of interest points need use the affine transformation matrix to determine the corresponding points.
- *Geometrically constrained registration.* For each interest point given in reference image, the position range of corresponding points in target image can be estimated by making use of the exterior orientation parameters of the Chang'E-1 orbiter. Then, the registration can be implemented by searching the corresponding points in the estimated small region.
- *Affine transformation matrix estimation.* Affine transformation matrix measured the relationship of the stereo images can be estimated using the corresponding points obtained by the geometrically constrained registration method.

All these steps can be illustrated by flow chart in Fig. 24.1.

24.3.1 Geometrically Constrained Registration

By area-based registration method, two sub-images overlapping the same region of lunar surface have been obtained from nadir and backward image. If image triplet is acquired in ideal conditions, that is, the trajectory is assumed to be a straight line with the constant flying height, and all three images are assumed to be taken with perfect parallel projection to the standard sphere along the flight direction, then the corresponding points of nadir and backward image should be one to one in pixel coordinates. However, due to the exterior orientation of the orbiter changes and undulant terrain, the position offset between nadir and backward image are existed. So in this step, it attempts to determine the position range of corresponding points in backward or forward image for a given interest point in nadir image.

Consider an image triplet of Chang'E-1, shown as in Fig. 24.2. Given an interest point in the nadir image, an image ray that connects the instant perspective center and this image point can be determined. Given a lunar surface point A, and the projection of A is A_0. The height approximation h can be computed by interpolation from LAM data of Chang'E-1 lunar mission. Let a denote the

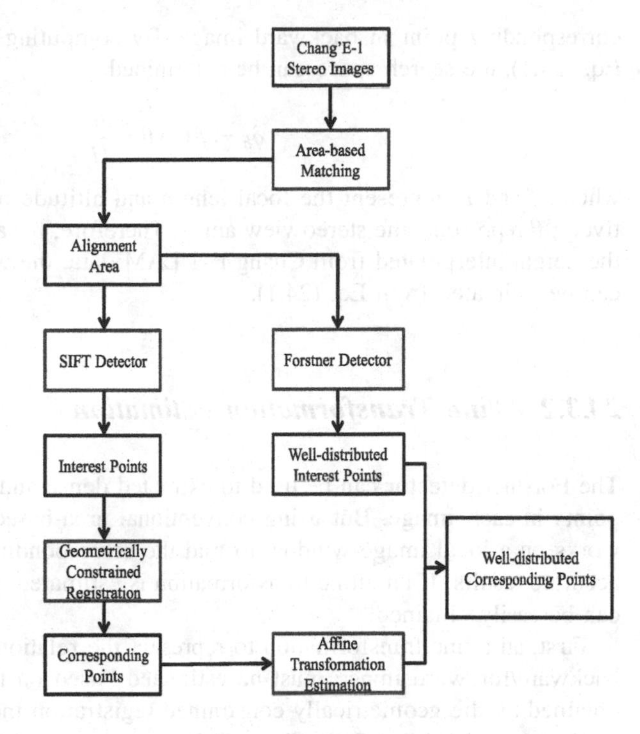

Fig. 24.1 Flowchart of the hierarchical registration method

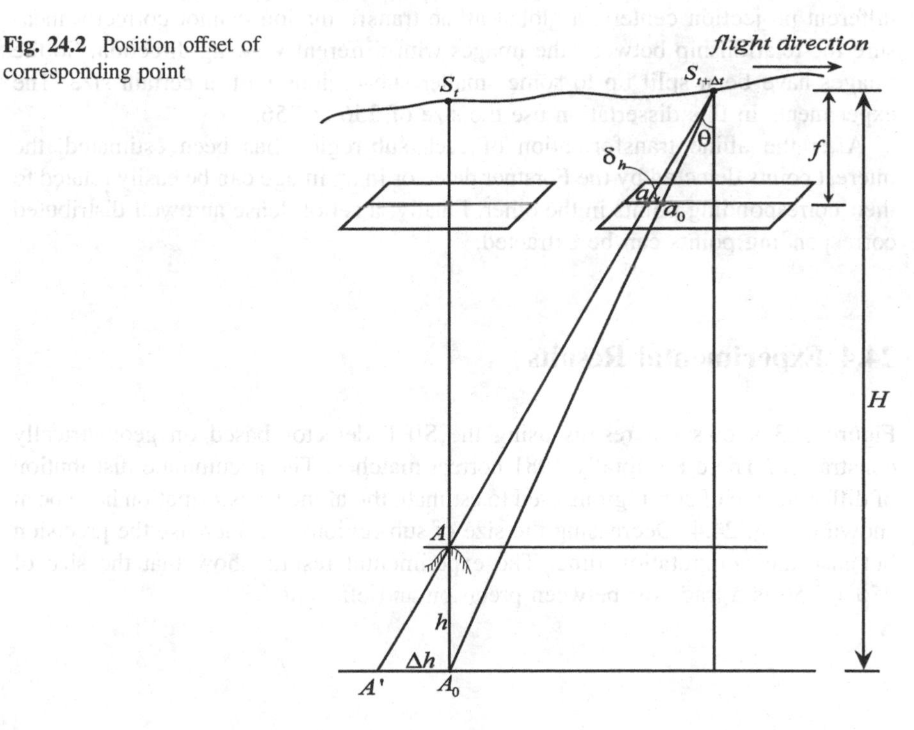

Fig. 24.2 Position offset of corresponding point

corresponding point in backward image. By computing the length of $\overline{aa_0}$ using Eq. (24.1), the search space can be determined.

$$\delta_h = f \tan\theta \times \frac{h}{H} \tag{24.1}$$

where, f and H represent the focal length and altitude of stereo camera, respectively; θ represents the stereo view angle. Therefore, for a given interest point and the height interpolated from Chang'E-1 LAM data, the width of the search space can be estimated from Eq. (24.1).

24.3.2 Affine Transformation Estimation

The Forstner detector can be used to extracted dense and well-distributed interest points in each image. But using conventional area-based method which directly works on a local image window to find the corresponding points cannot provide accurate results. If an affine transformation is estimated, the corresponding points can be easily obtained.

First, an affine transformation to represent the relationship of nadir image and backward/foreward image must be estimated based on the corresponding points obtained by the geometrically constrained registration method.

Because each line of the Chang'E-1 stereo images acquired by the sensor has different projection centers, a global affine transformation cannot correctly measure the relationship between the images with different viewing direction. So the images have been split up to some smaller sub-regions with a certain size. The experiments in this dissertation use the size of 256 × 256.

After the affine transformation of each sub-region has been estimated, the interest points detected by the Forstner detector in an image can be easily related to their corresponding points in the other. Finally, a set of dense and well-distributed corresponding points can be extracted.

24.4 Experimental Results

Figure 24.3 shows the results using the SIFT detector based on geometrically constrained. There are totally 1981 correct matches. The accumulate distribution of different size of sub-regions used to estimate the affine transformation have been shown as Fig. 24.4. Decreasing the size of sub-regions can increase the precision but also the computation time. The experimental results show that the size of 256 × 256 is a trade-off between precision and efficient.

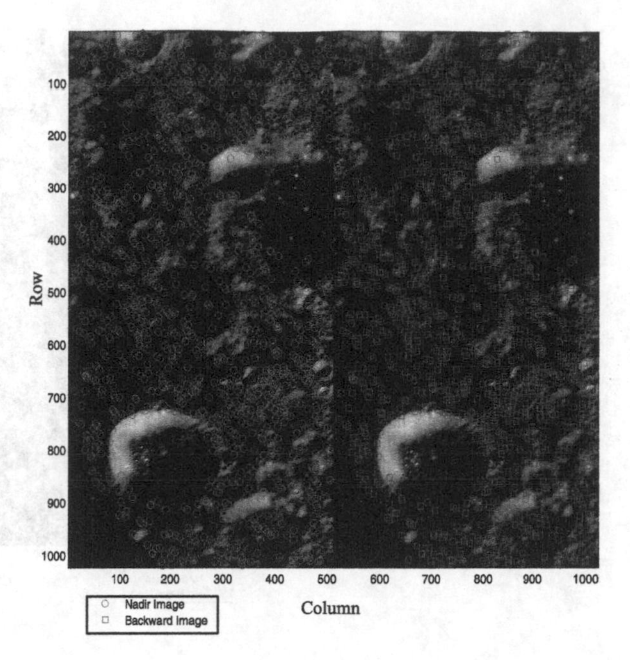

Fig. 24.3 Result of geometrically constrained matching (1981 matches)

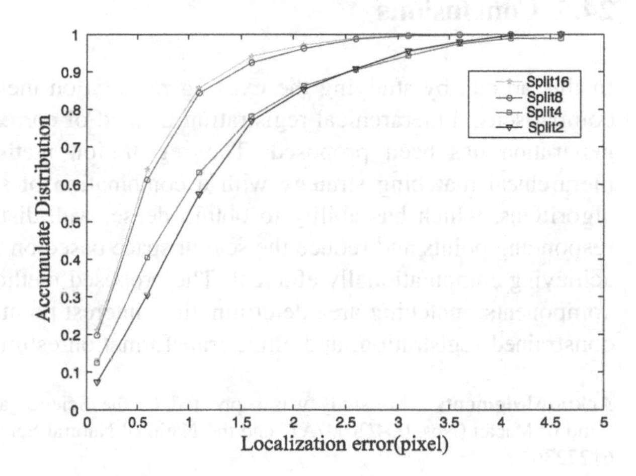

Fig. 24.4 The accumulate distribution between different split numbers

After the affine transformations of each sub-region have been estimated, the corresponding points of the interest points extracted by the Forstner detector can be obtained. Total 11362 matches are collected and shown as Fig. 24.5.

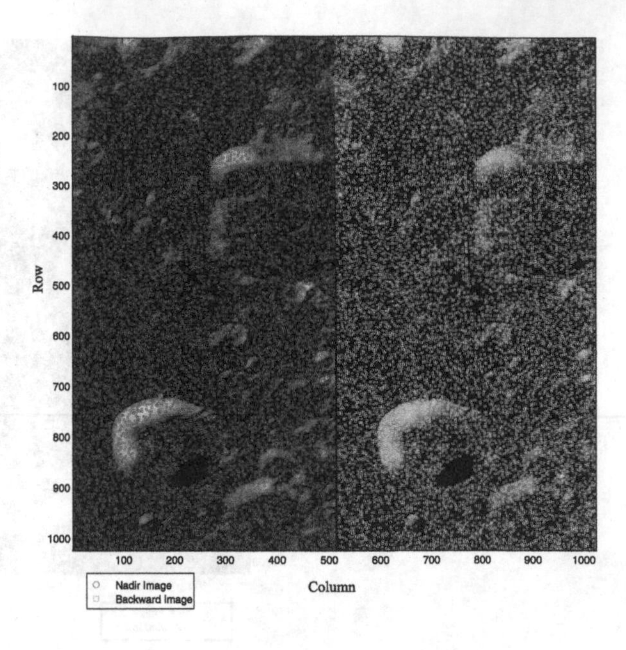

Fig. 24.5 The final results of corresponding points (11362 matches)

24.5 Conclusions

In this article, by studying the existing registration methods and from published comparisons, a hierarchical registration method of corresponding points for DEM generation has been proposed. The registration method uses a coarse-to-fine hierarchical matching strategy with a combination of several image registration algorithms, which has ability to obtain dense, well-distributed, and reliable corresponding points and reduce the search space based on geometrically constrained achieving computationally efficient. The proposed method mainly consists of four components: matching area determination, interest point extraction, geometrically constrained registration, and affine transformation estimation.

Acknowledgments This study was supported by the Science and Technology Development Fund of Macau (Nos. 004/2011/A1) and the National Natural Science Foundation of China (No. 61272364).

References

1. Babaud J, Baudin M, Witkin A et al (1986) Uniqueness of the Gaussian kernel for scale-space filtering. IEEE Trans Pattern Anal Mach Intel 8(1):26–33
2. Bay H, Tinne AE, Tuytelaars T, Van Gool L (2008) SURF—Speeded Up Robust Features. Comput Vis Image Underst (CVIU) 110(3):346–359
3. Brown LG (1992) A survey of image registration techniques. ACM Comput Surv 24(4):325–376

4. Burt PJ (1981) Fast filter transforms for image processing. Comput Vis Graph Image Process 16:20–51
5. Förstner W, Gülch, E (1987) A fast operator for detection and precise location of distinct points, corners and circular features. In: Intercommission conference on fast processing of photogrammetric data, pp 281–305
6. Harris C, Stephens M (1998) A combined corner and edge detector. In: Proceedings of the alvey vision conference, pp 147–151
7. Ke Y, Sukthankar R (2004) PCA-SIFT—A more distinctive representation for local image descriptors. Comput Vis Pattern Recogn 2:506–513
8. Koenderink J (1984) The structure of images. Biol Cybern 50(5):363–370
9. Li CL et al (2010) The global image of the Moon obtained by the Chang'E-1- Data processing and lunar cartography. Sci China Earth Sci 53:1091–1102
10. Li CL et al (2010) Laser altimetry data of Chang'E-1 and the global lunar DEM model. Sci China Earth Sci 53:1582–1593
11. Lindeberg T (1994) Scale-space theory—a basic tool for analyzing structures at different scales. J Appl Stat 21(2):225–270
12. Lindeberg T (1998) Feature detection with automatic scale selection. Int J Comput Vis 30(2):79–116
13. Liu JJ et al (2009) Automatic DEM generation from CE-1's CCD stereo camera images. Lunar and planetary science conference, pp 23–27
14. Lowe D (1999) Object recognition from local scale-invariant features. In: International conference on computer vision, Corfu, Greece, pp 1150–1157
15. Lowe D (2004) Distinctive image features from scale-invariant keypoints. Int J Comput Vis 60:91–110
16. Mikolajczyk K, Schmid C (2003) A performance evaluation of local descriptors. Comput Vis Pattern Recogn 2:257–263
17. Mikolajczyk K, Schmid C (2004) Scale and affine invariant interest point detectors. Int J Comput Vis 60(1):63–86
18. Moravec H (1981) Rover visual obstacle avoidance. In: Proceedings of the 7th International Conference on Artificial Intelligence, Vancouver, pp 785–790
19. Noble A (1989) Descriptions of image surfaces. (Ph.D.) Department of Engineering Science, Oxford University
20. Tomasi C, Kanade T (2004) Detection and tracking of point features. Pattern Recogn 37:165–168
21. Witkin AP (1984) Scale-space filtering: a new approach to multi-scale description. Acoustics, Speech, and Signal Processing. In: IEEE international conference on ICASSP, pp 150–153
22. Ye M, Tang Z (2013) Registration of correspondent points in the stereo-pairs of Chang'E-1 lunar mission using SIFT algorithm. J Earth Sci 24(3):371–381
23. Zhang L (2005) Automatic Digital Surface Model (DSM) generation from linear array images. PhD dissertation, Report no.88, Institute of Geodesy and Photogrammetry, ETH Zurich, Switzerland

Chapter 25
Efficient Adaptive Window Matching Algorithm Based on Cross Search

Qian Liang, Yingyun Yang and Bo Liu

Abstract The prominent problem with area-based algorithm is that the window sizes and shapes are hard to select. For the common problem, this paper presents a new adaptive window matching algorithm which contributes an outstanding performance on building adaptive window. The three main ideas of the algorithm are combining SAD, GRAD, CENSUS as the cost function, using cross search to determine the range of window sizes, and evaluating the given window cost to update window sizes/shapes. Through discarding the poor pixels found by given window cost, the support window has no limit on sizes or shapes. Moreover, the merging cross search with window cost function strengthens the relationships of points in one adaptive window and prevents window boundary from excessive growth. Experimental results show that the algorithm is superior to that of similar adaptive-window methods.

Keywords Stereo matching · Cross search · Widow cost · Adaptive-window

25.1 Introduction

Area-based matching is a widely used method for stereo matching which is one of the most active research areas in computer vision. Usually, area-based algorithms focus on the selection of support window such as constant window and adaptive window. Since constant window algorithms do not perform well, there has been some excellent work on adaptive window algorithms. For example, [1] proposed

Q. Liang (✉) · Y. Yang
College of Information Engineering, Communication University of China, Beijing, China
e-mail: 937663993@qq.com

B. Liu
State Radio Regulation of China, Beijing, China

A. A. Farag et al. (eds.), *Proceedings of the 3rd International Conference on Multimedia Technology (ICMT 2013)*, Lecture Notes in Electrical Engineering 278,
DOI: 10.1007/978-3-642-41407-7_25,

the multiple-window model. For each pixel, the approach performed the correlation with a small number of different windows and retained the one with the best cost. Multiple-window [1] could get better results by increasing the number of given windows. But the calculation time would grow at the same time [2] presented an adaptive window model based on regional growth. For each pixel, this algorithm increased the size of the support window and evaluated the corresponding window cost, then retained the appropriate window which had the lowest window cost. However, the algorithm had a limit on the shape of the windows [3] presented a cross-based algorithm. For each pixel, this algorithm found the four cross-arms and construct an adaptive window by merging the neighbor pixels with cross-arms constraint. In [4] an adaptive support-weight algorithm was proposed. The method adjusted the support-weights of the pixels in a given constant window based on geometric proximity and color similarity. So the algorithm had no need to define the support window sizes for each pixel. The algorithm [4] performed effectively in discontinuous and less textured areas.

In this paper, we propose a new adaptive-window algorithm. Compared with [2], our approach introduces the cross search and integrates it with window cost. The cross search can help to find the largest grow-range of each pixel in vertical and horizontal direction based on color similarity and geometric proximity. The range that we get from cross search defines the maximum size of the initial adaptive window. Then, we use the combination of three (SAD, GRAD and CENSUS) as the cost function. After that, we evaluate window cost to update the support window sizes and shapes by discarding the poor pixels. We found that the proposed has higher matching accuracy-rate compared with [2, 3]. It may be the result of merging cross search with window cost. Compared with [2], the proposed has no limit on the window shape, which determines the outperformance of our algorithm on discontinuous disparity areas. While instead of the common cross smoothing method [3], the proposed uses window cost to make points in one adaptive window have direct constraint relationships, which prevents an excessive growth in window boundary. The experimental results prove that our algorithm is effective.

The paper is organized as follows: in Sect. 25.2, the choice of matching cost function is introduced; in Sect. 25.3, our adaptive window algorithm based on cross search is presented; in Sect. 25.4, experimental results are given and finally in Sect. 25.5, conclusions and the future direction of the work are presented.

25.2 Matching Cost

Stereo matching algorithms rely on matching costs for computing the similarity of corresponding points. Thus, the choice of matching cost function has great effect on the final depth map.

Sum of absolute or squared differences (SAD/SSD), sum of gradient absolute difference (GRAD), and rank or census transforms (RANK/CENSUS) are the

Table 25.1 Matching cost performed by constant window

Cost	Tsukuba (%)	Cones (%)	Teddy (%)
SAD + CENSUS [5]	9.00	14.6	18.7
SAD + GRAD + CEN	8.72	14.1	18.6

common window-based matching costs. SAD uses the intensity information and performs effectively in computational simplicity. GRAD uses the gradient values between adjacent pixels and performs well in object boundary where gradient values are high. CENSUS uses hamming distance over census transform and has good anti-noise performance. In order to take full advantage of different matching costs, many algorithms adopt the linear mixed cost [5–7]. For example, [5] used a linear combination of SAD and CENSUS. By contrast, we combine and normalize the three (SAD, GRAD, CENSUS) as the matching cost function according to Eq. 25.5. Table 25.1 shows that the addition of gradient information has better experimental results.

$$C_d^{\mathrm{SAD}}(x,y) = \min\left\{\sum_W |I_l(x,y) - I_r(x+d,v)|, \lambda_{\mathrm{SAD}}\right\} \tag{25.1}$$

$$C_d^{\mathrm{GRAD}}(x,y) = \min\left\{\sum_W \left|\nabla_x, \nabla_y I_l(x,y) - \nabla_x, \nabla_y I_r(x+d,y)\right|\right\} \tag{25.2}$$

$$C_d^{\mathrm{CEN}}(x,y) = \min(\mathrm{Ham}(CT_l(x,y), CT_r(x+d,y)), \lambda_{\mathrm{CEN}}) \tag{25.3}$$

$$\rho(C,\lambda) = 1 - \exp(-C/\lambda) \tag{25.4}$$

$$C_d(x,y) = \rho(C_d^{\mathrm{SAD}}, \lambda_{\mathrm{SAD}}) + \rho(C_d^{\mathrm{GRAD}}, \lambda_{\mathrm{GRAD}}) + \rho(C_d^{\mathrm{CEN}}, \lambda_{\mathrm{CEN}}) \tag{25.5}$$

In Eq. 25.1, $C_d^{\mathrm{SAD}}(x,y)$ is the SAD cost computed from the left image to right in the point (*x*, *y*) with disparity *d*, window W; the maximum cost value of SAD is kept below a level λ_{SAD}, which prevents high cost measures for occluded pixels. Similarly, $C_d^{\mathrm{GRAD}}(x,y)$ corresponds to the gradient values calculating in both horizontal and vertical direction. And for $C_d^{\mathrm{CEN}}(x,y)$, we calculates the hamming distance between the bit streams of the corresponding pixels for disparity *d*; the bit stream comes from the census transform, which calculates the difference between the center pixel and the other ones in the census-window and transforms these distances to bits. We use ϱ to normalize the three cost values and sum them to be the final matching cost function. The experimental results of Table 25.1 in Sect. 25.4 show that our matching cost function has higher accuracy than that used in [5].

25.3 The Algorithm

Compared with [2], which constrained the shape of a support window to a rectangle, we propose an adaptive window algorithm which has no limit on the window size and shape. The proposed involves three main steps: first, the cross search; second, the calculation of window cost; third, disparity optimization.

25.3.1 Cross Search

The main idea of cross search [3] is finding the cross-arms. For each pixel, we compute to get the four cross-arms (up, down, right, left) according to color similarity (Eq. 25.6) and connectivity constraints (Eq. 25.8). *Dc* (*P*, *P1*) in Eq. 25.7 corresponds to color difference between the pixel *P* and *P1*. The equation calculates in three-channel (RGB) and keeps the maximum one.

$$D_c(P, P1) < \tau \tag{25.6}$$

$$D_c(P, P1) = \max_{i=R,G,B} |I_i(P) - I_i(P1)| \tag{25.7}$$

$$D_s(P, P1) < L \tag{25.8}$$

$$D_s(P, P1) = |P - P1| \tag{25.9}$$

The value of *Ds* (*P*, *P1*) in Eq. 25.9 is the spatial distance between pixel *P* and *P1*. The max value of *Dc* (*P*, *P1*) and *Ds* (*P*, *P1*) must below threshold values τ and *L*. Aimed at pixel *P*, the cross search first starts in left and right direction along the scanning line. Then, the search is going on along the vertical directions including up and down. When the process is over, the four cross-arms are memoried. The specific process is shown in Fig. 25.1.

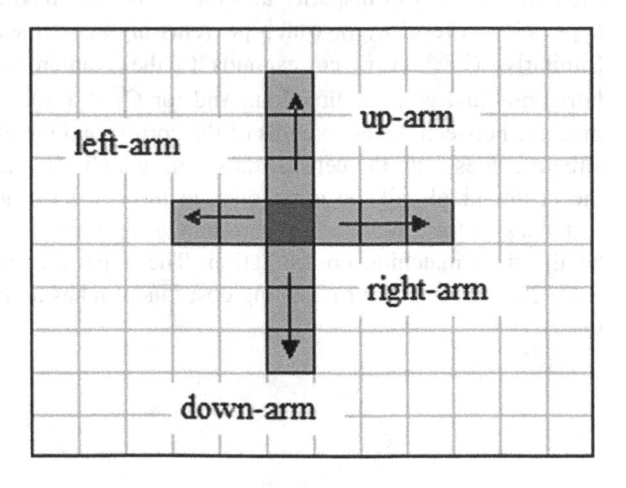

Fig. 25.1 Cross search

The information we can easily get is that the value of τ and L determine the length of arms; the greater the value, the longer the arms. However, the real result is not ideal for different areas calling for variable standard. For example, less textured areas need a larger window, so the distance threshold L should be larger; and discontinuous disparity areas need smaller window that calls for smaller L. The value of τ also varies in different cases. In order to establish a unified measure, which can take the problems mentioned above into account, the cross search do the following improvements shown in Eqs. 25.10–25.12.

$$D_c(P,P1) < \tau_1 \text{ and } D_c(P,P1+(1,0)) < \tau_1 \tag{25.10}$$

$$D_s(P,P1) < L_1 \tag{25.11}$$

$$\text{if } L_2 \leq D_s(P,P1) < L_1 \text{ then } D_c(P,P1) < \tau_2 \tag{25.12}$$

Equation 25.10 shows that the color difference not only between pixel P and $P1$, but also between P and the pixel followed by $P1$ should below the value τ_1, which ensures that the window boundaries include no discontinuous disparity areas. To balance less textured and discontinuous disparity areas, two values of L (larger L_1 and smaller L_2) are set. Equation 25.12 shows that if the spatial distance between P and $P1$ is not smaller than L_2, the color threshold will be replaced by smaller τ_2, which prevents an excessive growth in window boundary. And the larger L_1 in Eq. 25.11 can also meet the requirements of size for less textured areas. In this study, we use the improved constraint Eqs. 25.10–25.12.

25.3.2 Calculation of Window Cost

Conventional cross-based algorithms [3] had two main steps: finding the four cross-arms for each pixel and constructing an adaptive window by merging the neighbor pixels with cross-arms constraint. However, the support window of [3] were construct by smoothing method; the pixels in one window had no direct constraint relationships which might result in an excessive growth in window boundary. Instead of the traditional smoothing method, the proposed uses the combination of cross search and window cost. The window cost is used to exclude poor pixels and optimize the window. Compared with [3], the proposed makes window points have direct relation. The window strategy has two main steps, showing as follows:

a. For each pixel to be matched, a maximum window is construct with cross-arms obtained by the above step. The maximum window will be used in the following step as the initial window.
b. Using the Eq. 25.13 to calculate the window cost of initial window. For each pixel in the initial window, Eq. 25.14 is used to determine the pixels that participate in calculation. And for each participating pixel, Eq. 25.13 will be

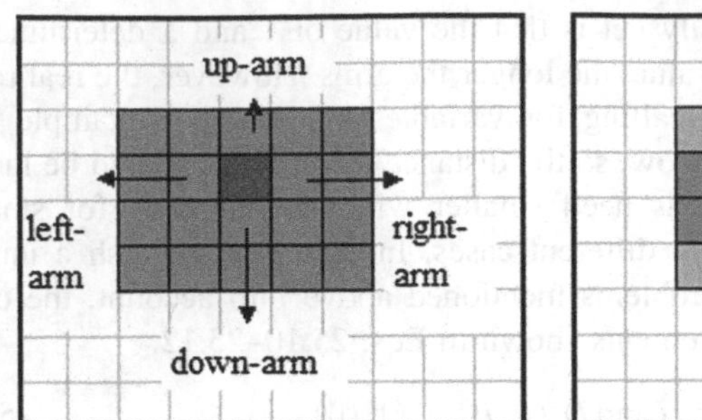

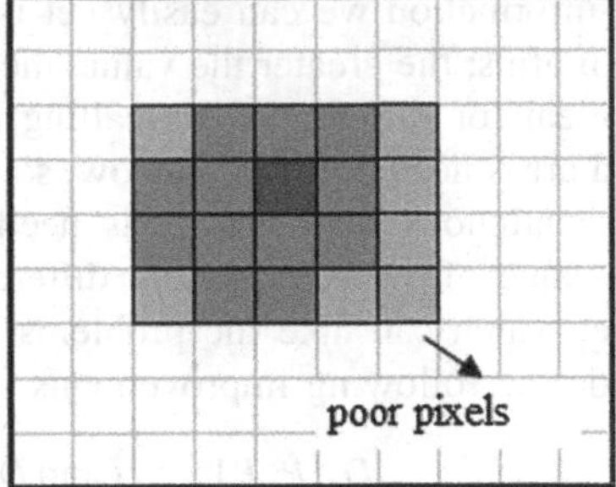

Fig. 25.2 Window strategy

performed again to calculate the window cost without this pixel; if the new window cost is smaller, then the calculated pixel is labeled to be poor and excluded. After that the window cost and window size will be updated. The specific process is shown in Fig. 25.2. The window cost is shown in Eq. 25.13.

$$C_d(W) = DSI/W + \beta\Big/\left(\sqrt{W} + r\right) \tag{25.13}$$

$$|P - P1| < L \tag{25.14}$$

In Eq. 25.13, DSI represents the initial matching cost calculation performed by Eq. 25.5 for each disparity; W is the number of pixels in the window. The first item of the Eq. 25.13 refers to the average error rate of the window. Obviously, the smaller the value of item is, the more similar given disparity d and realistic disparity is. The W is used to normalize the window cost for comparability of different sizes window. When the window is larger, the second item is smaller. That item is used to get larger window when the first item is same, which has an important effect on less textured areas [2]. The value of L determines the number of pixels that participate in the calculation; for different pixel to be matched, the value should be self-adaptive. In this study, we choose half of the largest length of four cross-arms as the value of L for poor pixels usually far from center. To have enough information, the minimum size of window is 3×3.

25.3.3 Disparity Optimization

In actual operation, the disparity obtained by the above steps needs further error detection and correction (disparity optimization) for the presence of smooth areas, the periodic pattern and the occluded. In this paper, we use the method of vote window as the optimization strategy. First, constructing dmax + 1 ballot boxes (dmax is the maximum disparity); for each pixel, every ballot box represents a disparity level. Then performing the voting process in a given vote window using color similarity. For the target pixel p, if the color difference between the voting pixel q in the vote window and target pixel p is smaller than the threshold value t, the votes in the corresponding ballot boxes will increase by one. After the end of the voting process, getting a statistic on the total number of votes in each ballot box. If the total number of votes in all ballot boxes is greater than T, then the voting process is efficient and the disparity of target p will be replaced with the corresponding disparity that has the highest number of votes.

25.4 Experimental Results

To compare different adaptive window algorithms, all algorithms are simulated on the platform of 3.10 GHZ Intel (R) Core (TM) i5-2400. In addition, we adopt MATLAB as the experimental platform. All the experiments are performed using the benchmark Middlebury stereo database. Results of evaluations and all the test images can be found on the web via [8].

In the experiments, the truncation threshold of SAD, GRAD and CENSUS is 20, 15 and 18, respectively; the census-window size is 5×4. Parameters τ_1, τ_2, L_1, L_2, β, α, t, T are set to 20, 6, 34, 17, 1, –1, 20, 190 respectively.

In Table 25.1, we compare the proposed matching cost with that mentioned in [5] on 7×7 constant window. The error rates of the proposed are lower than the one used in [5] for the three test images [8] (Table 25.1). The results prove the effectiveness of our matching cost function.

In Fig. 25.3, the stereo test images and disparity maps that estimated by the proposed algorithm are illustrated in the first and second row.

The first column in Table 25.2 lists names of the four evaluated adaptive-window algorithms: multi-window [1], var-window [2], conventional cross-based algorithm [3] and the proposed algorithm. The next four columns give error rates the algorithm makes in all areas (all), non-occluded areas (nonc), and

Fig. 25.3 The tsukuba, venus, teddy, cones stereo test images and disparity images of the proposed

Table 25.2 Error rate of adaptive-window algorithm

Algorithm	Tsukuba (%)			Venus (%)			Teddy (%)			Cones (%)			
	non	all	dis	non	all	dis	non	all	dis	non	all	dis	Average
Multiwindow [1]	5.94	6.28	7.85	12.0	12.8	9.77	19.8	27.1	27.9	17.1	24.8	22.9	16.2
Varwindow [2]	4.87	5.66	14.2	6.48	7.60	11.0	18.8	26.7	32.9	14.8	23.5	23.2	15.8
Conventional Cross-based [3]	2.54	2.85	8.35	0.33	0.72	3.01	7.21	14.8	18.4	3.97	12.4	10.5	7.08
The proposed	1.77	2.29	7.86	0.31	0.72	3.00	6.93	14.4	18.2	3.39	11.9	8.91	6.64

PUTv3 [63]	66.3	1.77 68	3.86 89	9.42 86	0.42 68	0.95 78	5.72 86	7.02 65	14.2 89	18.3 84	2.40 11	9.11 54	6.56 8	6.64
YOUR METHOD	67.4	1.77 69	2.29 66	7.86 68	0.31 52	0.72 59	3.00 48	6.93 64	14.4 95	18.2 81	3.39 53	11.9 96	8.91 58	6.64
GradAdaptWgt [60]	67.7	2.26 90	2.63 76	8.99 81	0.99 93	1.39 91	4.92 83	8.00 88	13.1 71	18.6 89	2.61 19	7.67 12	7.43 21	6.55
MultiCue [51]	68.0	1.20 26	1.81 35	6.31 33	0.43 69	0.69 55	3.36 61	7.09 68	14.0 87	17.2 71	5.42 109	12.6 101	12.5 101	6.89
SegTreeDP [22]	68.9	2.21 87	2.76 79	10.3 94	0.46 72	0.60 45	2.44 34	9.58 106	15.2 103	18.4 85	3.23 50	7.86 18	8.83 54	6.82
RT-ColorAW [106]	69.2	1.40 46	3.08 86	5.81 23	0.72 85	1.71 96	3.80 70	6.69 58	14.0 86	15.3 39	4.03 78	11.9 95	10.2 71	6.55
AdaptWeight [12]	69.4	1.38 43	1.85 45	6.90 49	0.71 84	1.19 88	6.13 87	7.88 82	13.3 76	18.6 90	3.97 76	9.79 71	8.26 42	6.67

Fig. 25.4 Middlebury online test results

discontinuous disparity areas (disc) for four test images. The final column is the average error rates, respectively. From Table 25.2 we can see that the proposed has higher matching accuracy-rate than [1] and [2]. And compared with [3], the results of the algorithm are better too. Expecially for Tsukuba test image, the proposed has a bigger superiority.

Figure 25.4 shows the online test results of the proposed on the Middlebury College website [8], we rank 67.4 in all algorithms including global method. The proposed outperforms AdaptWeight method [4] on Venus test image.

25.5 Conclusions

This paper proposes a novel area-based adaptive window matching algorithm. The main idea of this algorithm is the combination of cross search and window cost, which works efficiently on building the adaptive window. Compared with conventional cross-based algorithm, the proposed has better results. Moreover, the usage of combined matching cost (SAD, GRAD, CENSUS) has an important effect on final results.

Every method shows clear improvements, but also weaknesses. Although our algorithm has high accuracy, there is certain weakness in discontinuous disparity areas. There is plenty room for improvement.

Acknowledgments This work is supported by the National Science & Technology Pillar Program of China (No. 2012BAH39F02).

References

1. Fusiello A, Roberto V (1997) Efficient stereo with multiple windowing. Paper presented at the IEEE conference on computer vision and pattern recognition, San Juan, 858–863
2. Veksler O (2003) Fast variable window for stereo correspondence using integral images. In: IEEE computer society conference on computer vision and pattern recognition, Madision
3. Zhang K et al (2009) Cross-based local stereo matching using orthogonal integral images. IEEE Trans Circuits syst Video Technol 19(7):1073–1079
4. Yoon KJ, Kweon IS (2006) Locally adaptive support-weight approach for visual correspondence search. IEEE Trans Pattern Anal Mach Intell 28(4):650–0656
5. Cigla C, Alatan AA (2011) Efficient edge-preserving stereo matching. In: IEEE International Conference on Computer Vision Workshops (ICCV Workshops)
6. Veksler O (1999) Efficient graph-based energy minimization methods in computer vision, Cornell University, New York
7. Hirschmuller H (2008) Accurate and efficient stereo processing by semi-global matching and mutual information, CVPR
8. http://vision.middlebury.edu/stereo/

Chapter 26
Super-Resolution Reconstruction for Mixed Resolution Videos Using Key Frames and Adaptive Detail Warping

Yun-Jhen Chen, Jin-Jang Leou and Han-Hui Hsiao

Abstract In the mixed resolution coding approach, one type of scalable video coding, a mixed resolution video sequence contains low-resolution (LR) nonkey video frames at high frame rate and periodic high-resolution (HR) key video frames at low frame rate in order to reduce bit rate and encoding complexity. At the decoder, LR nonkey video frames can be upsampled to the corresponding HR video frames by neighboring HR key video frames via video super-resolution. In this study, a video super-resolution reconstruction approach for mixed resolution videos using key frames and adaptive detail warping is proposed. Resolution enhancement is realized by incorporating the high-frequency details into the HR nonkey video frames that are directly upsampled by same conventional spatial interpolation method. The corresponding high-frequency details of each nonkey frame are derived by adaptive detail warping of the high-frequency details extracted from two (forward and backward) neighboring HR key frames, using the non-local-means concept and motion estimation. Based on the experimental results obtained in this study, the performance of the proposed approach is better than those of three comparison approaches.

Keywords Super-resolution reconstruction · Mixed resolution video · Key frames · Nonkey frames · Nonlocal means · Adaptive detail warping

26.1 Introduction

In the mixed resolution coding approach, one type of scalable video coding, a mixed resolution video sequence contains low-resolution (LR) nonkey video frames at high frame rate and periodic high-resolution (HR) key video frames at

Y.-J. Chen · J.-J. Leou (✉) · H.-H. Hsiao
Department of Computer Science and Information Engineering, National Chung Cheng University, Chiayi, Taiwan 621, Republic of China
e-mail: jjleou@cs.ccu.edu.tw

A. A. Farag et al. (eds.), *Proceedings of the 3rd International Conference on Multimedia Technology (ICMT 2013)*, Lecture Notes in Electrical Engineering 278,
DOI: 10.1007/978-3-642-41407-7_26,

low frame rate in order to reduce bit rate and encoding complexity [1]. At the decoder, LR nonkey video frames can be upsampled to the corresponding HR nonkey video frames by neighboring HR key video frames via video super-resolution.

At the decoder, the simple way to upsample the LR nonkey video frames to the corresponding HR video frames is spatial interpolation [2], i.e., each LR nonkey video frame is directly upsampled to the corresponding HR video frame by some spatial interpolation algorithm, such as linear interpolation, bicubic interpolation, …, without using temporal correlation information within neighboring key and nonkey frames. This type of approaches usually suffer from several types of visual degradation, such as jagging and artifacts.

At the decoder, to up-sample the LR nonkey video frames to the corresponding HR video frames using both spational and temporal information in neighboring key and nonkey video frames, several video super-resolution (SR) reconstruction approaches are proposed, which can be classified into two categories: example-based SR algorithms [3–5] and reconstruction-based SR algorithms [6–8]. For example-based SR algorithms [3–5], correspondences between LR and HR image patches are learned from a database composed of LR and HR image patch pairs. Each LR patch of the input video frame finds the best matching example patch in the training database, and the corresponding HR example patch will be used to reconstruct the HR patch in the reconstructed HR video frame. Usually, the sum of absolute differences (SAD) is employed as the distortion measure for patch matching.

Based on the assumption that the movements between the LR video frames are locally translational, an image process is modeled by establishing a relationship between the unknown HR video frame and the LR image observation, which is used to estimate the HR video frame. Most reconstruction-based SR algorithms [6–8] use a camera model for downsampling the HR video frames.

For mixed resolution coding, Watanabe et al. [9] proposed a learning-based video super-resolution algorithm for mixed resolution video coding. Ancuti et al. [10] proposed a video super-resolution approach using high quality photographs and maximum a posteriori (MAP) estimation. Brandi et al. [11] proposed a video super-resolution approach using key frames and motion estimation. Basavaraja et al. [12] proposed a video super-resolution approach based on the nonlocal means (NLM) concept. Spatio-temporal warping of the HR details extracted from HR still images is used to estimate the missing HR details of video frames. Hung et al. [1] proposed a video super-resolution approach using multiple overlapped variable-block-size codebooks to reconstruct HR non-key video frames. Song et al. [13] proposed a video super-resolution algorithm using bi-directional overlapped block motion compensation and on-the-fly dictionary training.

26.2 Proposed Adaptive Detail Warping Video SR Reconstruction Approach

NLM [14] is one method for video denoising by replacing each pixel with a weighted average of its neighboring pixels. The weights are determined by using block matching measures between patches centered on the pixel and neighboring pixels. Recently, some video super-resolution reconstruction approaches [13, 14] based on NLM framework are proposed. It is based on the assumption that an image patch may repeat in some spatio-temporal neighborhood, and hence each pixel centered at one patch of an image (video frame) can be estimated from regions that are similar to the region centered around it. If $y(i,j)$ denotes the input pixel value and $x(k,l)$ denotes the output estimated pixel value, the NLM filter can be described as

$$x(k,l) = \frac{\sum_{i,j\in N(k,l)} w(k,l,i,j)y(i,j)}{\sum_{i,j\in N(k,l)} w(k,l,i,j)}, \tag{26.1}$$

where $N(k,l)$ denotes the neighborhood centered at pixel (k,l), $y(i,j)$ is the input pixel value of the neighborhood centered at pixel (i,j), and the NLM weight for the (i,j)th neighbor pixel is given as

$$w(k,l,i,j) = \exp\left\{-\frac{\text{dist}\left(R_{k,l}y, R_{i,j}y\right)^2}{2\sigma^2}\right\}, \tag{26.2}$$

where $R_{k,l}$ is an operator that extracts a patch of a fixed size centered at pixel (k,l) from an image y, σ is a tuning parameter standing for the variance of data, and $\text{dist}\left(R_{k,l}y, R_{i,j}y\right)$ is the distance between two patches.

As shown in Fig. 26.1, in the mixed resolution coding approach, each input video sequence contains LR nonkey video (NKFs) at high frame rate and periodic HR key frames (KFs) at low frame rate in order to reduce bit rate and encoding complexity. Each HR video frame can be divided into low-frequency and high-frequency parts, which are treated as a blurred frame and a detail frame, respectively. To reconstruct the HR version of each LR NFK, the HR blurred frame can be obtained by directly upsampling the LR NFK by some conventional spatial interpolation method, and the HR detail frame is obtained by using the high-frequency information from neighboring HR KFs in the video sequence via the proposed video SR reconstruction approach.

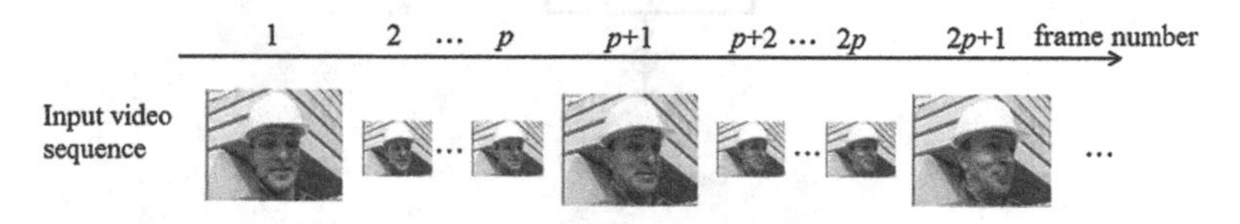

Fig. 26.1 An input mixed resolution video sequence including LR nonkey frames and periodic HR key frames

Basavaraja et al. [12] proposed a detail warping approach for super-resolution without explicit motion vector. However, the corresponding video reconstruction results are not good when the input video sequence contains global motions or high-speed moving object(s). In the proposed video SR reconstruction approach, a block-based motion estimation procedure is employed before detail warping to obtain the better video reconstruction results.

Assume that the input mixed resolution video sequence contains the LR video frames and periodic HR key frames. The objective is to obtain the corresponding HR video sequence containing details $X = \{X_m : m \in (1,2,\ldots,M)\}$ of $y = \{y_m : m \in (1,2,\ldots,M)\}$.

As shown in Fig. 26.2, assume that the neighboring forward and backward HR key frames of an LR video frame y_i are Z_k and Z_{k+p}. First, the LR video frame y_i is directly upsampled by bicubic interpolation to the HR blurred frame and the two neighboring key frames, Z_k and Z_{k+p}, of Y_i are downsampled and then upsampled by bicubic smooth and bicubic interpolation to obtain Y_k and Y_{k+p}, respectively. The high-frequency details, D_k and D_{k+p}, of the two key frames, Z_k and Z_{k+p}, are obtained by

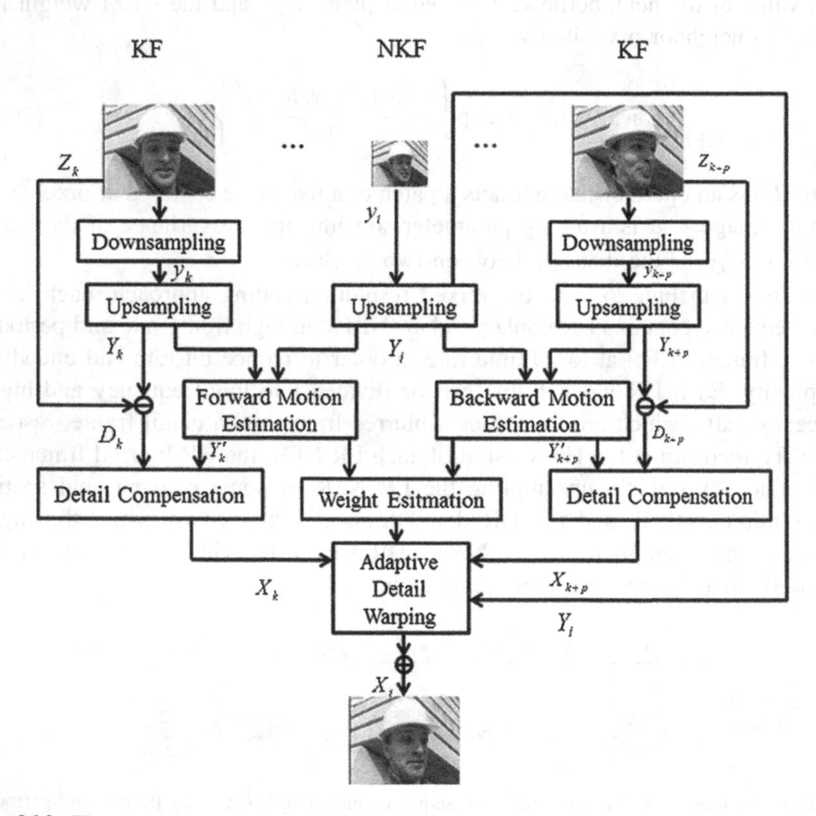

Fig. 26.2 The overview of the proposed video SR reconstruction approach

$$D_k = Z_k - Y_k \text{ and } D_{k+p} = Z_{k+p} - Y_{k+p}. \tag{26.3}$$

Additionally, both forward motion estimation between Y_k and Y_i and backward motion estimation between Y_i and Y_{k+p} are performed to get motion vectors MVs using block (patch) matching with block size $M \times M$.

The updated details, X_k and X_{k+p}, are obtained by applying detail compensation to the high-frequency details, D_k and D_{k+p}, based on forward and backward MVs, respectively. The "updated" smooth key frames, Y_k' and Y_{k+p}', are obtained by applying motion compensation to Y_k and Y_{k+p} using forward and backward MVs, respectively. The pixel-wise weight matrix W_k between Y_i and Y_k' is computed as

$$w(a,b,c,d) = \exp\left(\frac{-\|R_{a,b}Y_i - R_{c,d}Y_k'\|}{2\sigma_2^2}\right), \tag{26.4}$$

where $\|.\|$ denotes the Euclidean distance between two patches. The weights are used to estimate the high-frequency details of Y_i using adaptive detail warping.

Adaptive detail warping includes detail fusion and detail warping. As shown in Fig. 26.3a, based on two weight patches, w_k and w_{k+p}, a new weight patch w_i will be generated by selecting the maxima weight in the corresponding position in w_k and w_{k+p}. As shown in Fig. 26.3b, the detail fusion patch x_i' is obtained in the similar way as the detail weight patch w_i. Then, the final pixel $x_i(a,b)$ in the reconstructed HR nonkey video frame X_i is estimated as

$$x_i(a,b) = \frac{\sum_{c,d \in N(a,b)} w_i(a,b,c,d)x_i'(c,d)}{\sum_{c,d \in N(a,b)} w_i(a,b,c,d)}. \tag{26.5}$$

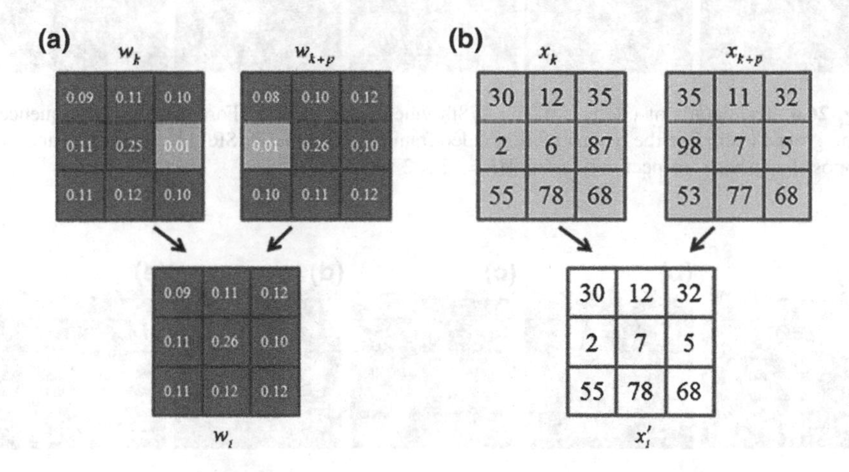

Fig. 26.3 An illustrated example patch of detail fusion

26.3 Experimental Results

The proposed approach has been implemented on an Intel Core i7-2600 CPU 3.40 GHz PC with 4 GB main memory using visual C++ of version 2010 software develop tool. In this study, seven test video sequences, "News," "Speaking," "Foreman," "Tennis," "Bus," "Dog," and "Stefan," are used to evaluate the performance of the proposed approach. "News" and "Speaking" contain small-motion objects; "Foreman" and "Tennis" contain moderate-motion objects; and "Bus," "Dog," and "Stefan" contain large-motion objects and global motions. Three comparison approaches, bicubic interpolation (Bicubic), motion-based SR (MSR) [11], and detail warping SR (DSR) [12] are also implemented.

In this study, the magnification factor (*MF*) is set to 2×2, the patch size $S \times S$, the search neighborhood N, and σ for DSR [12] and the proposed approach are set to 7×7, 7×7, and 10, respectively. The block size and the search range for motion estimation in MSR [11] and the proposed approach are set to 7×7 and 16×16, respectively. Additionally, the peak-signal-to-noise ratio (PSNR) is used as the objective performance measure.

Figure 26.4 shows the details of one part of the 178th video frame of the "Foreman" sequence, including the ground truth and the reconstruction video frames by Bicubic, MSR [11], DSR [12] and the proposed approach with

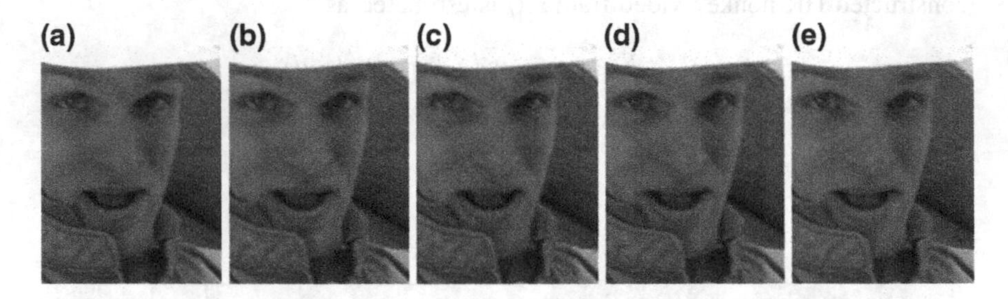

Fig. 26.4 The details of one part of the 178th video frame of the "Foreman" video sequence: **a** the ground truth; **b–e** the reconstruction video frames by Bicubic, MSR [11], DSR [12], and the proposed approach, respectively, with $MF = 2 \times 2$ and $p = 3$

Fig. 26.5 The details of one part of the 88th video frame of the "Stefan" video sequence: **a** the ground truth; **b–e** the reconstruction video frames by Bicubic, MSR [11], DSR [12], and the proposed approach, respectively, with $MF = 2 \times 2$ and $p = 3$

Table 26.1 The average PSNR values (dB) of the "Speaking" video sequence by Bicubic, MSR [11], DSR [12], and the proposed approach with $MF = 2 \times 2$ and different key frame intervals p

p	Bicubic	MSR [11]	DSR [12]	Proposed
2	37.03	40.19	45.28	45.88
5	37.03	39.43	44.78	45.1
10	37.03	39.21	43.88	44.57
15	37.04	38.94	43.01	44.07
20	37.03	38.44	42.36	43.44

Table 26.2 The average PSNR values (dB) of the "Tennis" video sequence by Bicubic, MSR [11], DSR [12], and the proposed approach with $MF = 2 \times 2$ and different key frame intervals p

p	Bicubic	MSR [11]	DSR [12]	Proposed
2	41.08	40.81	41.6	44.53
5	41.05	40.74	41.39	43.56
10	41.17	40.40	41.31	42.99
15	41.25	40.88	41.22	42.65
20	41.35	41.01	41.18	42.79

$MF = 2 \times 2$ and key frame interval $p = 3$. Figure 26.5 shows the details of one part of the 88th video frame of the "Stefan" video sequence, including the ground truth, and the reconstruction video frames by Bicubic, MSR [11], DSR [12], and the proposed approach with $MF = 2 \times 2$ and key frame interval $p = 3$. Table 26.1 lists the average PSNR values (dB) of the "Speaking" video sequence by Bicubic, MSR [11], DSR [12], and the proposed approach with $MF = 2 \times 2$ and different key frame intervals p. Table 26.2 lists the average PSNR values (dB) of the "Tennis" video sequence by Bicubic, MSR [11], DSR [12], and the proposed approach with $MF = 2 \times 2$ and different key frame intervals p.

26.4 Concluding Remark

In this study, a video super-resolution reconstruction approach for mixed resolution video coding using key frames and adaptive detail warping is proposed. Resolution enhancement is realized by incorporating the high-frequency details into the HR nonkey video frames that are directly upsampled by some conventional spatial interpolation method. The corresponding high-frequency details of each nonkey frame are derived by adaptive detail warping of the high-frequency details extracted from two (forward and backward) neighboring HR key frames, using the nonlocal-means concept and motion estimation. Based on the experimental results obtained in this study, the performance of the proposed approach is better than those of three comparison approaches.

Acknowledgments This work was supported in part by National Science Council, Taiwan, Republic of China under Grants NSC 102-2221-E-194-028-MY2 and NSC 102-2221-E-194-041-MY3.

References

1. Hung EM et al. (2012) Video super-resolution using codebooks derived from key-frames. IEEE Trans on Circuits Syst Video Technol 22(9):1321–1331
2. Lehmann T, Gonner C, Spitzer K (1999) Servey: interpolation methods in medical image processing. IEEE Trans Med Imaging 18(11):1049–1075
3. Freeman WT, Jones TR, Pasztor EC (2002) Example-based super-resolution. Comput Graphics Appl 22(2):56–65
4. Yang J, Wright J, Huang T, Ma Y (2008) Image super-resolution as sparse representation of raw image patches. In: 2008 IEEE international conference on computer vision and pattern recognition, pp 1–8
5. Yang X, Su GD, Chen J, Moon YS (2010) Restoration of low resolution car plate images using PCA based image super-resolution. In: 17th IEEE international conference on image processing, pp 2789–2792
6. Mudenagudi U, Banerjee S, Kalra PK (2011) Space-time super-resolution using graph-cut optimization. IEEE Trans Pattern Anal Mach Intell 33(5):995–1008
7. Hsieh CC, Huang YP, Chen YY, Fuh CS (2011) Video super-resolution by motion compensated iterative back-projection approach. J Inf Sci Eng 27(3):1107–1122
8. Keller SH, Lauze F, Nielsen M (2011) Video super-resolution using simultaneous motion and intensity calculations. IEEE Trans Image Proc 20(7):1870–1884
9. Watanabe K, Iwai Y, Haga T, Yachida M (2008) A fast algorithm of video super-resolution using dimensionality reduction by DCT and example selection. In: 19th international conference on pattern recognition, pp 1–5
10. Ancuti C, Ancuti CO, Bekaert P (2010) Video super-resolution using high quality photographs. In: 2010 IEEE international conference on acoustics, speech, and signal processing, pp 862–865
11. Brandi F, Queiroz R, Mukherjee D (2008) Super-resolution of video using key frames and motion estimation. In: 15th IEEE international conference on image processing, pp 321–324
12. Basavaraja SV, Bopardikar AS, Velusamy S (2010) Detail warping based video super-resolution using image guides. In: 17th IEEE international conference on image processing, 2009–2012
13. Song BC, Jeong SC, Choi Y (2011) Video super-resolution algorithm using bi-directional overlapped block motion compensation and on-the-fly dictionary training. IEEE Trans Circuits Syst Video Technol 21(3):274–285
14. Protter M et al (2009) Generalizing the nonlocal-means to super-resolution reconstruction. IEEE Trans Image Proc 18(1):36–51

Chapter 27
Segmentation Based on Spiking Neural Network Using Color Edge Gradient for Extraction of Corridor Floor

XiaoWei Wang, QingXiang Wu, Zhenming Zhang, Zhiqiang Zhuo and Liuping Huang

Abstract In this paper, for the purpose of obstacle avoidance for blind men in the environment of indoor corridor, a corridor ground segmentation algorithm is proposed using image processing mechanism of the human visual system combined with the existing segmentation algorithms in robot visual navigation techniques. The segmentation algorithm is based on a spiking neural network. First, three color image gradient maps are generated utilizing a spiking neural network. The best gradient map is generated from three color components to extract the effective and useful image edges. Then threshold segmentation method is used to eliminate unwanted gradient to identify the boundary of floor. Finally, the corridor ground is extracted. The experimental results show that the algorithm works efficiently and the boundary of floor can be extracted accurately for corridor images with certain noise textured and nontextured. The algorithm has the practicality and robustness for identification of ground floor in blind navigation.

Keywords Color edge detection · Spiking neural networks · Receptive field · Visual system

27.1 Introduction

Intelligent vehicle is to achieve independence for driving, requiring being able to initiatively perform overtaking, avoiding in the case of no intervention of human drivers. The visual navigation techniques enable mobile robots to automatically

X. Wang · Q. Wu (✉) · Z. Zhang · Z. Zhuo · L. Huang
College of Photonic and Electronic Engineering, Fujian Normal University, Fuzhou, China
e-mail: qxwu@fjnu.edu.cn

X. Wang
e-mail: shuiguoalnzi397@fjnu.edu.cn

A. A. Farag et al. (eds.), *Proceedings of the 3rd International Conference on Multimedia Technology (ICMT 2013)*, Lecture Notes in Electrical Engineering 278,
DOI: 10.1007/978-3-642-41407-7_27,

avoid obstacles in indoor and outdoor environments. So these techniques can also be used to help the blind men who lose sight.

Eyes are the windows of the soul, through which people come to understand and perceive the external world. The humans obtain 80 % of information by the eye. So living in the dark world, the blind men cannot get the colorful world, and cannot describe and understand it, and their daily lives have been a considerable degree of restrictions, especially in traveling, and thus blind navigation brings themselves or their family the inconvenience to some degree. And how to help the blind more conducive to safe travel is bound to become one of the main aspects. In view of the navigation algorithm better being used to robots and vehicles, the concept of navigation is also widely used in the blind in recent years.

Currently research is of blind navigation system which is still infant in the world, and computer vision technology is widely used in the blind road detection. CAO Yu-zhen [1] proposed a blind navigation system to detect the crosswalk. The target area is selected using the method of bipolar gray scale contrast estimation and then the transform domain is used to extract the crosswalk features such as number, length, stripes, angle, etc. Zheng nanning [2] proposed a road detection algorithm based on statistical features and principal component analysis, which has good robustness. Kantan wong [3] proposed a road traffic signs which can be detected in the navigation system, using the color and shape different from natural world to extract the characteristics of the traffic signs. Among them, the blind detection is the hotspot of blind navigation system. Jinglei Tang [4] proposed a method of detecting and identifying blind road in the natural scene using dual-threshold segmentation and the algorithm of blind road localization and tracking.

In this paper, for the blind navigation in the corridor, the main research is about the algorithm of detecting road boundary from the interior hallway floor area. Literature [5] proposed a score model for evaluating the likelihood of line segments belonging to the wall-floor boundary and has done some processing appropriately for the specular reflection, but has not very good adaptability for the ground and the wall with a certain texture. The literature [6] mainly uses the information of texture and color to segment an indoor obstacle region in order to facilitate robot navigation. But it has only been applied to areas of the ground with no obstructions which is fully different from around obstacles and walls. Although Corridor environment has not so complicated and harsh as external environment, it still bring some difficulties in segmentation because of light, shadows, and reflections, etc. Especially some corridors are very dark due to backlight, or a strong light suddenly shines from the end of the corridor, or shines from windows, or there will be large quantities of goods heap or ground is cluttered because some corridor left unattended. If the corridor environments are very complicated, there is still no algorithm applicable to all kinds of such complicated corridors. Current segmentation procedures are difficult to handle the situation where wall and floor are similar or have various texture, lighting, shadow, reflection, and water stains, etc. The proposed algorithm in this paper can improve the accuracy of the above cases, and the algorithm has robustness, so that it can be applied to most natural scenes with less cost of computations.

In this article we first present how to obtain the respective edge of the gradient for three components of the color image using a spiking neural network, and then compare them to select the Maximum gradient which is regarded as the original color gradient map. The multithreshold segmentation and watershed segmentation of the gradient map are used to identify the boundaries. The floor region is merged using the result graph marked by watershed algorithm. Finally two seeds are used to filling algorithm to further prevent the phenomenon of over-segmentation, and accuracy boundaries are extracted.

Experimental results show that this algorithm is effective computing speed and can be well suited for indoor corridor identification to provide services for the blind navigation in real-time with robustness. For example, it can be implemented in a smart phone.

27.2 The Proposed Algorithm Framework Based Spiking Neural Network

27.2.1 Edge Gradient Information of a Color Image

Luguan Ming [7] proposed a multiscale morphological gradient operator. Let $B_i(0 \leq i \leq n)$ be a square structure elements and n is the structure factor. The size of B_i is $(2i+1) \times (2i+1)$, such multiscale gradient operator is defined as follows:

$$MG(f) = \frac{\sum_{i=1}^{n} \{[(f \oplus B_i) - (f \odot B_i)] \odot B_{i-1}\}}{n} \tag{27.1}$$

where $\oplus$ and $\odot$ represent dilate and erode of morphological operations respectively. Due to the averaging operation, if structure factor n is bigger, ability of anti-noise is stronger, but the operation takes more time.

Gongtian Xu [8] defines a gradient operator using image color component to calculate gradient information:

$$\begin{aligned} \|\Delta f_{c_1c_2c_3}\|^2 = \Big(& \left(\|\Delta f^{c_1}_{c_1c_2c_3}(x)\| + \|\Delta f^{c_1}_{c_1c_2c_3}(x)\| + \|\Delta f^{c_1}_{c_1c_2c_3}(x)\|\right)^2 \\ & + \left(\|\Delta f^{c_1}_{c_1c_2c_3}(y)\| + \|\Delta f^{c_1}_{c_1c_2c_3}(y)\| + \|\Delta f^{c_1}_{c_1c_2c_3}(y)\|\right)^2 \end{aligned} \tag{27.2}$$

where $f_{c_1c_2c_3}$ represents the color image of the color space $c_1c_2c_3$, $\triangle$ represents the gradient operator, and $f^{c_1}_{c_1c_2c_3}(x)$ means that c_1 components of the image $f_{c_1c_2c_3}$. The gradient templates for calculating the gradient of X and Y directions are shown as follows:

$$\Delta f(x) = \begin{pmatrix} 1 & 4 & 1 \\ 0 & 0 & 0 \\ -1 & -4 & -1 \end{pmatrix} \quad \Delta f(y) = \begin{pmatrix} 1 & 0 & -1 \\ 4 & 0 & -4 \\ 1 & 0 & -1 \end{pmatrix} \tag{27.3}$$

Daoqing Lin [9] proposed a single-scale morphological gradient operator to the component of the color image, such as gray gradient map of the R component diagram can be obtained using the following formula

$$G_r = \frac{F_r \oplus S - F_r \odot S}{2} \tag{27.4}$$

wherein G_r is the required R gradient map, F_r is the original R component diagram, and S is a circle template of radius 1.

$$G(i,j) = \mathrm{MAX}\big(G_r, (i,j), G_g, (i,j), G_b, (i,j)\big) \tag{27.5}$$

However, this algorithm is sensitive to noise of the boundary and makes the edge position change, which can sometimes lead to errors of corridor floor segmentation. Therefore, in this paper we propose an edge gradient [10] extraction algorithm based on spiking neural network with the use of the human visual mechanisms and biological receptive field activities. It is recognized that the brain is a complex nervous system consisting of thousands of neurons. The human cerebral cortex has more than 10 billion neurons, about 10,000/mm^3, which combine with each other to form the neural network. And each neuron cell is composed of the cell body, dendrites and axons. Artificial neural network is put forward for the simulation of the human brain structure. Inspired by the principles from the human visual system, a spiking neural network is proposed to obtain edge gradients, as shown in Fig. 27.1. More details can referred to literature [10].

In the network, the neurons are connected by excitatory(X) and inhibitory(Δ) synapses. The neuronal connection weights in the receptive field $\mathrm{FR}_{\mathrm{rept}}(x, y)$ are represented as follows.

$$w_{x,y}^{up\_ex} = \begin{cases} 0 & \text{if } (y - y_c) \le 0 \\ w_{e\max}\, e^{-\frac{(x-x_c)^2}{\delta_x^2} - \frac{(y-y_c)^2}{\delta_y^2}} & \text{if } (y - y_c) > 0 \end{cases} \tag{27.6}$$

$$w_{x,y}^{up\_ih} = \begin{cases} 0 & \text{if } (y - y_c) > 0 \\ w_{i\max}\, e^{-\frac{(x-x_c)^2}{\delta_x} - \frac{(y-y_c)^2}{\delta_y}} & \text{if } (y - y_c) \le 0 \end{cases} \tag{27.7}$$

where (X_c, Y_c) is center of receptive field $\mathrm{RF}_{\mathrm{rcpt}}$, δ_x and δ_y are constants, $w_{e\max}$ and $w_{i\max}$ are the maximum weight of excitatory and inhibitory synapse.

Shown in Fig. 27.1, an image present in the input layer. The information is processed in the intermediate layer through corresponding four parallel arrays of neurons with different weight matrix. And finally an edge map can be obtained in the output layer. These four weight matrix detect the edge direction of down, left, and right directions, achieving by the excitatory and inhibitory synapses.

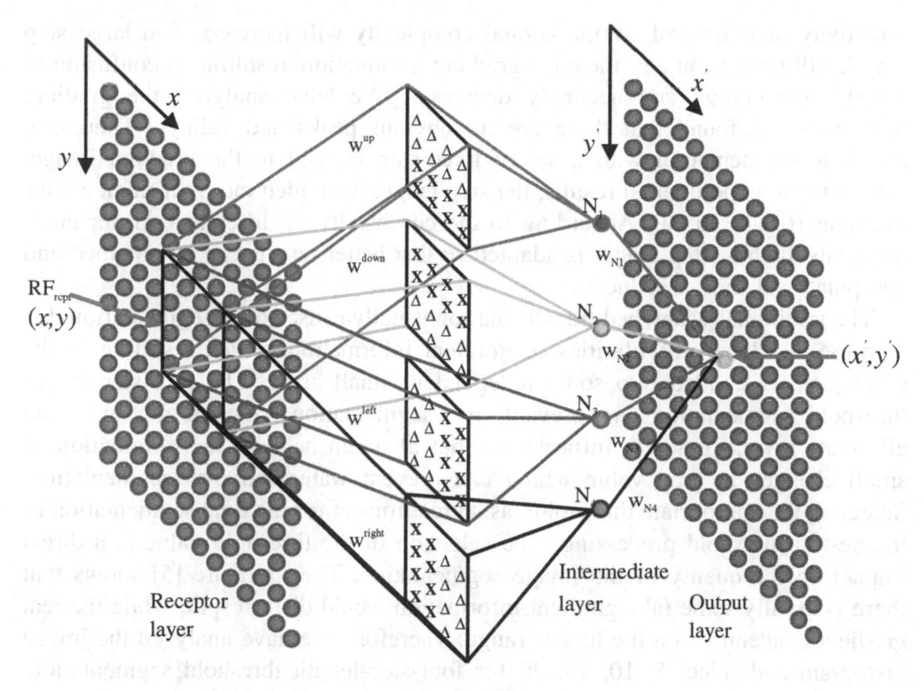

Fig. 27.1 Spiking neural network model for edge detecting where (X_c, Y_c) is center of receptive field RF_{rcpt}, δ_x and δ_y are constants, w_{emax} and w_{imax} are the maximum weight of excitatory and inhibitory synapses

First, the spiking neural network in Fig. 27.1 is used to compute the gradient of components of RGB color image, and then equation (27.5) is used to generate a gradient image for identify boundary of floor.

27.2.2 Gradient Map of the Multithreshold Segmentation

Threshold segmentation can be divided into following categories: point-based global threshold method, regional and global threshold method or local threshold method, and multithreshold method. Among them, the point-based global threshold method includes p fraction, Otsu, and entropy method. Region-based global threshold method has simple statistical and histogram method. Local threshold and the multithreshold method refer to threshold difference method, waterline threshold algorithm and multithreshold method, and so on. A multithreshold segmentation with the fixed step length is used to the proposed algorithm.

When the multithreshold segmentation algorithm is applied to gradient image, it is very important to determine the step length. If the step length is too small, the sensitivity of noise and computational complexity will increase. Too large step length will ignore some of the edge gradient information, resulting in confusion of target segmentation and accuracy decreases. We have analyzed the gradient histogram and found that there are no obvious peaks and valleys. Therefore, threshold segmentation with a set of fixed-step is used to the gradient image. According to experimental results, the step length is divided into four numbers for example 10, 20, 30, 40. According to the complexity of different corridor environments the best step length is adapted so that better results can be obtained and computational time is reduced.

The traditional watershed transformation usually causes over-segmentation due to noise and local irregularities of gradient information, which resulting in the target area is divided into some independent small areas. The four-step-length threshold segmentation can prevent over-segmentation, since it not only can eliminate invalid gradient information, but also can achieve the elimination of small changes in gray value which can prevent watershed over-segmentation. Selecting the appropriate threshold has a great impact for the final segmentation of images on threshold processing. The selection of the threshold value is a direct impact on the quality of the image segmentation. The literature [5] shows that there is usually some fake gradient information would disturb split, while the real gradient concentrate on the higher range. Therefore, we have analyzed the image histogram and select 5, 10, 15, 20 for four-step-length threshold segmentation. Experimental results have validated that these thresholds can achieve good results for the segmentation algorithm to generate the best gradient map.

27.2.3 Regional Merger and Seed Filling Algorithm

A new gradient graph can be generated using the watershed segmentation [11] from the gradient map, and in order to further prevent over-segmentation problem, the regions in the divided image can be marked, and then the marks are used to merge each regions appropriately according to the similarity criteria of the gray values.

Let $D(i, j)$ represents gray value of pixel (i, j) in D region $A*B$, the average gray value of the image can be defined as:

$$M = \frac{\sum_{i}\sum_{j} D(i, j)}{A \times B} \tag{27.8}$$

The similarity criteria between regions D_x and D_y is defined as flows:

$$mp_x = \frac{D(M_x)}{n}, \quad mp_y = \frac{D(M_y)}{n} \quad x = 1, 2, \ldots \mathrm{n} \tag{27.9}$$

$$S(D_x, D_y) = |mp_x - mp_y| \tag{27.10}$$

where $D(M_x)$ and $D(M_y)$ represent the mean values of Region D_x and D_y respectively, n is the total marks in the regions. When $S(D_x, D_y)$ is less than a certain value, then Region D_x and D_y are combined.

Then the seed filling algorithm is used to fill the same mark to the corridor floor, and finally the Moore boundary tracking algorithm [12] is used to extract the boundary contours.

27.2.4 Summary of the Proposed Algorithm

The main steps of the algorithm are as follows:

1. First, calculate the respective gradient information of three components of the color image respectively using the spiking neural network model, and then compute best gradient using the formula (27.5).
2. Using the four-fixed step length multithreshold segmentation to the resulting gradient map, to remove the impact from noises and invalid gradients.
3. Then use the watershed segmentation to produce a new gradient graph and mark the separated regions in the image.
4. Finally, region merging and Moore boundary tracking algorithm are used to extract the wall—floor boundaries.

27.3 Experimental Results

To validate the proposed algorithm, an image database of more than one hundred images was taken in different corridors built with diverse materials. The experimental results indicate that the proposed algorithm efficiently obtain the boundary which match real boundary very well, as shown in Fig. 27.3. Experiments show that the algorithm has applicability and is able to meet real-time requirements (Fig. 27.2).

In the image database corridors include indoor or outdoor, and materials of corridor are tile, cement, and wood. Corridor's environment includes reflections, shadows, and different illumination. It is worth mentioning that, there is a baseboard sometimes between walls or doors and floor, and its colors and textures are different from the wall or door. As the baseboard area is significantly less than the area of floor ground, and the real edge is in the middle of the baseboard, so it is always ignored. The proposed algorithm has obtained accuracy boundaries in most situations shown in Fig. 27.3.

Viewing from the first row of Fig. 27.4, Ling's algorithm cannot deal with a small gradient of the wall-floor situation due to dust or water, while the proposed algorithm can handle it well. It is worth mentioning that there are actually water

Fig. 27.2 Successful segmentation examples of the corridor floor in our image database which contains corridor pictures with the different material, contours, as well as different building environments

Fig. 27.3 Results of the proposed algorithm on images download from the paper [5]

spots in the upper right side area of the floor. From the second row, Lee's algorithm can segment a wall from the floor and ceiling using geometric reasoning and it is time-consuming. The proposed algorithm can detect the floor. And the last row show that Li's algorithm can accurately segment the ground area but apparently ignored the edge information of the wall on both sides, while the proposed algorithm can deal with it.

Fig. 27.4 The first column shows the original images, the second one shows results from Daoqing Ling, Lee [13], and Yinxiao's algorithms, and the third column shows results obtained by the proposed algorithm

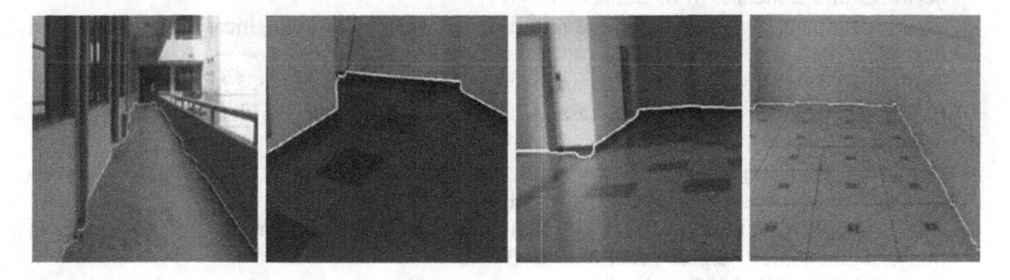

Fig. 27.5 Results of the proposed algorithm in case that is strong texture on the floor or the wall

Viewing from Fig. 27.5, the proposed algorithm can deal with some situation, such a textured floor or wall, while Li's algorithm is difficult to deal with these situations.

However, the proposed algorithm also has some limitation. It is difficult to deal with very dark corridors in a building, or a very bright light at the end of corridors, or the ground and walls with very complex and dense texture.

27.4 Conclusion

In this paper, we present a corridor ground segmentation algorithm using color edge gradient information to watershed segmentation in which color edge gradient is obtained by a spiking neural network inspired from the human visual system. The algorithm achieves good results in most cases with robustness and adaptability, whether the floor and wall's textures are similar or completely different. However, we still want to improve the algorithm to deal with particularly complex and harsh environment of the corridors. The future work will focus on the extraction of space area of the corridor for example, length and width of the corridor, and implement the algorithm in a start phone as a flexible "eye" for the blind navigation.

Acknowledgments The authors gratefully acknowledge the fund from the Natural Science Foundation of China (Grant No.61179011) and Science and Technology Major Projects for Industry-academic Cooperation of Universities in Fujian Province (Grant No.2013H6008).

References

1. Zhen CY, Gang LIU (2008) Research on the detection of pedestrian crossing in blind man navigation system. Comput Eng Appl 176–198
2. Zheng NN, Cheng H (2003) Road recognition algorithm using principal component neural networks and k-means. SPIE 5286:77–80
3. Wong KT, Phanprasit S (2007) Road traffic signs detection and classification for blind man. ICCAS 82:847–852
4. Tang JL (2009) blind-road location and recognition in natural scene. 82(3):284–301
5. Li YX, Stanley YB (2010) Image-based segmentation of indoor corridor floors for a mobile robot. Intell Robots Syst 837–843
6. Hernandez JF, Campos C (2010) Detecting obstacle-free regions for visual robot navigation by inferring scene horizons. Robotic Sens Environ 21:1–6
7. Lu GM, Li SH, Multiscal (2001) Morphological gradient algorithm and its application in image segmentation. Signal Process 17(2):37–40
8. Gong TX, Peng JX (2003) Color image segmentation based on watershed transform. Comput Vis Image Underst 77(3):317–370
9. Qing D, Gao ZY (2007) Segmentation for images of VCH-F1 based on improved watershed algorithm. Comput Eng Appl 43(35):355–364

10. Wu QX, McGinnity TM, Maguire LP, Belatreche A, Glackin B (2007) Edge detection based on spiking neural network model. Springer LNAI 4682:26–34
11. Vincent L (1993) Morphological grayscale reconstruction in image analysis: applications and efficient algorithms. IEEE Trans Image Proc 2(2):176–201
12. Moore GA (1968) Automatic scanning and computer processes for the quantitative analysis of micrographs and equivalent subjects. In: Cheng GC et al (eds) Pictorial pattern recognition. Thomson, Washington, D.C, pp 275–326
13. Lee DC, Hebert M, Kanada T (2009) Geometric reasoning for single image structure recovery. In: IEEE conference vision and pattern recognition (CVPR), 2136–2143

Chapter 28
Interactive Scene Text Detection on Mobile Devices

Jinlong Hu, Baihua Xiao, Chunheng Wang, Cunzhao Shi and Song Gao

Abstract With the increasing resolution and availability of digital cameras, text detection in natural scene images receives a growing attention. When taking pictures using a mobile device, people generally only concerned with interesting texts instead of all of the text in the image. In this paper, we propose an interactive method to detect and extract interesting text in natural scene images. We first draw a line to label a region which contains the texts we want to detect. Then a coarse-to-fine strategy is adopted to detect texts in this label region. For coarse detection, we apply Canny edge detection and connected component (CC)-based approach to extract coarse region from the label region. For fine detection, some heuristic rules are specially designed to eliminate some non-text CCs and then to merge the remaining CCs in the coarse region. To better evaluate our algorithm, we collect a new dataset, which includes various texts in diverse real-world scenarios. Experimental results on the proposed dataset demonstrate very promising performance on detecting text in complex natural scenes.

Keywords Text detection · Interactive · Coarse-to-fine · Connected component analysis

J. Hu · B. Xiao (✉) · C. Wang · C. Shi · S. Gao
The State Key Laboratory of Management and Control for Complex Systems,
Institute of Automation, Chinese Academy of Sciences,
Beijing, China
e-mail: baihua.xiao@ia.ac.cn

J. Hu
e-mail: jinlong.hu@ia.ac.cn

C. Wang
e-mail: chunheng.wang@ia.ac.cn

C. Shi
e-mail: chunzhao@ia.ac.cn

S. Gao
e-mail: Song.gao@ia.ac.cn

A. A. Farag et al. (eds.), *Proceedings of the 3rd International Conference on Multimedia Technology (ICMT 2013)*, Lecture Notes in Electrical Engineering 278,
DOI: 10.1007/978-3-642-41407-7_28,

28.1 Introduction

Text detecting in natural scene images plays a very important role in content-based image analysis. However, this is a challenging task due to the wide variety of text appearances, such as variations in font and style, geometric and photometric distortions, partial occlusions, and different lighting conditions.

Text detection has been considered in many recent studies and numerous methods are reported in the literature [1–6]. Most of the existing methods of text detection could be roughly classified into two categories: region-based and CC-based. Region-based methods need to scan the image at multiple scales and use a text/non-text classifier to find the potential text regions. Chen et al. [1] proposed a fast text detector base on a cascade AdaBoost classifier. As opposed to region-based method, CC-based methods first use various approaches such as edge detection, color clustering or stroke width transform to get the CCs, and heuristic rules or classifiers are used to remove non-text CCs. Pan et al. [7] adopted region-based classifier to get the initial CCs and use the CRF to filter non-text components.

Most of previous methods focus on detecting all of the text in the image. However, when taking pictures using a mobile device, we are only interested in certain text in the image. Moreover, detecting all of the text in the image requires more computation and storage capacity. Therefore, most of these methods are not suitable for use in real-time applications and on mobile devices. In this paper, we propose an interactive method to detect texts in natural scene images. The overall process of the text detection is illustrated in Fig. 28.1. First, Canny edge detection and CC-based approach are applied to quickly extract coarse region from the label region (Fig. 28.1c, d). Then specially designed heuristic rules are used to eliminate the non-text CCs [8]. Finally, the remaining CCs in the coarse region are merged to get the fine region (Fig. 28.1e, f).

In comparison to previous text detection approaches, our algorithm offers the following major advantages. First, interactive text detection only concerned with the region we want to detect. Further, our method provides a reliable binarization for the detected text, which can be directly passed to OCR for text recognition.

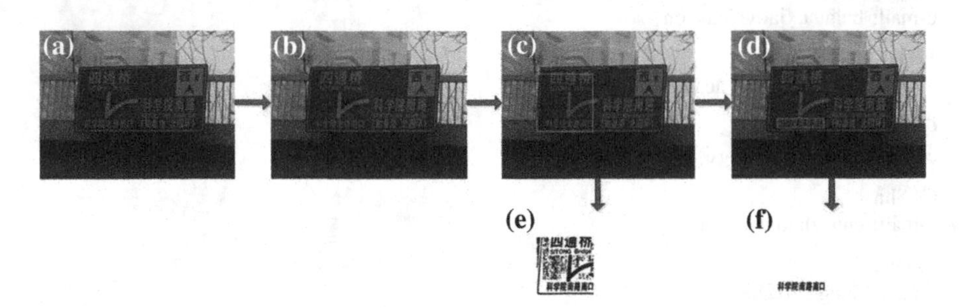

Fig. 28.1 Overview of our method

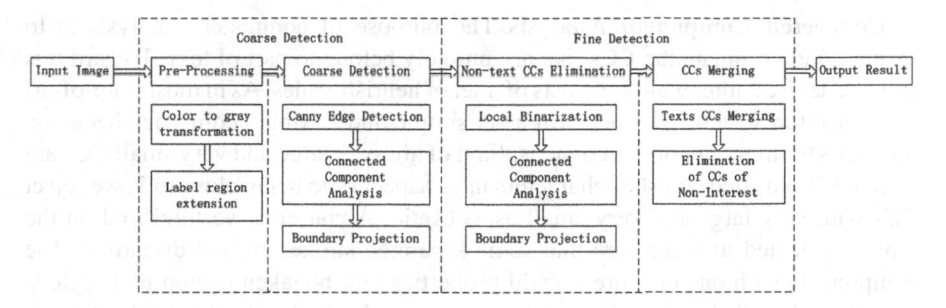

Fig. 28.2 The flowchart of the proposed method

28.2 Text Detection Method

The flowchart of our text detection algorithm is shown in Fig. 28.2. The algorithm works on a gray scale image and can be separated into three main steps: (1) Extract coarse region from the label region; (2) Eliminate the non-text CCs using heuristic rules; (3) Merge the rest CCs to get the fine region.

28.2.1 Coarse Region Extraction

The aim of this step is to extracts coarse region from the label region.

Connected Component Extraction To extract CCs from the image, we use Canny edge detector [9] to produce an edge map (Fig. 28.3c) from the extended image (Fig. 28.3b). This edge detector is efficient and to provide accurate results which makes it suitable for our purpose. With the result of Canny detection, we obtain a binary image where the foreground CCs are considered as text candidates.

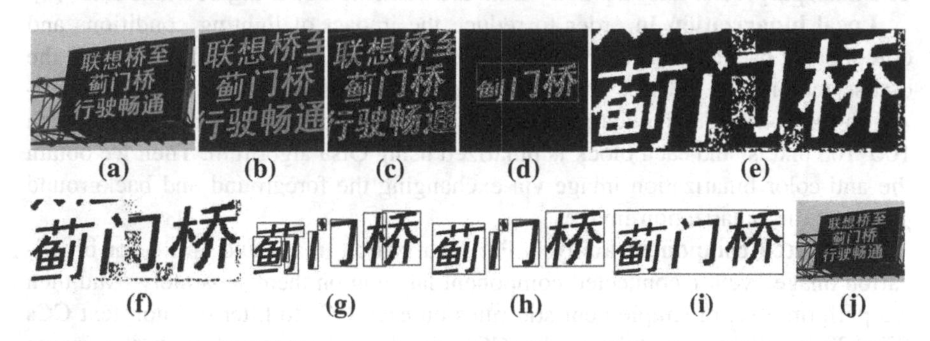

Fig. 28.3 Text detection process. **a** Original image with line labeled by user. **b** Extended image. **c** Edge map. **d** Boundaries of coarse region. **e** Coarse region. **f** Component filtering result. **g** CCs before merged. **h** CCs after merged. **i** Fine region. **j** Detected result

Connected Component Analysis The purpose of component analysis is to identify and eliminate the CCs that are unlikely belong to part of text. Toward this end, we devise a filter which consists of a set of heuristic rules. As in most state-of-art text detection systems, we perform a set simple and flexible geometric checks on each CCs to filter out non-text objects. First of all, very large and very small CCs are rejected. Then, since most of characters have aspect ratio being close to 1, we reject CCs with very large and very small aspect ratio. A conservative threshold on the scale is selected to make sure that some separated strokes are not discarded. The components with one or more invalid properties will be taken as non-text regions and discarded. This preliminary filter proves to be both effective and efficient. A large portion of obvious non-text regions are eliminated after this step.

Boundary projection According to the results of connected component analysis, we apply the projection method to find the boundaries [10]. To find height boundaries, we scan pixels along horizontal line with lengths equals to the width of label line. We scan these lines and calculate the ratio between foreground pixels and background pixels. The procedure is applied until the ratio is less than a certain threshold. As an outcome of this procedure we obtain the top and bottom boundaries. To find the width boundaries, we scan pixels along vertical line with heights equals to the height of top and bottom boundaries. We scan these lines following the same pixel criteria used earlier. The algorithm moves the lines toward left and right until this criteria are fulfilled (Fig. 28.3d).

The combination of these three procedures computes a rectangular bounding box that encloses the text (Fig. 28.3e). However, the produced bounding box may be slightly larger than the minimum bounding box due to noise present in the image. This bounding box is a coarse region, we will extract fine region in this smaller region.

28.2.2 Non-text CCs Elimination

The main purpose of this step is to eliminate non-text CCs using heuristic rules [8].

Local binarization In order to reduce the impact of lighting conditions and complicated background, we propose a local binarization approach to binarize the coarse region (Fig. 28.3e). Our technique is essentially based on Otsu's binarization method [11]. The coarse region is divided in nonoverlapping blocks of in 100*100 pixels, and each block is binarized using Otsu algorithm. Then we obtain the anti-color binarization image via exchanging the foreground and background pixels of the binarization image.

Connected component analysis For binarization image and anti-color binarization image, we use connected component labeling on them separately. And then we perform a set of simple heuristic rules on each CCs to filter out non-text CCs (Fig. 28.3f). First, we remove the CCs which are connected with the image boundary. Then, very large and very small CCs are rejected. Finally, we remove the isolated CCs which are far away from its surrounding CCs. We select the threshold following the same criteria used earlier to make sure that some separated strokes are not discarded.

According to the number of remaining CCs in two kinds of binarization images, we can determine the polarity of the image and choose the appropriate one as the binarization image.

28.2.3 Rest CCs Merging

We found there are still some narrow non-text CCs in the remaining CCs because these CCs are very similar to separated strokes of some characters. Therefore, we propose an algorithm to merge that separated strokes which belong to the same character.

For every remaining CCs (Fig. 28.3g), we scan each CCs around it and decide whether these two CCs are merged based on the heights, widths, and position of their bounding boxes. If one CCs contains another, or two components have overlapping region, or two components are close enough, they are merged into one component (Fig. 28.3h). Each rectangle represents a CC. This step also serves as a filtering step because the CCs which very small and cannot be merged are taken as components casually formed by noises or background clutters, and thus are discard. Once the non-text CCS are eliminated and text CCs are merged, the CCs close to the label line are preserved as the final text region and we obtain the fine region (Fig. 28.3i, j).

28.3 Experimental Results

28.3.1 Dataset and Experiment Setting

For evaluating the performance of the proposed methods, we introduce a dataset for evaluating text detection algorithm, which contains images of real-world complexity.

Although widely used in the community, the ICDAR dataset [12, 13] has two major drawbacks. First, most of the text lines (or single characters) in the ICDAR dataset are horizontal. In real scenarios, however, text may appear in any orientation. The second drawback is that all the text lines or characters in this dataset are in English. These two shortcomings are also pointed out in [7, 14]. In this work, we generate a new multilingual image dataset with horizontal as well as skewed and slant texts. This dataset contains 250 natural images in total. These images are taken from indoor (office and mall) and outdoor (street) scenes using a mobile camera. Some typical images from this dataset are shown in Fig. 28.4a.

The dataset is very challenging because of both the diversity of the texts and the complexity of the backgrounds in the images. The texts may be in different languages (Chinese, English, or mixture of both), fonts, sizes, colors, and orientations.

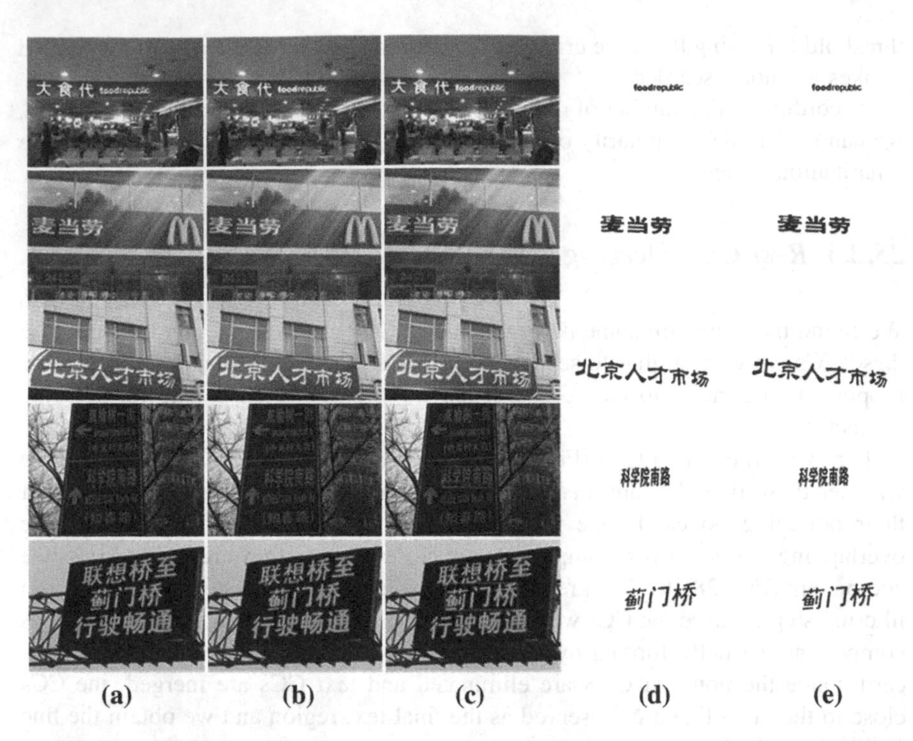

Fig. 28.4 Selected results of our coarse-to-fine method on the proposed dataset. **a** Original image with the line labeled by user. **b** Manually labeled text areas (*red rectangles*). **c** Detected text by our method (*green rectangles*). **d** Manually extracted text. **e** Extracted text by our method

Our evaluation method is inspired from the one used in the ICDAR2003 competition, but it is much simpler. The definitions of precision and recall are:

$$\text{precision} = |\text{TP}|/|E| \quad \text{recall} = |\text{TP}|/|T| \tag{28.1}$$

For text detection, E and T are the sets of estimated rectangles and ground truth rectangles. For text extraction, E is the text pixels extracted by our algorithm, T is the manually labeled text pixels. Where TP is their intersection. There is usually a trade-off between precision and recall for a given algorithm. It is therefore necessary to combine them into a single final measure of quality f:

$$f = 2\text{pr}/(p + r) \tag{28.2}$$

28.3.2 Results and Analysis

To evaluate our coarse-to-fine method for text detection, according to Eq. (28.1) we can compute precision and recall using image areas. Some text detection examples of the proposed algorithm are presented in Fig. 28.4d. The algorithm can

Table 28.1 Performances of text detection method evaluated on the proposed dataset

Step	Precision	Recall	F-measure
1st step	73.80	81.37	77.40
2nd step	86.78	77.20	81.71
3rd step	93.24	76.74	**84.19**

Table 28.2 Performances of text extraction method evaluated on the proposed dataset

Step	Precision	Recall	F-measure
1st step	49.33	70.03	57.88
2nd step	61.82	69.58	65.47
3rd step	91.05	66.20	**76.66**

handle several types of challenging scenarios, e.g., variations in text font, color and size, as well as repeated patterns and background clutters. The results of this experiment are reported in Table 28.1.

To evaluate our coarse-to-fine method for text extraction, according to Eq. (28.1) we can compute precision and recall using image areas expressed in terms of number of pixels. Examples of our algorithm on this dataset are shown in Fig. 28.4e. The results of this experiment are reported in Table 28.2.

From Tables 28.1 and 28.2, we observe that our algorithm achieves significantly enhanced performance when detecting texts of arbitrary orientations. It demonstrates the effectiveness of the proposed method. The images in Fig. 28.5 are some typical cases where our algorithm failed to detect the texts or gave false positives. The misses are mainly due to strong highlights, blur, and low resolution.

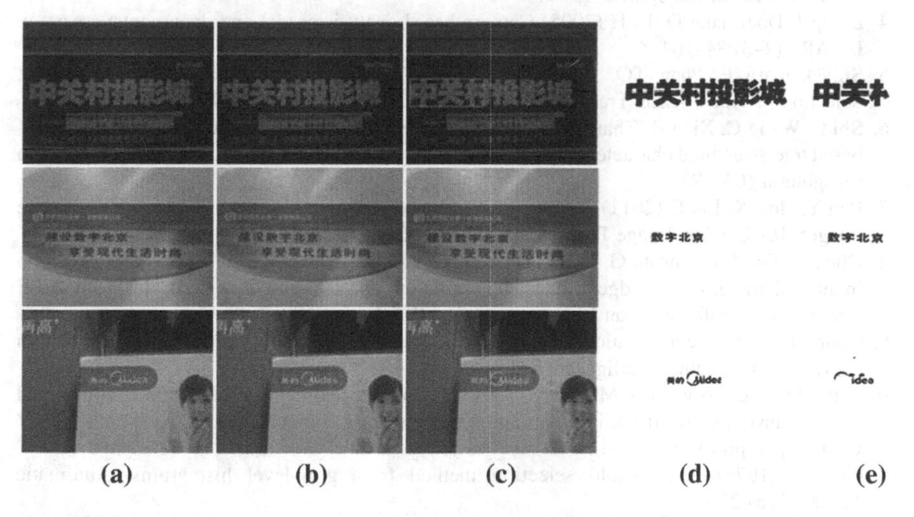

Fig. 28.5 Examples of failure cases. **a** Original image with the line labeled by user. **b** Manually labeled text areas (*red rectangles*). **c** Detected text by our method (*green rectangles*). **d** Manually extracted text. **e** Extracted text by our method

28.4 Conclusions

In this paper, an interactive method was proposed to detect and extract interesting text in natural scene images. We first draw a line to label a region which contains the texts we want to detect. Canny edge detection and CC-based approach are applied to quickly extract coarse region from the label region, then some specially designed heuristic rules are used to eliminate the non-text CCs and the remaining CCs in the coarse region are merged to get the fine region. Then use the high complexity precise approach to detect the small amount of candidate regions can greatly accelerate the speed of text detection and localization. Experimental results on the proposed dataset demonstrate very promising performance on detecting text in complex natural scenes.

Acknowledgments This work was supported by the National Natural Science Foundation of China (NSFC) under Grants No.60933010, No.61172103 and No.61271429.

References

1. Chen X, Yuille AL (2004) Detecting and reading text in natural scenes. In: Proceedings of the 2004 IEEE computer society conference on computer vision and pattern recognition (CVPR), vol 2, pp 366–373
2. Epshtein B, Ofek E, Wexler Y (2010) Detecting text in natural scenes with stroke width transform. In: IEEE conference on computer vision and pattern recognition (CVPR), pp 2963–2970
3. Jung K., Kim K. I., Jain A. K (2004) Text information extraction in images and video: a survey. Pattern Recognit 37(5):977–997
4. Liang J, Doermann D, Li H (2005) Camera-based analysis of text and documents: a survey. IJDAR 7(2–3):84–104
5. Shivakumara P, Phan TQ, Tan CL (2011) A laplacian approach to multioriented text detection in video. IEEE Trans Pattern Anal Mach Intell 33(2):412–419
6. Shi C, Wang C, Xiao B, Zhang Y, Gao S, Zhang Z (2013) Scene text recognition using part-based tree-structured character detection. In: IEEE conference on computer vision and pattern recognition (CVPR)
7. Pan Y, Hou X, Liu C (2011) A hybrid approach to detect and localize texts in natural scene images. IEEE Trans Image Process 20(3):800–813
8. Chen H, Tsai SS, Schroth G, Chen DM, Grzeszczuk R, Girod B (2011) Robust text detection in natural images with edge-enhanced maximally stable extremal regions. In: 18th IEEE international conferences on Image processing (ICIP), pp 2609–2612
9. Canny J (1986) A computational approach to edge detection. In: IEEE transaction on pattern analysis and machine intelligence, vol 6, pp 679–698
10. Petter M, Fragoso V, Turk M, Baur C (2011) Automatic text detection for mobile augmented reality translation. In: IEEE international conferences on computer vision workshops (ICCV workshops), pp 48–55
11. Otsu N (1975) A threshold selection method from gray-level histograms. Automatic 11(285–296):23–27
12. Lucas SM (2005) Icdar 2005 text locating competition results. In: Proceedings of the eighth international conference on document analysis and recognition, IEEE, pp 80–84

13. Sosa LP, Lucas SM, Panaretos A, Sosa L, Tang A, Wong S, Yound R (2003) Icdar 2003 robust reading competitions. In: Proceedings of the seventh international conference on document analysis and recognition, Citeseer
14. Yi C, Tian Y (2011) Text string detection from natural scenes by structure-based partition and grouping. IEEE Trans Image Process 20(9):2594–2605

Chapter 29
A New Edge Detection Algorithm Using FFT Procedure

Tongfeng Yang, Jun Ma, Shaomang Huang and Qian Zhao

Abstract Edge detection is the basic computation in image segmentation, feature extraction and image matching. Most of the existing detection algorithms extract edges through a convolution operation in the spatial domain, which may loose the overall characteristics of the image and make the resulted edges uncompleted and unbalanced. In this paper we proposed a novel edge detection algorithm using the FFT procedure. A simple edge model and a systemic extraction steps are described. In the experiments, we compared the edges extracted by our method with those by three classical algorithms of Sobel, Prewitt and Roberts, where we use Canny's results as benchmarks. The experiments illustrate that edges produced by our algorithm are accurate, complete and balanced.

Keywords Image processing · Edge detection · FFT · Edge growing

29.1 Introduction

The image is the sampling of the real-world scene. It is a kind of 2D discrete signal. Image edges are the parts where the sample values change significantly.

Lots of edge detection methods have been developed since now, such as methods based on morphological operation [1], methods based on the principal component analysis algorithm [2], methods based on the fuzzy logic technique [3] and so on. Each of these methods has its unique advantages, but the edge detection methods based on the differential operators are thought to be better taking into account the efficiency, the accuracy and the completeness. The first-order differential edge detection operators, also known as the gradient edge detection

T. Yang · J. Ma (✉) · S. Huang · Q. Zhao
Department of Computer Science and Technology, Shandong University, Jinan, China
e-mail: majun@sdu.edu.cn

A. A. Farag et al. (eds.), *Proceedings of the 3rd International Conference on Multimedia Technology (ICMT 2013)*, Lecture Notes in Electrical Engineering 278,
DOI: 10.1007/978-3-642-41407-7_29,

operators, use the property that image gradient obtains the local maxima value at the edge location to detect edges, such as Sobel [4], Prewitt [5], Kirsch [6], and Canny [7]. The second-order differential edge detection operators use the property that the second derivative value of the image equals zero at the edge location to detect edges, such as Marr [8], Lindeberg and LoG and so on. Since they are based on the pixel-level operations of the spatial domain, they cannot grasp the overall characteristics of the image. So the edge detection results of these derivative-based operators are often different from what are desired by human. Operators based on operations in a frequency domain can handle the characteristics of the whole image. Examples are mostly the edge detectors using a wavelet [9]. But here are very few methods based on a FFT procedure, for it is thought difficult to locate the exact positions of edge pixels. In this paper an edge detection method using a FFT procedure is proposed, who can get a complete and accurate edge result.

The main contributions of this paper can be summarized as: First, we explored a novel effective edge detection method. Second, the pixel-growing phase which we use to locate the exact edge pixels may be used in other edge algorithms.

29.2 Algorithm

To facilitate the analysis we consider one-dimensional edge profiles. That is, we will assume that two-dimensional edges locally have a constant cross-section in some direction. As shown in Fig. 29.1a shows a ridge edge model and Fig. 29.1c shows a step edge model. Though the original edge models have different shapes, both of them become a local maximum point in their corresponding high-pass reconstructed image i.e. Fig. 29.1b, d. So both edge models can be detected in a same way in the reconstructed image. However, because of the interference of ringing artifacts, it is not very easy to get a complete and accurate edge from a reconstructed image, especially in a two-dimensional image.

The edge detection explored in this paper is abbreviated as FFT Edge Detector (FED) for simple. In our procedure, a Butterworth filter [10] is used to get the high-frequency parts of the image which are corresponding to the image edges. The ringing artifact (ripple response) [11–13] occurs due to the sudden change of the signal. We use a ratio threshold to eliminate the ripple response. Inevitably, some edge pixels with small-scale steps are deleted at the same time. In order to

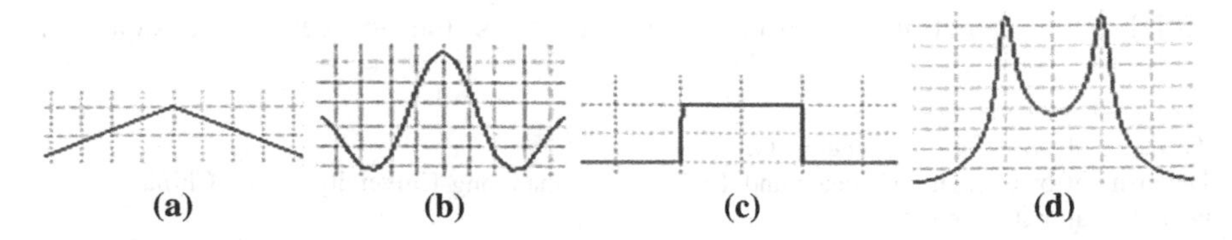

Fig. 29.1 Edge models and corresponding high-pass reconstructed image

find the lost edge pixels back and make the gotten edges straight and complete, a pixel-growing phase is implemented based on the exact edge pixels which can be obtained through a non-maximal suppression step.

29.2.1 The Acquisition of the High-Frequency Reconstructed Image

An image I is a 2D discrete signal, can be expressed as: $\{(x,y)|x \in (0,M); y \in (0,N)\}$. The image spectrum $F(u,v)$ can be computed using the Fast Fourier Transformation procedure:

$$F(u,v) = \frac{1}{\sqrt{MN}} \sum_{x=0}^{M-1} \sum_{y=0}^{N-1} f(x,y) e^{-j2\pi\left[\frac{ux}{M}+\frac{vy}{N}\right]}, \quad u \in \{0,1,\ldots,M-1\}; v \in \{0,1,\ldots,N-1\} \tag{29.1}$$

The high frequency part of the image spectrum is obtained using a Butterworth filter [10, 14]. Given a spectrum with the size is$M \times N$and the center is $(u_c, u_c) = (M/2, N/2)$, $D(u,v)$ is used to reprensent the distance between a point on the spectrum (u,v) and the center(u_c, u_c). So the transfer function of Butterworth filter $H(u,v)$ can be computed as follows:

$$D(u,v) = \sqrt{\left[\left(u - \frac{M}{2}\right)^2 + \left(v - \frac{N}{2}\right)^2\right]}, \quad H(u,v) = \frac{1}{1 + \alpha\left[\frac{D_0}{D(u,v)}\right]^{2n}}, \quad u \in \{0,1,\ldots,M-1\}; v \in \{0,1,\ldots,N-1\} \tag{29.2}$$

where, α is used to control how much should the low-frequency part be suppressed, and D_0 is the truncation frequency.

The high-frequency part of the image spectrum $F_h(u,v)$ is computed by

$$F_h(u,v) = F(u,v) * H(u,v) = F(u,v) * \frac{1}{1 + \alpha\left[\frac{D_0}{D(u,v)}\right]^{2n}}, \quad u \in \{0,1,\ldots,M-1\}; v \in \{0,1,\ldots,N-1\} \tag{29.3}$$

A reconstructed image $f_h(x,y)$ can be obtained using an inverse Fourier Transformation procedure based on $F_h(u,v)$:

$$f_h(x,y) = \frac{1}{\sqrt{MN}} \sum_{u=0}^{M-1} \sum_{v=0}^{N-1} F_h(u,v) e^{j2\pi\left(\frac{ux}{M}+\frac{vy}{N}\right)}, \quad u \in \{0,1,\ldots,M-1\}; v \in \{0,1,\ldots,N-1\} \tag{29.4}$$

29.2.2 The Suppression of the Ringing Artifacts

In signal processing, particularly digital image processing, ringing artifacts are artifacts that appear as spurious signals near sharp transitions in a signal. They occur when a sudden change exists in the input signal [11–13]. The ringing artifacts occur in FED procedure due to the unnatural suppression of the low-frequency waveforms in the Butterworth high-pass filtering phase. Considering the energy of minor lobe is usually 15–20 % of the energy of main lobe in the FED procedure, we use a ratio threshold γ to eliminate the minor lobe which is noise actually.

$$f_{\mathrm{dh}}(x,y) = \begin{cases} f_h(x,y) & \text{if}\left(f_h(x,y) \geq \gamma * \max_{u,v} f_h(u,v)\right) \\ 0 & \text{otherwise} \end{cases} \tag{29.5}$$

Where, $f_{\mathrm{dh}}(x, y)$ is the result image without ringing artifacts, γ is the ratio threshold and $\gamma = 0.2$ in our experiments. Since in the reconstructed image $f_h(x, y)$, it exists that the energy of minor lobe of one edge may be bigger than the energy of main lobe of another edge, some edge pixels are deleted in the ringing artifacts suppression phase.

29.2.3 Non-Maximal Suppression

In f_{dh} a few ringing artifact pixels with high energy remained while some edge pixels are deleted. Both the retrieve of the deleted edge pixels and the deletion of the remained ringing artifact pixels are based on the non-maximal suppression. Because the pixels on a main lobe are edge pixels and the maximum pixel of a local region is surely on a main lobe, we can find out these exact edge pixels using a non-maximal suppression operation.

We use a *Cover* of size $w \times w$, whose center is at (0,0). In order to decide whither a pixel on $f_{\mathrm{dh}}(x, y)$ is an exact edge pixel or not, we move the *Cover* to make its center aligned at (x, y). Then a f_c is obtained by multiplying the values on the *Cover* and the corresponding covered values on f_{dh}. The maximum value of f_c is selected into the E_p which represents the exact edge pixel set.

$$f_c(u,v) = f_{\mathrm{dh}}(u,v) * \mathrm{Cover}(u-x, v-y) \tag{29.6}$$

$$E_p(x,y) = \begin{cases} 1 & \text{if}\left(f_c(x,y) = \max\limits_{u,v} f_c(u,v)\right) \\ 0 & \text{otherwise} \end{cases} \tag{29.7}$$

29.2.4 Edge Growing Using Exact Edge Pixels as Seeds

The goal of edge growing is to find out as much edge pixels as possible from f_{dh} and make sure the edges are precise, straight and complete. The edge growing operations are carried out on local regions of size $s \times s$. $N_s(u, v)$ is used to represent a local region having a top-left corner at (u,v). Its center location can be computed to be $\left(C_{N,s,u}, C_{N,s,v}\right) = \left(u+\frac{s-1}{2}, v+\frac{s-1}{2}\right)$. We use E to represent the edge image. $E(x, y)$ is assigned **1** if it is identified as an edge pixel, otherwise it is assigned **0**.

Because the exact edge pixels are on the main lobes and the pixels on the main lobes have higher energy than their minor lobes, we can find out pixels on main lobes take the exact edge pixels and the pixel values as clues. In order to be sure that edge growing makes sense, there should be at least one exact edge pixels in $N_s(u, v)$ and a pixel should not be identified to be a new exact edge pixel unless that its pixel value is big enough. According to Canny's good edge criteria [3], an edge should be single (i.e. only one response to a single edge). So in $N_s(u, v)$, the exact edge pixels numbers should be less than s (s is the region's width or length). Based on the above principles, several determination conditions are defined as follows.

T_1:The pixel at $\left(C_{N,s,u}, C_{N,s,v}\right)$is not an edge pixel until now.

$$0 = E\left(u + \frac{s-1}{2}, v + \frac{s-1}{2}\right) \tag{29.8}$$

T_2: At least one exact edge pixels should exist in $N_s(u, v)$.

$$\sum\nolimits_{\Delta y=0}^{s-1} \sum\nolimits_{\Delta x=0}^{s-1} E(u + \Delta x, v + \Delta y) \geq 1 \tag{29.9}$$

T_3: The edge pixels number should be less than s.

$$\sum\nolimits_{\Delta y=0}^{s-1} \sum\nolimits_{\Delta x=0}^{s-1} E(u + \Delta x, v + \Delta y) \leq s \tag{29.10}$$

T_4: The pixel value at $\left(C_{N,s,u}, C_{N,s,v}\right)$ should in top s within$N_s(u, v)$.

$$\text{Rank}_{N_s(u,v)}\left(fh\left(u + \frac{s-1}{2}, v + \frac{s-1}{2}\right)\right) \leq s \tag{29.11}$$

T_5: The pixel value at $\left(C_{N,s,u}, C_{N,s,v}\right)$ should be bigger than μ percent of the maximum pixel value within $N_s(u, v)$.

$$fh\left(u + \frac{s-1}{2}, v + \frac{s-1}{2}\right) \geq \mu * \max_{0 \leq \Delta x, \Delta y < s} fh(u + \Delta x, v + \Delta y) \tag{29.12}$$

The steps of pixel growing described below.

(1) Initially $E = E_p$.

(2) For each local region$N_s(u, v)$ determine the center pixel to be an edge pixel if all the conditions from T_1 to T_5 are true.

$$E\left(u+\frac{s-1}{2}, v+\frac{s-1}{2}\right)=\begin{cases}1 & \text{if}(T_1\&T_2\&T_3\&T_4\&T_5)\\ E\left(u+\frac{s-1}{2}, v+\frac{s-1}{2}\right) & \text{otherwise}\end{cases} \quad (29.13)$$

(3) Repeat step (2) until the number of edge pixels $N(E)$ do not change.

$$N(E)=\sum\nolimits_{x=0}^{M-1}\sum\nolimits_{y=0}^{N-1}E(x,y),\quad x\in\{0,1,\ldots,M-1\};y\in\{0,1,\ldots,N-1\} \quad (29.14)$$

(4) The edge image E is the output result of FED method.

29.3 Experiments

In order to prove the effectiveness of our method, we compare it with three other edge detection methods which are Sobel, Prewitt and Roberts. The edges detected using Canny algorithm are selected to be benchmarks. All the experiments are carried out in Matlab R2011b. We use the library function of Matlab for the implementation of the methods Canny, Sobel, Prewitt and Roberts Fig. 29.2.

We use an evaluation criterion similar to the one used by Zhao [2] and Martin [15]. Detection Correct Ratio (DCR) and Detection Error Ratio (DER) are computed in the experiments.

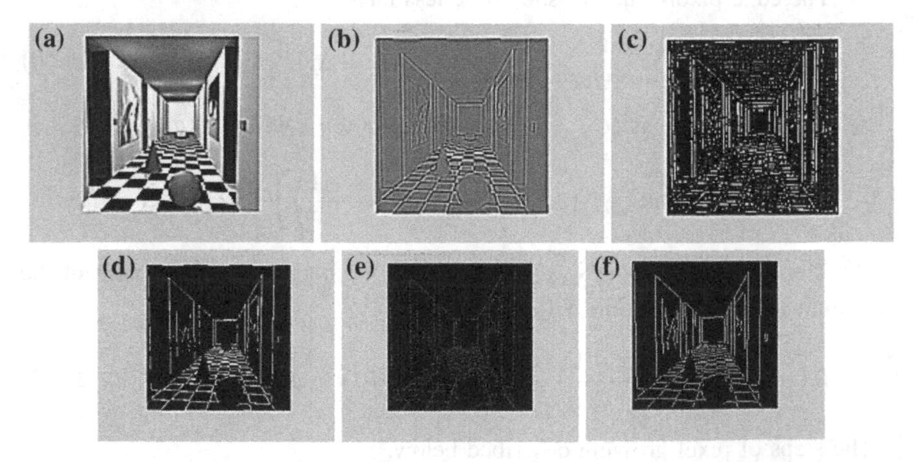

Fig. 29.2 Major steps of FED method, **a** Input image, **b** High-pass reconstructed image, **c** Ringing artifacts, **d** Ringing suppression, **e** Exact edge pixels, **f** Output edge image

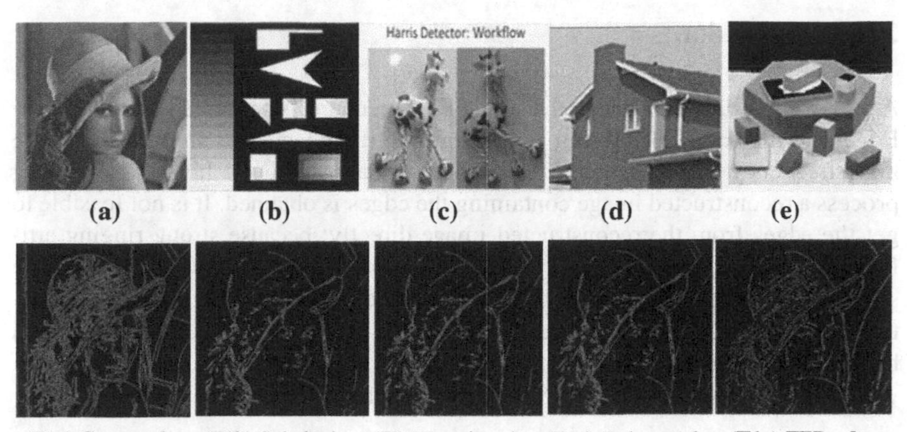

(T1a) Canny edge (T1b) Sobel edge (T1c)Prewitt edge (T1d) Roberts edge **(T1e) FED edge**

Fig. 29.3 Edge detection results of methods, **a** Image 1, **b** Image 2, **c** Image 3, **d** Image 4, **e** Image 5, **T1a** Canny edge **T1b** Sobel edge **T1c** Prewitt edge **T1d** Roberts edge **T1e** FED edge

As shown in Fig. 29.3, the results of these methods on 1 images are demonstrated. It is easy to find out that the edges obtained by FED method are much complete and balanced. To be complete means that more edge pixels are detected. To be balanced means that the edge pixels are not concentrated in one or several contain regions but have a balanced distribution. Both the two properties are illustrated particularly prominent in the comparison between the results of the third image.

In order to make the comparison between each algorithm quantitative, the DCR and DER of each edge result are computed and demonstrated in Table 29.1. The edge results obtained using Canny algorithm are selected to be the benchmarks. The DCRs of FED method are bigger than that of other three edge detection methods, which illustrates the completeness and accuracy of the detection results of FED methods. It is not occasional that the DERs of each method results are nearly zero. Because Canny algorithm can find out edges caused by even very small changes and the results of Sobel, Prewitt, Roberts and FED are subsets of the results of Canny.

Table 29.1 DCR/DER of methods

Test images	DCR(bigger is better)				DER(smaller is better)			
	Sobel	Prewitt	Roberts	FED	Sobel	Prewitt	Roberts	FED
Image 1	0.3486	0.3450	0.2919	0.3509	0	0	0	0
Image 2	0.5625	0.5550	0.6280	0.6864	0	0	0	0
Image 3	0.4075	0.4033	0.4005	0.9153	0	0	0	0
Image 4	0.7370	0.7285	0.7031	0.8632	0	0	0.0109	0.0329
Image 5	0.7069	0.6910	0.6097	0.8995	0	0	0	0

29.4 Conclusion

A novel edge detection method using FFT procedure is presented in this paper. Edges are the parts of changing significantly in the image, and are reflected to be high frequency portions in the frequency domain. Through high-pass filtering process a reconstructed image containing the edges is obtained. It is not feasible to get the edges from the reconstructed image directly; because strong ringing artifacts are caused by the truncation of spectrum in the frequency domain. The subsequent operations of ringing artifacts suppression, exact edge pixels acquisition and pixel growing make it become reality to get edges complete, accuracy and balanced. Experiments illustrate that our method can match or exceed the effect of widely spread methods such as Sobel, Prewitt and Roberts.

Acknowledgments This work was supported by Natural Science Foundation of China (61272240,60970047, 61103151), the Doctoral Fund of Ministry of Education of China (20110131110028) and the Natural Science foundation of Shandong province (ZR2012FM037).

References

1. Lee J, Haralick R, Shapiro L (1987) Morphologic edge detection. IEEE J Robotics Autom 3(2):142–156
2. Hua J, Wang J, Yang J (2009) A novel approach to edge detection based on PCA. J Imag Gr 14(5):912–919
3. Alshennawy AA, Aly AA (2009) Edge detection in digital images using fuzzy logic technique. World Acad Sci Eng Technol 27 14(15):16
4. Vincent O, Folorunso O (2009) A descriptive algorithm for sobel image edge detection. In: Proceedings of informing science and IT education conference (InSITE)
5. Prewitt JMS (1970) Object enhancement and extraction. In: Lipkin B, Rosenfeld A (eds) Picture processing and psychopictorics. Academic Press, New York, pp 75–149
6. Castleman KR (1998) Digital image processing. Tsinghua University Press, Beijing
7. Jahne B (2002) Digital image processing. Springer, New York
8. Marr D, Hildreth E (1980) Theory of edge detection. In: Proceedings of the royal society of London, Series B 207
9. Zhang L, Paul B (2002) Edge detection by scale multiplication in wavelet domain. Pattern Recognit Lett 23(14):1771–1784
10. Selesnick IW, Burrus CS (1998) Generalized digital butterworth filter design. IEEE Trans Signal Process 46(6):1688–1694
11. Bankman IN (2000) Handbook of medical imaging. Academic Press, San Diego, Section I.6, p 16
12. Chitode JS (2008) Digital signal processing. Technical Publications, Pune, pp 4–70
13. Glassner AS (2004) Principles of digital image synthesis, 2nd edn. Morgan Kaufmann, San Francisco, p 518
14. Butterworth S (1930) On the theory of filter amplifiers wireless engineer. Radio Eng 7:536–541
15. Martin D, Fowlkes C, Tal D, Malik J (2001) A database of human segmented natural images and its application to evaluating segmentation algorithms and measuring ecological statistics. In: Proceedings of 8th international conference on computer vision, vol 2, pp 416–423

Chapter 30
A Novel Error Concealment Method Based on Adaptive Ordering of Block Match

Fan Zhou, Weiwei Xu and Yaowu Chen

Abstract Error concealment methods can reduce the quality degradation induced by packet loss under the unstable condition of wireless network at the decoder side. In this paper, a new exemplar-based spatial error concealment algorithm is proposed utilizing adaptive ordering block match. To achieve a better subjective video quality, structural information like the edge is first restored, and then the texture is concealed with an adaptive ordering block match algorithm. First, the local similarity is generated from the edge information and multiscale pictures. Next, the edges around the loss area are paired and restored under the constraint of local similarity. A block-based spatial error concealment procedure is then performed to conceal the loss area. The priority of concealment is decided by the intensity, reliability, and the similarity of the nearby areas. From experimental results, the proposed algorithm gains 1.6 to 1.8 db on the objective quality comparing with the algorithm implemented in JM14.0.

Keywords Error concealment · Adaptive ordering block matching · Local similarity

30.1 Introduction

With the rapid growth of video terminal unit, multimedia applications over wireless networks have been fast developing. Due to the picture structure of bitstream, the loss of the reference frame causes the error propagation. It makes the good quality delivery a very challenging work [1, 2]. To solve this, error resilience and error concealment (EC) techniques have been developed [3, 4]. Bilinear interpolation (BI) was commonly used for spatial error concealment (SEC), but it

F. Zhou (✉) · W. Xu · Y. Chen
The Institute of Advanced Digital Technology and Instrument
and Zhejiang Provincial Key Laboratory for Network Multimedia Technologies,
Zhejiang University, Hangzhou, China
e-mail: fanzhou@mail.bme.zju.edu.cn

A. A. Farag et al. (eds.), *Proceedings of the 3rd International Conference on Multimedia Technology (ICMT 2013)*, Lecture Notes in Electrical Engineering 278,
DOI: 10.1007/978-3-642-41407-7_30,

could not recover prominent image edges [5, 6, 8]. In [7, 9, 10, 12], the directional entropy of neighboring edges was used to ensure better edge preservation. In [11], the support vector machine (SVM) and distributed genetic algorithm (DGA) were used. The authors in [13] proposed a spatial shape EC technique. In [14], edge-directed error concealment (EDEC) algorithm was used. Criminisi et al. [15] proposed an exemplar-based image inpainting method.

In this paper, an adaptive ordering block match algorithm is proposed for SEC. The isophote of the picture in different scale level is utilized to obtain the structure attributes of the image, then prominent image edges around the loss area are paired and the edge is recovered as being the initial edge. A block match algorithm is performed under the constraint of structure attributes. The priority of EC is based on the intensity and similarity of blocks nearby the lost block. The proposed method offers satisfying robustness to preserve important edge features using local available information while maintaining the natural texture for loss areas.

30.2 The Proposed Adaptive Ordering Block Match Method

In the situation of losing continual macro-blocks (MBs), the edges may pass through many MBs. Unreasonable restoration order may lead to false edges which degrade the visual quality. Figure 30.1b illustrates the restoration order using the line scan and Fig. 30.1c illustrates a better result using the adaptive restoration order. The proposed concealment procedure is as follows: first, the local similarity is calculated and the constraint of loss area is established. Next, the edges in the loss areas are concealed. Finally, the adaptive ordering restoration is performed.

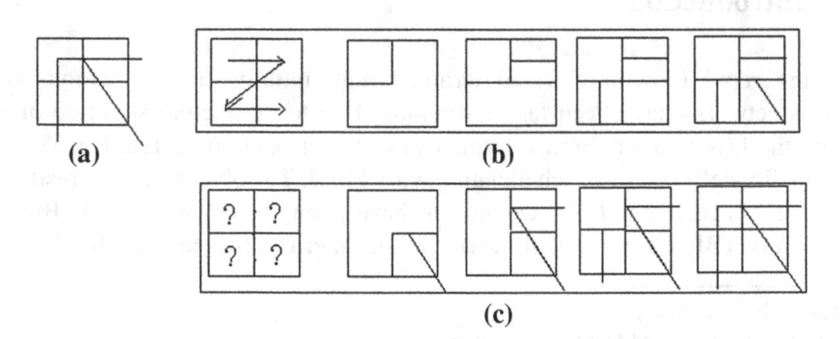

Fig. 30.1 **a** The Right Result for Concealment, **b** Result using line scan order, **c** Result using adaptive order

30.3 Calculating Local Similarity and Constraint Property

All connected lost MBs are one loss area; all pixels in the current loss area are defined as Θ, and current frames as F(n). All MBs around Θ are B; B∈F(n)-Θ. All the lost MBs nearby B that belong to Θ are BI; BI∈Θ. All the received MBs around B are BN; BN∈F(n)-Θ-B. All the block around Θ are B4; B4∈F(n)-Θ. All the blocks around B4 are BN4; BN4∈ BN-B4. Structure attributes vector (Y_p, U_p, V_p, G_p, $G16_p$, S_p, V_p, A_p, AP_p) is utilized. Pixel value attributes Y_p, U_p, and V_p are the Y, U, and V component. G_p is the gradient of the 4 × 4 area. $G16_p$ is the 16 × 16 level gradients. S_p is the structural extending attribute. V_p is the structural extending vector attribute. A_p is the area label attribute. AP_p is the area paired attribute and it is equal to the A_p of the paired area.

The attribute of local similarity is calculated using a distortion function $D_a(p, q)$. $D_{pixel}(p, q)$ means the sum absolute difference (SAD) between Y, U, and V component, and $D_g(p, q)$ means a different surrounding gradient structure; $D_g(p, q)$ is related to conformation;

$$D_a(p, q) = D_{pixel}(p, q) + a \times D_g(p, q) \tag{30.1}$$

$$D_g(p, q) = \sum\nolimits_{p \in Gp, q \in Gq} (|p - q|) \tag{30.2}$$

Scaled pyramid picture is used for calculating the spatial vector for propagation of the gray value and the texture shape in the order from coarseness to fine. Sobel descriptor is used to calculate the gradient. The attributes of $D_{pixel}(p, q)$ in 16 × 16 level pixels are used to measure the low frequency part of local area. The local gradient structure is used for describing the details of the texture as shown in Fig. 30.2.

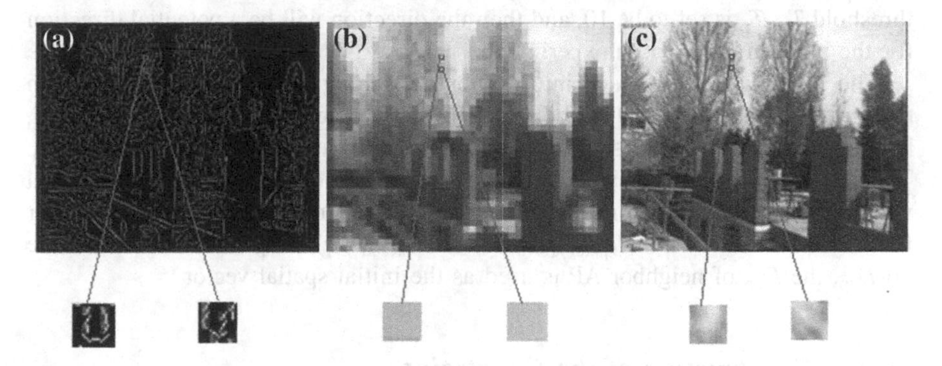

Fig. 30.2 **a** The gradient of 4 × 4 level picture. **b** The gray value of 16 × 16 level picture. **c** The actual texture of original picture

30.3.1 Local Gradient Similarity

Current patch is defined as P_p; $P_p \in F(n)\text{-}\Theta$. It is a 4*4 matrix of 4 × 4 level pixels, and (1, 1) of P_p is defined as central point; marking it as p, and 4 × 4 level pixel position as (x_p, y_p). The search area is SW_p; $SW_p = \{q{:}q \in F(n)\text{-}\Theta$, where q is a 4 × 4 level pixel; $(-w-1) < (x_q - x_p) < (w+2)$, $(-w-1) < (y_q - y_p) < (w+2)\}$, and w is set to be 4. 16 × 16 and 4 × 4 level pixels are utilized for calculating gradient information;

$$D_{\text{mask}}(\text{MP}_p, \text{MP}_q) = \frac{\left(\sum_{p\in \text{MPp},\, q\in \text{MPq}} (p \times q)\right)}{\left(\sum_{p\in \text{MPp},\, q\in \text{MPq}} (p + q)\right)} \tag{30.3}$$

The 4 × 4 mask matrix MP_p of P_p is created; if a 4 × 4 level pixel having its gradient strength bigger than T_e, its mask will be set as 1 and as 0 otherwise. The difference between two 4 × 4 masks MP_p and MP_q is defined as $D\text{mask}(\text{MP}_p, \text{MP}_q)$. Patch with $\min(D_{\text{mask}}(\text{MP}_p, \text{MP}_q))$ would be chosen as the prediction, and the spatial vector V_p is then calculated.

30.3.2 Local Gray Value Similarity

For every 16 × 16 level pixel p in BN, coordinate in 16 × 16 level is (x_p, y_p), and the spatial vector in B is calculated. The set of initial spatial vectors is defined as D_p. Search area is defined as S_d; $S_d = \{q\in F(n)\text{-}\Theta$; q is a 16 × 16 level pixel with $\max(|x_p - x_q|, |y_p - y_q|) \le 1\}$. For $q\in S_d$, if its gradient strength is greater than T_e, its gradient direction will be included in D_p. For every direction in D_p, calculate the $D_{\text{pixel}}(p, q)$ in 16 × 16 level for p. If $D_{\text{pixel}}(p, q)$ in certain direction is under threshold T_p, T_p is set to be 10, and then the direction will be a potential direction for the pixel in B based on experiments.

The MBs in B are divided into groups depending on the difference between the gray value and the distance. For 16 × 16 pixel p and q, $\max(|y_p - y_q|, |x_p - x_q|) = 1$, $p\in$B, and $q\in$B. If $D_{\text{pixel}}(p, q) < T_p$, p and q will have the same A_p. The set of pixels with the same A_p are defined as AP. For $p\in$AP, the actual spatial vector D_{sp} is calculated from the maximum and minimum potential spatial vector for AP which has the $\min(D_{\text{pixel}}(p, q))$ and $D_{\text{pixel}}(p, q) < T_p$. For APs which have no D_{sp}, the D_{sp} of neighbor AP is used as the initial spatial vector.

30.3.3 Construction of the Constraint

Constraints from pairing areas are first calculated. For 16 × 16 level pixel $p \in$ B, the paired pixel will be found across the loss area. The diffusing procedure to the

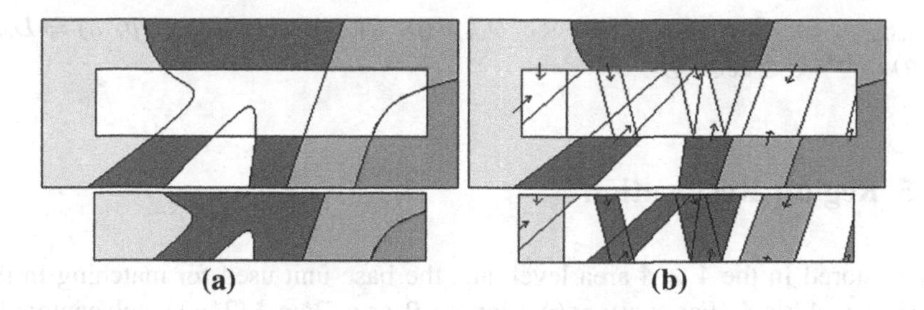

Fig. 30.3 **a** The lossless whole picture (*up*) and the original central area. **b** The picture with lost (*up*) and the constraint built in loss area

loss area should be performed under the guide of connection. The extending process is started from p in the direction of D_{sp} in the loss area until another pixel in B is met, and set it to be q. If $D_{pixel}(p, q) < T_p$, then p and q is supposed to be paired; For the 4 × 4 level pixel $p \in B4$, using D_{sp} in D, and find the spatial vector $D4_{sp}$ in 4 × 4 level across the loss area until the 4 × 4 level pixel in B is met, and set it to be q. If $D_{pixel}(p, q) < T_p$, then p and q is paired, and $D = D_{sp}$ of p. The 16 × 16 and 4 × 4 level constraints are constructed first using the max and min Y value in pixel level of the paired 16 × 16 and 4 × 4 pixel.

Constraints from neighbor areas are also calculated. For a 16 × 16 level area $p \in B$, its feature should be analyzed to see if it were an "open" area. If the gradient direction of most pixels on the boundary is perpendicular to the direction of the boundary line, this area is not an "open" area; it is "closed." After excluding "closed" area, the edge is extended from BN4 to B4. The final constraint is shown in Fig. 30.3.

30.4 Edge Pair

The edges around the loss area are paired to build up the initial structure of lost area. The order of pairing starts from the strong edge to the poor edge. Areas are paired first. Get the left, central, and right 16 × 16 pixel level area attributes A_L, A_C, and A_R for p and $A1_L$, $A1_C$, and $A1_R$ for q, and then use the paired area to constrain the edge pair. Define $E(A_p, A_q)$: If A_p equal to A_q, $E(A_p, A_q) = 1$, otherwise $E(A_p, A_q) = 0$. The total number of areas that are paired is calculated. Number of areas paired in the area of edge pair will be calculated as NP_p. If NP_p is bigger than 0, p and q is marked as being associated, then go to the next step.

Edges are then paired. Use the Y, U, and V component value of the edge area, as well as pixel on the left side and right side of the edge area to calculate the difference. Use the distance between the current edge and the paired edges for the further constraint. If $D_{edge}(p, q) < \min(De_{dge}(p, q'))$ and $NP_p > \min(NP_p)$, then this step will be executed. The distance between the two edges is defined to be

$D_{distance}(p, q)$. Then the difference D_{total} (p, q) becomes D_{total} $(p, q) = D_{edge}$ $(p, q) + b*D_{distance}(p, q)$.

30.5 Region Restoration

Θ is restored in the 4 × 4 area level, and the base unit used for matching in this step is 4 × 4 block. For every $p \in \Theta$, suppose P_j as a $(2*w)*$ $(2*w)$ patch centered at p, and PE_j as a $(2*(w + 1))*$ $(2*(w + 1))$ patch centered at p. w is set according to the local property. The attributes' vector (C_p, E_p, S_p, V_p, N_p, and AM_p) is used for pixel $q \in F(n)$. C_p is the confidence attribute, E_p is the edge attribute, S_p is the structural extendable attribute, V_p is the structural extending vector attribute, N_p is the copying time attribute that this pixel has been used, and AM_p is the amercement attribute for the missing match.

30.5.1 Concealing Order and Patch Size Decision

Search in Θ to find the P_j with the maximum priority, the priority of the p is defined as below with A is a weighted factor set to be 100;

$$P(p)=C(p)+D(p)+S(p) \tag{30.4}$$

$$C(p)=\left(\sum\nolimits_{q \in Pj \cap F(n)} Cp\right) \Big/ \|Pj\| \tag{30.5}$$

$$D(p)=\left(\sum\nolimits_{q \in Pj \cap F(n)} Ep\right) \Big/ \|Pj\| \tag{30.6}$$

$$S(p) = \left(\sum\nolimits_{q \in Pj \cap F(n)} Sp\right) \times A \tag{30.7}$$

In [15], P(p) was defined to be C(p)*D(p), which emphasized that the strong gradient area had higher priority for extending to loss area. It had the tendency to make strong edge grow faster. In [16], gradient threshold of whole image was used to divide the point, and (C(p) + D(p)) was used. The proposed algorithm utilizes the structural information of the scaled picture which includes information of both local and whole. The local similarity S(p) is also added, which is calculated from the scaled picture. This is designed for describing the extension of isophote and texture on the overall structure for balancing the extending speed and controlling the extending direction. Both the intensity and the local similarity of P_j are used to decide the patch size w for EC. For punishing P_j which could not find a match in (F(n)-Θ), smaller size is used in the next round. Then w is set to be 1.

30.5.2 The Best Match

The spatial search area is defined as S; $S \in F(n)-\Theta$. The spatial search area center is set using V_p and zero MV. Q_j is $(2*w)*(2*w)$ patch centered at q, and $Q_j \in S$. The difference between patches set as $DP_a(p, q)$, $D_{pixel}(p, q)$ means the SAD between Y, U, and V component.

$$DP_a(p, q) = DP_{pixel}(p, q) + DP_g(p, q) - N_q \tag{30.8}$$

$$DP_g(p, q) = \sum_{p \in Pj,\, q \in Qj} \left| G_p - G_q \right| \tag{30.9}$$

$$Nq = \sum_{q \in Qj} N_p \tag{30.10}$$

If the $DP_a(p, q)$ is larger than T_p, another difference function $DP1_{pixel}(p, q)$ is used. The average of pixel difference AY, AU, and AV are also used; $AY = (\sum(p \in P_j,\ q \in Q_j)\ (Y_p - Y_q)/(\|P_j\|)$, $AU = 4*(\sum(p \in P_j,\ q \in Q_j)\ (U_p - U_q)/(\|P_j\|)$, $AV = 4*(\sum(p \in P_j, q \in Q_j)\ (V_p - V_q)/(\|P_j\|)$. If AV is below T_a, T_a is defined to be 20, then AY, AU, and AV are used for calculating $DP1_{pixel}(p, q)$;

$$\begin{aligned} DP1_{pixel}(p, q) = & \left|Y_p - Y_q - AY\right| + \left|U_p - U_q - AU\right| \\ & + \left|V_p - V_q - AV\right| \end{aligned} \tag{30.11}$$

$$DP1_a(p, q) = DP1_{pixel}(p, q) + DP_g(p, q) - N_q \tag{30.12}$$

The most similar patch PM_j is found by $\min(Q_j \in S)\ DP_a(P_j, Q_j)$ for P_j. If the difference between PM_j and P_j is below a threshold T_p for the early stop, this patch is used for concealment of the lost parts of P_j. The attributes of lost parts in P_j will be set too. The confidence attribute C_p is updated as been calculated in step 2 as $C(p)$, E_p is updated using the sobel descriptor, N_p is copied from the N_p of PM_j, and S_p is set (if the new vector between PM_j and P_j is similar to V_p), the new vector will be set to V_p and P_j will be eliminated from Θ. If P_j cannot find a matching PM_j and the difference between them is below T_p, AM_p will be increased to punish this patch and it will be given a lower priority in the following step.

30.6 Simulation Results

The proposed algorithm is simulated using H.264 codec. The video sequence is encoded by JM14.0 encoder using different configurations. The video sequence "foreman" compressed at 30 frames per second is used for testing (Baseline profile is used, the interval of I-pictures is set to 0 and 30, and 5 reference frames are used). In the simulation, multislice mode is used; the slice size is set to be 700 and 1,400 bytes. Two kinds of bit-streams are used; bit-streams that only have I-frames and bit-streams with P-frames. Slices of these bit-streams are dropped in

Table 30.1 Testing results of foreman under different slice size and loss rate

Slice size	Method	Loss rate		
		10 %	20 %	30 %
700	JM	35.43	32.23	30.67
	The proposed	38.30	33.02	31.78
1400	JM	34.41	31.23	26.27
	The proposed	36.27	32.28	28.74

Table 30.2 Testing results of foreman under different slice size and loss rate for bit-stream with Intra frames only under 5 % loss rate

Slice size	Method	
	JM	The proposed
700	36.24	36.76
1400	36.92	37.48

a random pattern to simulate the loss in transmission errors on the I-frame only, and the loss rate is set to be 5, 10, and 20 %. The peak signal-to-noise ratio (PSNR) is used to quantitatively evaluate the recovered video quality, and the average results for the same network and bit-stream configuration are illustrated. The proposed EC algorithm is compared with the EC algorithm implemented in the JM14.0 decoder.

As shown in Tables 30.1 and 30.2, when the GOP structure is normally IPPP, an improvement of average 1.6 db is achieved for the slice size of 700 bytes, and an improvement of average 1.79 db is achieved for the slice size of 1,400 bytes. To compare the subjective qualities, simulation results on a slice size of 1,400 and 700 bytes in the "Forman" sequence are shown in Figs. 30.4 and 30.5, respectively. Observation shows that the proposed method has achieved noticeable improvements than JM14.0.

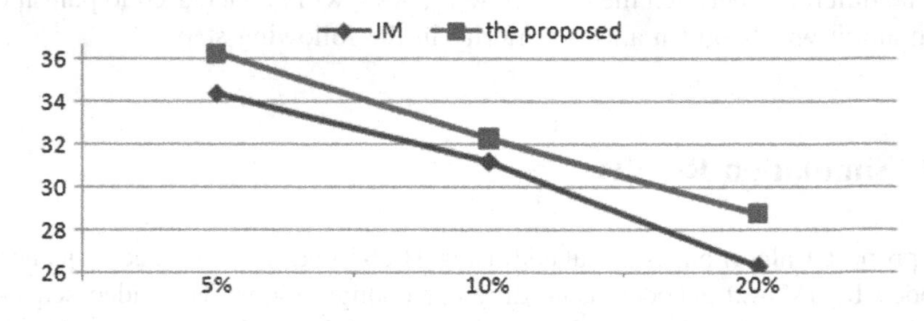

Fig. 30.4 The Result of Forman IP30 for Different Loss Rate, Slice Size 700

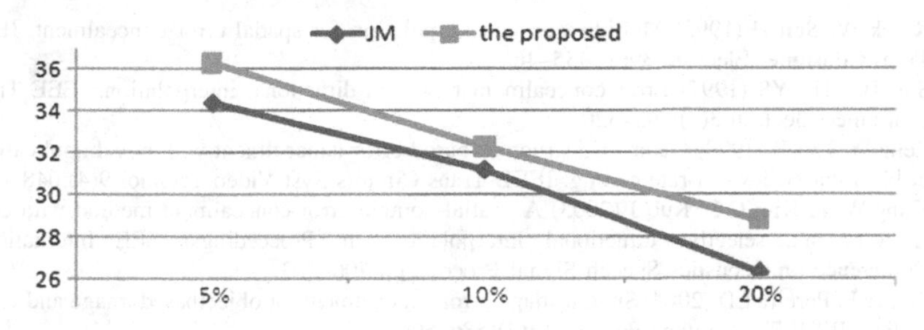

Fig. 30.5 The Result of Forman IP30, Slice Size 1,400

30.7 Conclusion

This paper presents a new robust SEC method for block loss recovery. The proposed method is based on modified block match under the framework of adaptive order arrangement model. The proposed algorithm can deal with loss areas in different sizes, fully utilize the structural information, and adaptively apply it on a patch-based restoration procedure. The results show that the proposed EC technique performs better comparing to previous techniques on both the subjective and objective qualities of the reconstructed images.

Acknowledgments This work is supported by National Natural Science Foundation of China (Grant No. 40927001), the project of Key Scientific and Technological Innovation Team of Zhejiang Province, (Grant No. 2011R09021-06), the Fundamental Research Funds for the Central Universities.

References

1. Wang Y, Wenger S, Wen J et al (2000) Review of error resilient coding techniques for real-time video communications. IEEE Signal Process Mag 17(4):61–82
2. Wang Y, Zhu QF (1998) Error control and concealment for video communication: a review. Proc IEEE 86(5):974–997
3. Kumar S, Xu L, Mandal M K et al (2006) Error resiliency schemes in H.264/AVC standard. J Vis Commun Image Represent 17(2):425–450
4. Wang Y, Zhu QF (1998) Error control and concealment for video communication: a review. Proc IEEE 86(5):974–997
5. Hsia SC, Cheng SC, Chou SW (2005) Efficient adaptive error concealment technique for video decoding system. IEEE Trans Multimedia 7(5):860–868
6. Wang Y K, Hannuksela M M, Varsa V (2002) The error concealment feature in the H.26L test model. Proc IEEE ICIP 729–736
7. Agrafiotis D, Bull DR, Canagarajah CN (2006) Enhanced error concealment with mode selection. IEEE Trans Circuits Syst Video Technol 16(8):960–973
8. Hsia SC, Chou SW (2007) VLSI implementation of high-performance error concealment processor for TV broadcasting. IEEE Trans Circuits Syst Video Technol 17(8):1054–1064

9. Kwok W, Sun H (1993) Multidirectional interpolation for spatial error concealment. IEEE Trans Consumer Electron 39(3):455–460
10. Suh JW, Ho YS (1997) Error concealment based on directional interpolation. IEEE Trans Consumer Electron 3(3):295–320
11. Zeng W, Liu B (1999) Geometric-structure-based error concealment with novel applications in block-based low-bit-rate coding. IEEE Trans Circuits Syst Video Technol 9(4):648–665
12. Kung W Y, Kim C S, Kuo J (2003) A spatial-domain error concealment method with edge recovery and selective directional interpolation. In: Proceedings IEEE International Conference on Acoustics Speech Signal Process, pp 700–703
13. Soares F, Pereira LD (2004) Spatial shape error concealment for objectbased image and video coding. IEEE Trans Image Process 13(4):586–599
14. Ma MY, Oscar C, Au SH et al (2010) Edge-Directed Error Concealment. IEEE Trans Circuits Syst Video Technol 20(3):382–395
15. Criminisi A, Pierez P, Toyama K (2004) Region filling and object removal by exemplar-based image inpainting. IEEE Trans Image Process 13(9):1200–1212
16. Nie DH, Ma LZ, Xiao SJ (2006) Digital image inpainting by example-based image synthesis method. High Technol Lett 12(3):276–282

Chapter 31
Template Selection for Lookup Table Based on Genetic Algorithm

Zhiqiang Wen, Wenqiu Zhu and Zhigao Zeng

Abstract A template selection method based on genetic algorithm was proposed for inverse halftoning in this paper. A constrained mathematical model was built to describe the template optimization problem. Since the existing greedy method was said not to find the globally optimum template, we solved this optimization problem by using genetic algorithm. The details about genetic operators, such as encoding and decoding scheme, selection and reproduce, crossover, mutation, and fitness function, were discussed. In experiments, the convergence process of genetic algorithm was analyzed. Two comparison experiments were done on two classic data sets. Experimental results showed the template found by our method is superior to that of the greedy method.

Keywords Lookup table · Genetic algorithm · Template selection · Inverse halftoning

31.1 Introduction

Inverse halftoning (IH), which belongs to the image restoration, is the inverse process of digital halftoning. It has a wide range of applications, e.g., paper image digitization, digital publishing system, image compression, and printed image processing. Generally, IH algorithms can be grouped into three categories: the filtering-based IH algorithms [1, 2], the iteration-based IH algorithms [3], and the

Z. Wen (✉)
School of Computer and Communication, Hunan University of Technology, Zhuzhou 412007, China
e-mail: zhqwen20001@163.com

W. Zhu · Z. Zeng
School of Computer and Communication, Hunan University of Technology, Zhuzhou, China

A. A. Farag et al. (eds.), *Proceedings of the 3rd International Conference on Multimedia Technology (ICMT 2013)*, Lecture Notes in Electrical Engineering 278,
DOI: 10.1007/978-3-642-41407-7_31,

learning-based IH algorithms. The developed learning-based IH algorithms involve lookup table (LUT)-based IH [4, 5], the classified vector quantization [6], etc. The current research of LUT focuses on following:

- Computation of index. Chung et al. [7] divided the local edge information of halftone image into 39 categories to form the index by combined with the local neighborhood pixels. Huang et al. [5] classified the DCT textures of halftone images into four categories to constitute the index together with the local neighborhood pixels.
- Optimal template selection. Denecker et al. [8] proposed a suboptimal template selection scheme which adopts a reduced greedy search strategy to produce a close-to-optimal template. Mese et al. [4] proposed a recursive algorithm to choose the template.
- Integration with the other methods. Chang et al. [9] utilized LUT method to reconstruct the multilevel gray image from halftone image, but when an empty value is referred, the LMS method was used to reconstruct the gray-level value. Suetake et al. [10] integrated LUT with Gaussian filtering to improve the reconstruction image quality.

Our research focused on the optimal template selection. Although an iterative search optimal template methods were proposed by [4, 8], they are actually the greedy methods and do not find the globally optimum template. Therefore, it was necessary to present a more effective method for template selection. Since template optimization problem is not differential, the classic optimization methods, such as gradient descent method, do not work, but the intelligent optimization methods provide an effective approach. Based on these reasons, the genetic algorithm was used to solve template selection problem in this paper.

31.2 LUT Method

Suppose, we are given a set of training image pairs $\{(G_i(x, y), H_i(x, y)) | 1 \leq i \leq M, 1 \leq x \leq W, 1 \leq y \leq H\}$ where G_i denotes the ith original gray image and H_i denotes the corresponding halftone image of G_i. The training image H_i was obtained by applying the Floyd-Steinberg error diffusion algorithm to the image G_i. Symbol T denotes a template as shown in Fig. 31.1a (see [4]). Both "O" pixel and all "a" pixels were used to predict the contone gray value of pixel "O." In this template, $K = 16$ means the number of "a" and "O". The template T is used as a sliding window to build the lookup table. Let LUT[I] denotes a lookup table and $N[I]$ denotes the number of pixels in training image set. The method to build up the lookup table is

$$\mathrm{LUT}[I] = \sum_{(i,x,y)\in S} G_i(x, y) \Big/ N[I]$$

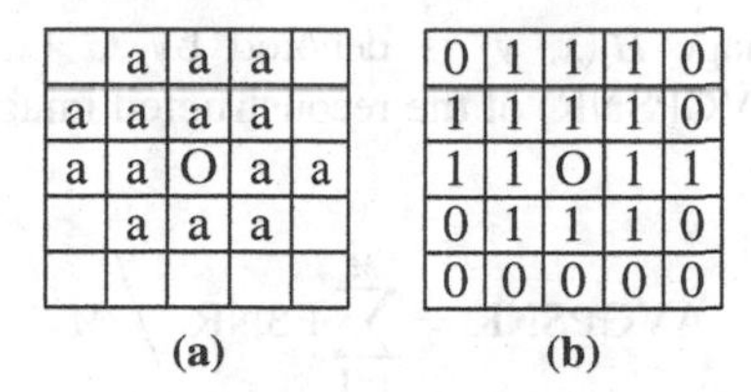

Fig. 31.1 Schematic diagram of a template. Size of template is 5 × 5 and this template involves 16 neighborhood pixels. In (**a**), "O" denotes the current estimated pixel and "a" is the neighborhood pixel of "O." In (**b**), "1" means that this pixel belongs to the neighborhood of "O" while "0" means this pixel is not included in the neighborhood of "O"

where I is an index satisfying $I = F(S_1, S_2,\ldots, S_K)$ and $F(\cdot)$ denotes a mapping function from K dimensional vector to index I. $S_1, S_2,\ldots, S_K$ is called the K dimensional binary vector that subimage of H_i covered by template T based on the raster scan order. $S_i \in \{0,1\}$, $1 \leq i \leq K$. S_p denotes $H_i(x, y)$ corresponding to "O" of T, $1 \leq p \leq K$. Other elements denote the neighborhood pixels. Set $S = \{(i, x, y)|I = F(S_1, S_2,\ldots, S_K) \wedge S_p = H_i(x, y)\}$ and $N[I] = |S|$ where $|S|$ denotes the size of set S.

The testing image set $\{TH_i(x, y)\}|1 \leq i \leq N_t, 1 \leq x \leq W, 1 \leq y \leq H\}$ denotes a series of halftone images. $IH_i(x, y)$, $1 \leq i \leq N_t$, denotes the reconstructed gray images of $TH_i(x, y)$ according to LUT. The reconstructed method is: let template T covers the local region of IH_i by a way of sliding window and get the K dimensional binary vector $S_1, S_2,\ldots, S_K$ based on the raster scan order. We can get the reconstructed pixels $IH_i(x, y) = LUT[I]$ where $I = F(S_1, S_2,\ldots,S_K)$ and $S_p = TH_i(x, y)$.

In [4, 9] the mapping function F is set as

$$F(S_1, \ldots, S_K) = \sum_{i=1}^{K} 2^{i-1} S_i$$

This means that there exist 2^K indexes in $S_1, S_2,\ldots, S_K$ while each index may be mapped to 2^l gray values (If the maximum gray value is 255 and $l = 8$). Therefore, there are 2^{K+l} mapping relations resulting in a one-to-many problem.

31.3 Template Selection Problem

Mese et al. [4] pointed out that different templates will lead to different inverse halftone image quality in LUT. In this section, we will discuss the optimal template selection problem. We assume the size of template T is $w \times h$ and T contains K neighborhood pixels (containing the estimated pixel, general $K < w \times h$). Assuming that template size and K are fixed, our goal is to choose a template for the best reconstruction image quality. Assuming the reconstructed grayscale images of halftone image $H_i(x, y)$ is denoted by $G'_i(x, y)$, the average peak signal-to-noise ratio (AVGPSNR) of the reconstructed image $G'_i(x, y)$, $1 \leq i \leq M$, is

$$\text{AVGPSNR} = \sum_{i=1}^{M} \text{PSNR}_i \Big/ M, \tag{31.1}$$

where PSNR_i is the peak signal-to-noise ratio (PSNR) of G'_i with G_i. The template optimization problem is to seek T such that the AVGPSNR be largest.

The size of the solution space of template optimization problem is $(wh)!/K!(wh-K)!$. The genetic algorithm is a heuristic search strategy based on the genetic processes of biological organisms. Over many generations, natural populations evolve according to the principles of natural selection and "survival if the fittest" [11]. The genetic algorithm may be used to solve the optimization problems, such as function optimization, automatic control, and image recognition. In next section, we try to use the genetic algorithm to search the optimal template.

31.4 Template Selection Based on Genetic Algorithm

The template optimization problem is a constrained optimization problem in this paper. The template in LUT can be converted to the form of Fig. 31.1b. Since the element "O" is regarded as the neighborhood pixel, it should satisfy

$$\sum_{i=1}^{w} \sum_{j=1}^{h} T(i, j) = K \tag{31.2}$$

The template optimization problem is converted into a constrained optimization problem: under the condition of (31.2), it is required that a template be found to let (31.1) as largest as possible.

The genetic algorithm can solve the multiobjective constrained optimization problem, but the template optimization problem of this paper is different from general constrained optimization problems. The constrained solution in this paper is not restricted to a certain range, but a set of discrete values. The genetic algorithm will be modified to satisfy the constrained condition (31.2) in this paper. Next, we will introduce the details about genetic operators: coding, selection, crossover, mutation, and fitness function.

- Coding

  The binary code is used in this paper. Since the template is a matrix, it is required to be converted to a one-dimensional vector according to the scan order way, as shown in Fig. 31.2. So the 24 dimension vector is formed. Decoding is the inverse process of encoding. Since template size is $w \times h$, the encoded vector is $w \times h-1$ dimension.

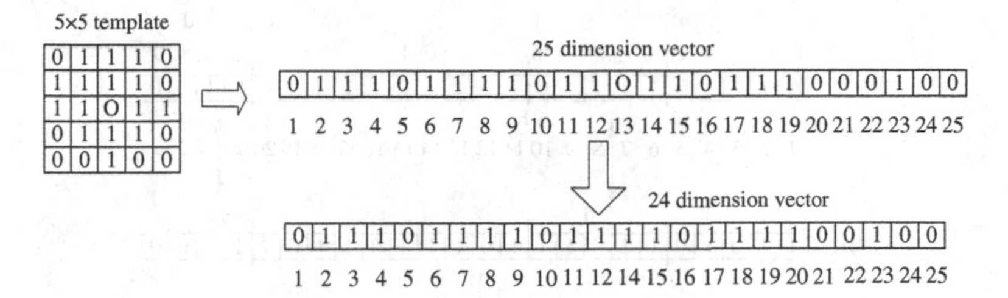

Fig. 31.2 Coding method. "O" is regarded as a fixed entry. Template size is 5 × 5 and the encoded vector is 24 dimensions

- Selection

  We may use roulette method to select individuals from the previous populations, but this method may repeatedly choose the same individuals resulting in the premature convergence. Our method is to sort the individuals in descending order according to their fitness values and select *n*/2 prior individuals. After selection, the individual number is *n*/2. *n* is the initial number of individuals.

- Cross

  The purpose of the cross operator is to randomly select two individuals for crossover operation from *n*/2 individuals. Based on the multipoint crossover operator, two individuals will produce four individuals after crossover operation. Our crossover operator requests that any substring, which is used to carry out crossover operation, contains one gene "0" at least. Therefore, for one of individuals, gene "0" are marked as 1, 2,..., S ($S = wh-K$) from left to right in turn and then four integer numbers are generated randomly denoting as a, b, c, d satisfying $1 \leq a<b<c<d \leq S$, respectively, corresponding to the above marks. The selected gene "0" correspond to a', b', c', d', respectively, satisfying $1 \leq a' < b' < c' < d' \leq wh-1$. a', b', c', and d' are regarded as the intersection to produce four individuals by crossing the selected substrings, respectively. Figure 31.3 shows an example of crossover operator.

- Mutation

  In general mutation operator, the value of a gene changes from "0" to "1" or "1" to "0" at a very small probability. In our mutation operator, some genes are required to change in three cases: a. The selected two individuals are the same completely; b. The two substrings, which will be crossed, are the same; c. An individual does not satisfy expression (31.3). In the first case, if A = B, A is fixed while a random sequence of "0" and "1" with the same length of A is generated as B satisfying A ≠ B. In the second case, if $A_{ab} = B_{ab}$, the A_{ab} is fixed while a random sequence of "0" and "1" with the same length of A_{ab} is generated as B_{ab} satisfying $A_{ab} \neq B_{ab}$; In the last case, the number of "1" in individual A is counted up and denoted as $\hat{K}$. Two different cases will be treated:

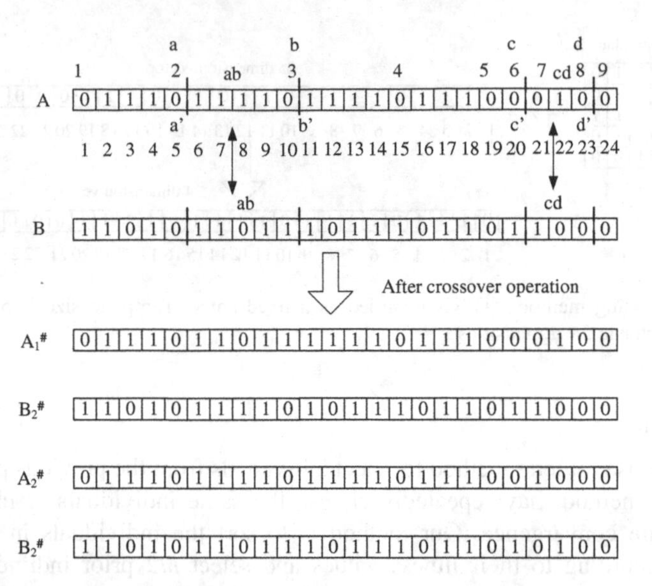

Fig. 31.3 Example of crossover operator. After crossover operation on A and B, four new individuals named as $A_1^{\#}$, $A_2^{\#}$, $B_1^{\#}$, $B_2^{\#}$ are produced

a. if $\hat{K} > K$, $\hat{K}$-K genes of “1” are randomly selected from individual A to turn into “0;” b. if $\hat{K} < K$, K-$\hat{K}$ genes of “0” are randomly selected from individual A to turn into “1.” After mutation operation, any individual should satisfy (31.3).

- Fitness function

  Expression (31.1) was regarded as the fitness function to evaluate the individuals. Calculation of expression (31.1) includes two steps: building lookup table and reconstruction. Since both reconstructed images and training images belong to the same dataset, so the halftone image reconstruction will not involve the null value problem.

31.5 Experimental Results

We test the proposed method on two image sets. The first set is the 512 × 512 pixel standard test image including airplane, Barbara, boat, camera, Goldhill, Lena, Mandrill, pepers, called Set1. The second set named Set2 has 60 images coming from a public dataset (http://www.systems.caltech.edu/mese/). All halftone images are produced by Floyd-Steinberg error diffusion algorithm. The initial population size is 8 and maximal iteration is set 50. We use the function rand() provided by matlab to produce the pseudorandom numbers in our experiments. All experiments are conducted under the environment of Windows XP, Matlab 7.0, an Intel Core Ci5-2500CPU, 3.3 GHz processor, and 4 GB of RAM. The convergence process of

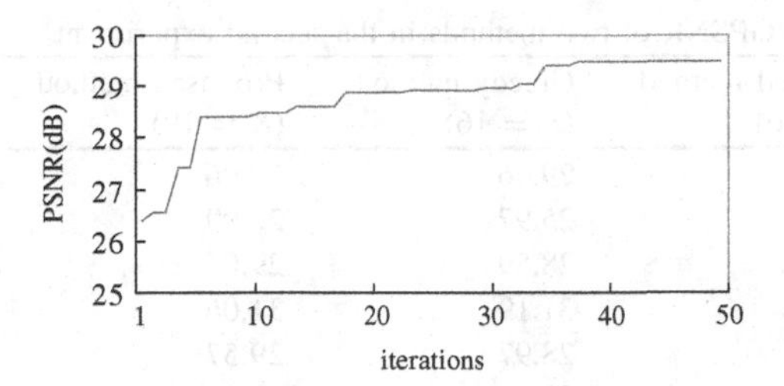

Fig. 31.4 Convergence process of genetic algorithm. The optimal PSNR is 29.50 dB

genetic algorithm is shown in Fig. 31.4 from which with the increasing of iterations, the AVGPSNR will reach to optimum.

In order to evaluate the performance of the proposed method, we present a comparison between the proposed method and the greedy method [4] on above two data sets. The size of T is fixed as 5×5. Parameter $K = 16$ or 19. Two experiments are done. In the first experiment, training set Set1 were used as testing set. The PSNR of test halftone images after reconstruction are shown in Table 31.1. In either case of $K = 16$ or 19, AVGPSNR of test halftone images in the proposed method is larger than that in the greedy method. The AVGPSNR of $K = 19$ is also larger than that of $K = 16$.

In the second experiment, training set is from Set2 while testing set is from Set1. The PSNR of test halftone images after reconstruction are shown in Table 31.2. The null values are estimated by the low-pass filtering. In either case of $K = 16$ or 19, AVGPSNR of the proposed method is larger than that of the greedy method. The AVGPSNR of reconstructed images in the second experiment is larger than that in the first experiment because of the inaccurate estimation of null values in the second experiment.

Table 31.1 The AVGPSNR of two methods in the first experiment

Method	Proposed method ($K = 16$)	Greedy method ($K = 16$)	Proposed method ($K = 19$)	Greedy method ($K = 19$)
Airplane	31.18	31.96	32.36	32.14
Barbara	26.51	26.48	27.41	27.37
Boat	29.46	29.29	31.28	31.10
Camera	31.72	31.49	33.07	32.73
Goldhill	29.46	29.26	31.20	29.98
Lena	31.24	31.87	32.27	32.01
Mandrill	25.82	25.74	26.7	26.65
Pepers	31.59	31.33	31.57	31.34
AVGPSNR	29.50	29.30	31.48	31.29

Values of column 2 are larger than that of column 3; Values of column 4 are larger than that of column 5

Table 31.2 The AVGPSNR of two methods in the second experiment

Method	Proposed method ($K = 16$)	Greedy method ($K = 16$)	Proposed method ($K = 19$)	Greedy method ($K = 19$)
Airplane	29.63	29.46	29.96	29.79
Barbara	26.03	25.97	25.99	25.94
Boat	28.78	28.59	29.07	28.91
Camera	31.40	31.19	31.06	31.89
Goldhill	29.18	28.97	29.57	29.33
Lena	31.56	31.19	31.23	31.95
Mandrill	25.30	25.23	25.09	25.03
Pepers	29.86	29.64	31.35	31.15
AVGPSNR	28.72	28.53	29.04	28.87

Values of column 2 are larger than that of column 3; Values of column 4 are larger than that of column 5

31.6 Conclusions

We proposed a solving strategy based on genetic algorithm for template optimization problem in LUT IH. A mathematical model of the constrained optimization for template selection was established. Many details about coding, selection, crossover, mutation, and fitness function were discussed. In experiments, the convergence process of the genetic algorithm was studied, and a lot of comparison experiments about halftone image reconstruction were done. The experimental results showed that the proposed method is better than the greedy method. The future research work will focus on the further performance improvement based on genetic algorithm.

Acknowledgments This study was supported in part by National Natural Science Foundation in China under Grant No. 61170102, Natural Science Fund of Hunan province in China under Grant No.11JJ3070, and Education Department Fund of Hunan province in China under Grant No.12A039.

References

1. Kite TD, Damera-Venkata N, Evans BL, Bovik AC (2000) A fast, high-quality inverse halftoning algorithm for error diffused halftones. IEEE Trans Image Process 9(9):1583–1592
2. Shen MY, Kuo CCJ (2001) A robust nonlinear filtering approach to inverse halftoning. J Vis Commun Image Represent 12(1):84–95
3. Wong PW (1995) Inverse halftoning and kernel estimation for error diffusion. IEEE Trans Image Process 4(4):486–498
4. Mese M, Vaidyanathan PP (2001) Look up table (LUT) method for inverse halftoning. IEEE Trans Image Process 10(10):1566–1578
5. Huang Y, Chung K, Dai B (2011) Improved inverse halftoning using vector and texture-lookup table-based learning approach. Expert Syst Appl 38(12):15573–15581

6. Lai JZC, Yen JY (1998) Inverse halftoning of color images using classified vector quantization. J Vis Commun Image Represent 9(3):223–233
7. Chung KL, Wu ST (2005) Inverse halftoning algorithm using edge-based lookup table approach. IEEE Trans Image Process 14(10):1589–1983
8. Denecker K, Assche SV, Neve PD et al (1999) Context-based lossless halftone image compression. J Electron Imaging 8(4):415–421
9. Chang PC, Yu CS, Lee TH (2001) Hybrid LMS-MMSE inverse halftoning technique. IEEE Trans Image Process 10(1):95–103
10. Suetake N, Tanaka G, Uchino E (2009) Look-up table and gaussian filter-based inverse halftoning method excellent in gray-scale reproducibility of details and flat regions. Opt Rev 16(6):594–600
11. Melanie M (1999) An Introduction to genetic algorithms. The MIT Press, Cambridge

Chapter 32
Color Preserved Image Compositing

Hao Wu and Dan Xu

Abstract Poisson cloning is an effective method for image compositing, but it suffers from color distortion when the hue of target image is different from that of source image. This paper proposes a method to correct the artifact of color distortion induced by Poisson cloning while keep seamless boundary. First, seamless cloning is applied to luminance component for keeping the local contrast of intensity. Then, the color belief that guide color correction is estimated based on general geodesic distance transform. Next, the color components of each pixel are corrected by minimizing an object function with corresponding color belief. Compared to other image compositing methods that only optimize boundary conditions, the proposed method keeps both the original color appropriately and the relative changes of colors, and using soft color belief to compositing provides a smooth transition from foreground to background, which cannot be done by hard constraints. Experiment results demonstrate the proposed method reduces the color distortion of composition effectively while keeps the seamless boundary, which requires only few interactions from user.

Keywords Image compositing · Poisson cloning · Color correction · Color belief

32.1 Introduction

Image compositing is a kind of methods that try to composite region of interest (ROI) of source image onto the target image seamlessly. The word "seamlessly" may be one of most important challenges to image compositing for an ideal

H. Wu (✉) · D. Xu
Department of Computer Science and Engineering, Yunnan University,
Kunming 650091, China
e-mail: wuhao19820311@163.com

D. Xu
e-mail: danxu@ynu.edu.cn

A. A. Farag et al. (eds.), *Proceedings of the 3rd International Conference on Multimedia Technology (ICMT 2013)*, Lecture Notes in Electrical Engineering 278,
DOI: 10.1007/978-3-642-41407-7_32,

composite that should be seamless in luminance, color, texture, boundary, noise pattern, size, etc. In a word, it should be seamless in perception without any noticeable artifacts. Satisfactory composites may be obtained by modifying per-pixel manually, but the work is time-consuming and tedious, and become unpractical when the ROI contains millions of pixels. Thus, large numbers of works have been developed to implement image compositing automatically or by few user interactions. Image segmentation [1, 2] and Image matting [3, 4] are the most common methods used for image compositing. However, matting is ill-posed problem, the additional assumption such as local smoothness has to be imposed to obtain a reasonable solution. Hence, image composting by segmentation or matting often fails when the assumption cannot be satisfied well, for example, foreground is not separated clearly from background, or the ROI contains highly textured regions [5]. Another kind of technique for image compositing is gradient domain based image cloning, such as Poisson cloning [6]. Although Poisson cloning can keep the local change of color well, it inflicts improper global change of color on composites so long as there is significant difference between source image and target image at boundary of ROI, which results in undesired color distortion on foreground object of composites, and the artifact is more obvious when the hue of ROI is highly different from target image.

To address the color distortion, we propose a method to composite the ROI of source image onto target image seamlessly while preserve the original color of foreground object contained in ROI. The contributions of our method are (1) A new model for correcting color distortion that occurs in image compositing, and a robust and accurate method for estimating color belief that is used to guide color correction, (2) An improved method for image compositing by combining the Poisson cloning with color correction, which aims to preserve the original color of foreground object as well as to keep the local change of intensity and seamless boundary in composites.

32.2 Related Work

Besides inducing color distortion, Poisson cloning [6] suffering from another artifact called "bleeding effect." Jiaya [7] alleviates bleeding effect by finding an optimal boundary between the boundary of ROI and a foreground object boundary extracted by segmentation algorithm. Sunkavalli [8] using a multi-scale technique to address the problem. The composites created by this multi-scale technique are consistent with target images in contrast, noise, and style, but it does not work well when the textures of source image and target image are not stochastic. To speed up the computation of Poisson cloning, a seamless cloning based on mean-value coordinates (MVC) was proposed [9], the MVC cloning spreads the difference between source image and target image at boundary to the entire ROI smoothly by a direct interpolation rather than solving the time-consuming Poisson equation, however, the quality of composites is not be improved for MVC cloning is an alternative of Poisson cloning in essence. The digital photomontage [10] stitches multiple ROIs from different images together to create a seamless composite. The best seam of each ROI is found by an optimization based on the graph-cut [11], and the artifacts of composite are reduced further by gradient domain operation. Although these works have improved Poisson cloning to a certain extent, they are not designed to address the color distortion problem. A work aimed to keep the original color of ROI in image composting can be found in literature [12]. The method first marks some pixels in ROI to fix their color to be same as source image or target image, which is equivalent to enforce hard constraints on marked pixels in addition to the boundary constraints, then they solve a weighted Poisson equation with these constraints to obtain the final composites. Their method reduces the color distortion of Poisson cloning, but often suffers from more "bleeding effect" due to the introduction of hard constraints. In contrast, our work in this paper avoids using the hard constraints, but employs a soft color belief and closed-form energy for color correction, and the composites obtained from our method have less color distortion and less "bleeding effect."

32.3 Overview

The main task in our method is generating composites that are seamless near the boundary, while keeping the original color of foreground object appropriately. We address the problem by three stages referred as luminance cloning, color belief estimation, and color correction. Figure 32.1 overviews the work flow of our method.

32.4 Algorithm

Given a target image I_t and a source image I_s with user specified ROI Ω_s, Our algorithm seamlessly composites Ω_s onto I_t while preserves the original color of foreground object contained in Ω_s appropriately. Here, we describe our key steps of our algorithm in details.

Luminance compositing. To operate on luminance and color component of image separately, the images are presented in CIE-Lab color space instead of RGB color space. The three channels of I_t and I_s are denoted as L_t, a_t, b_t, and L_s, a_s, b_s separately. When compositing Ω_s onto target image, it is necessary to keep the contrast of luminance in for human visual system is much more sensitive to local contrast than slow changes in the luminance [9]. In addition, the absolute luminance values near the boundary of Ω_s should consist with that of I_t, or a visible seam will arise in composites due to the luminance changes at boundary, and this artifact will be more obvious when the luminance L_s is highly different from L_t. For Poisson cloning [6] is an effective way to keep the original local contrast of Ω_s

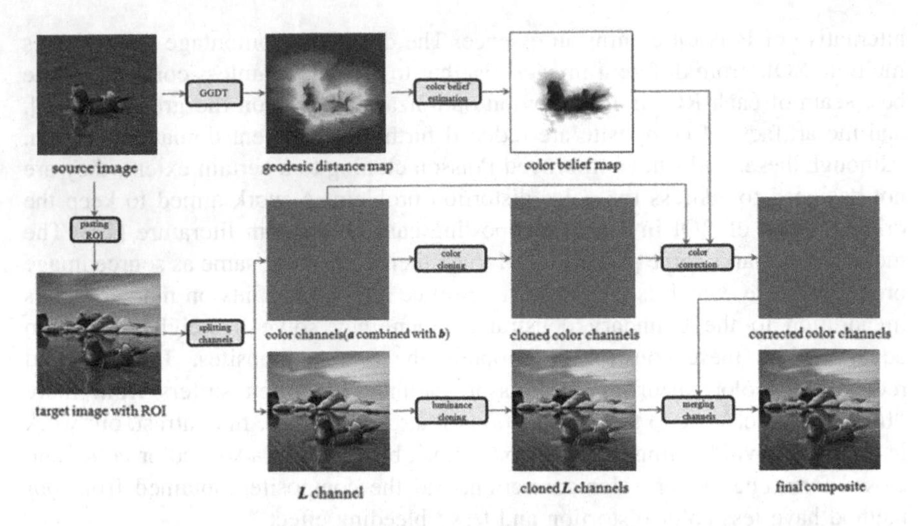

Fig. 32.1 The work flow of color preserved image compositing, while the color components were visualized by combining the color channels with a constant luminance (CIE L*a*b*)

and to generate a seamless boundary, we use it on luminance component directly to satisfy above constraints. Thus, the luminance cloning can be formulated as

$$\arg \min_{L_c} \iint_{\Omega_s} \|\nabla L_c - \nabla L_t\|^2 \tag{32.1}$$

with Dirichlet boundary condition

$$L_c|_{\partial\Omega_s} = L_t|_{\partial\Omega_s} \tag{32.2}$$

while L_c is the cloned luminance. Although Poisson cloning can be applied to each channel independently to obtain a chromatic composite, the color of composite is unnatural when the hue of Ω_s is far from that of I_t(Fig. 32.4c). Hence, we propose a method to correct the color distortion, which adjusts the color components of composites according to estimated color belief.

Color belief estimation. Color belief measures how much the color components obtained from Poisson cloning can be trusted, and we denote it as $B(X) \in [0,1]$ for each pixel location X. Intuitively, the color belief is always lower in foreground object and rises with the distance from foreground increase. Furthermore, color belief should be smooth to make the color components have natural transition from foreground to background. Thus, we model the color belief as a smooth interpolation of the distant function $D(X)$

$$B(X) = 3\left[\frac{D(X) - D_{\min}}{D_{\max} - D_{\min}}\right]^2 - 2\left[\frac{D(X) - D_{\min}}{D_{\max} - D_{\min}}\right]^3 \quad X \in \Omega_s \tag{32.3}$$

While $D_{\min}$, $D_{\max}$ are the minimal and maximal value of $D(X)$ in Ω_s separately. The precision of estimated color belief is highly dependent on the distance function $D(X)$, and the simple distance transform such as Euclidean distance provides pool results. Thus we resort to a more effective distance transform referred as general geodesic distance transform (GGDT) [13], which is built upon a soft, real-values mask $M_s(X) \in [0,1]$ that represents the probability of each pixel belongs to the background. It can be formulated as

$$D_{\text{GGDT}}(X, M_s, I_s) = \min_{X' \in \Omega_s}(d(X, X') + vM_s(X')) \tag{32.4}$$

while the free parameter v establishes the important degree of soft belief relative to the spatial distance. With

$$d(X, X') = \min_{\Gamma_{X,X'}} \int_0^1 \sqrt{\|\Gamma'(\mathrm{s})\|^2 + \gamma^2(\nabla I_s \cdot u)^2} \mathrm{ds} \tag{32.5}$$

with $P_{X,X'}$ is the set of all paths between the pixels X and X', and each path is described by parametric function $\Gamma(\mathrm{s})$, while $\mathrm{s} \in [0,1]$. $\Gamma'(\mathrm{s}) = \frac{\partial \Gamma(s)}{\partial \mathrm{s}}$, and $u = \frac{\Gamma'(\mathrm{S})}{\Gamma'(\mathrm{S})}$ is unit vector tangent to the direction of the path $\Gamma(\mathrm{s})$. The factor γ weighs the contribution of the image gradient versus the spatial distances. The soft mask M_s is modeled as

$$M_s(X) = g\left(\ln \frac{P_r(I(X)|\alpha(X) = \mathrm{BG})}{P_r(I(X)|\alpha(X) = \mathrm{FG})}\right) \tag{32.6}$$

While the $g(\cdot)$ is a sigmoid transformation, which is $g(x) = 1/(1 + \exp(-x/\sigma))$, and we set $\sigma = 4.5$ in this paper. $P_r(I(X)|\alpha(X) = BG)$ and $P_r(I(X)|\alpha(X) = FG)$ are pixel-wise foreground and background likelihood separately, and $\alpha(X)$ is binary per-pixel labeling that indicates pixel X to belong to foreground or background. Like the way used in Grabcut [2], we model the foreground and background likelihood as different Gaussian mixture models (GMM) separately, and the parameters of each of GMM are estimated from a few seed samples specified by user strokes (Fig. 32.3a). Because GMM is more robust to noise and to the case that background and foreground have similar color distribution, the likelihood obtained from GMM is more accurate than that from histograms accumulating [13].

From formulation (32.4), we know that GGDT tries to find a path with minimal energy from X to all of other pixels, and the energy of the path is computed by geodesic distance $d(X, X')$ and soft mask $M_s(X)$. In other words, GGDT is minimal distance from X to all of pixels of image, and it accounts for the spatial distance, distance in color space and probability distance. Now, we set $D(X) = D_{\text{GGDT}}(X, M_s, I_s)$ for each pixel and compute the color belief $B(X)$ directly by Eq. (32.3). It is worth noting that the image segmentation obtained by GDDT are more robust and precise for using soft probabilistic mask can convey more information, and our estimation of color belief is also benefit from GDDT for the same reason. Figure 32.2 gives an example of color belief estimation with a few user strokes.

Color correction. This step corrects the color distortion induced by Poisson cloning according the estimated color belief. We model the color correction as minimization of an energy function that combines color fidelity term and color propagation term

$$E = E_{\mathrm{cf}} + \lambda E_{\mathrm{cp}} \tag{32.7}$$

Color fidelity term E_{cf}. To preserve the original color of foreground, we constrain the corrected color being close to that of source image when the pixels have low color belief. If the color belief of pixels is high, the corresponding corrected color will be more close to the color components obtained by Poisson cloning. We defined E_{cf} in a weighted squared L2 Norm

$$E_{\mathrm{cf}} = \sum\nolimits_{X\in\Omega_s} B_1(X)\left\|\overrightarrow{C_r}(X) - \overrightarrow{C_s}(X)\right\|^2 + B_2(X)\left\|\overrightarrow{C_r}(X) - \overrightarrow{C_P}(X)\right\|^2 \tag{32.8}$$

while the color components of source image are combined as a vectors $\overrightarrow{C_s} = (a_s, b_s)^{\mathrm{T}}$, similarly, the $\overrightarrow{C_P} = (a_P, b_P)^{\mathrm{T}}$ and $\overrightarrow{C_r} = (a_r, b_r)^{\mathrm{T}}$ denote the color components obtained by Poisson cloning and corrected color components. The weight function $B_1(X) = 1 - B(X)$ and $B_2(X) = B(X)$.

Color propagation term E_{cp}. For a pixel that has intermediate color belief value, linear blending $\overrightarrow{C_s}(X)$ and $\overrightarrow{C_P}(X)$by color belief cannot produce a satisfactory result. A more subtle alternative is to be propagating the color of pixels that have extreme color belief value (lower or higher color belief) to that have intermediate color belief value. In other words, the $\overrightarrow{C_r}(X)$ will be estimated from the colors of its neighbors when the degree of a pixel X belongs to foreground is close to the degree of it belongs to background. We model the color propagation term as

$$E_{\mathrm{cp}} = B_3(X)\sum\nolimits_{X\in\Omega_s}\left\|\overrightarrow{C_r}(X) - \sum_{\widetilde{X}\in N(X)} w(X,\widetilde{X})\,\overrightarrow{C_r}(\widetilde{X})\right\|^2 \tag{32.9}$$

While $N(X)$ is the set of neighbor pixels of X. $w(X,\widetilde{X})$ is a weight function defined on the similarity between X and its neighbor $\widetilde{X}$, and we define it as an affinity function that take the spatial distance and color distance into consideration.

$$w(X,\widetilde{X}) = \exp\left(-\left\|I_s(X) - I_s(\widetilde{X})\right\|^2/\sigma_1^2\right)\exp\left(-\left\|X - \widetilde{X}\right\|^2/\sigma_2^2\right) \tag{32.10}$$

While σ_1 is computed from all pixels in $N(X)$, and we fix $\sigma_2 = 2.5$ in this paper. $B_3(X)$ is used for weighting the color propagation term, and we defined it as

$$B_3(X) = \exp\left(-|B_1(X) - B_2(X)|^2/\sigma_3^2\right) \tag{32.11}$$

σ_1 controls the relative importance of color propagation term, small σ_1 will make propagation term have larger weight. Obviously, if the pixels have intermediate color belief value ($|B_1(X) - B_2(X)|$ is smaller), color propagation term

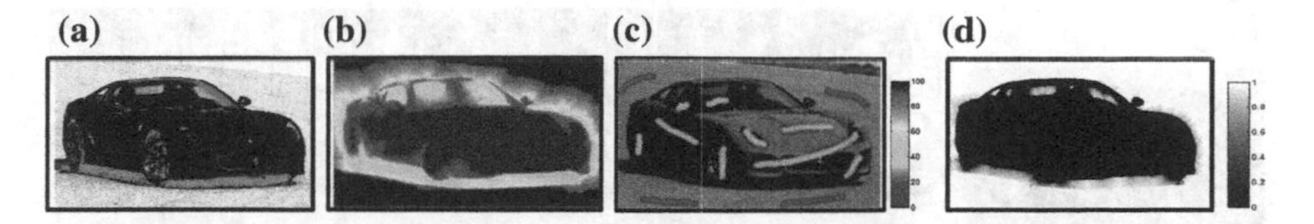

Fig. 32.2 Color belief estimation, **a** user interactions, foreground samples are specified by *green strokes*, background samples by *red strokes*. **b** Soft mask obtained from GMM, which uses **a** as seed samples. The likelihood belongs to background is lower to darker pixels. **c** General geodesic distance computed with the soft mask. **d** The estimated color belief from

will be more predominant in energy function. With the color fidelity term E_{cf} and color propagation term E_{cp}, we minimize the energy function E by solve a large sparse overdetermined linear system, The final composite I_f is constituted of the luminance components obtained by Poisson cloning and the color components corrected by our method.

32.5 Results

With few interactions from user, our method succeed in generate composite with correct color of foreground and seamless boundary. A comparison of our method with that of other compositing is shown in Figs. 32.3, 32.4, 32.5, 32.6, 32.7. It is worth noticing that we only use few strokes on foreground and on background separately besides specifying ROI in all examples.

Fig. 32.3 Comparison of our method with other compositing methods, **a** source image, **b** Poisson cloning [6], **c** cloning with optimal boundary [7], **d** Composting with α matte, **e** our method

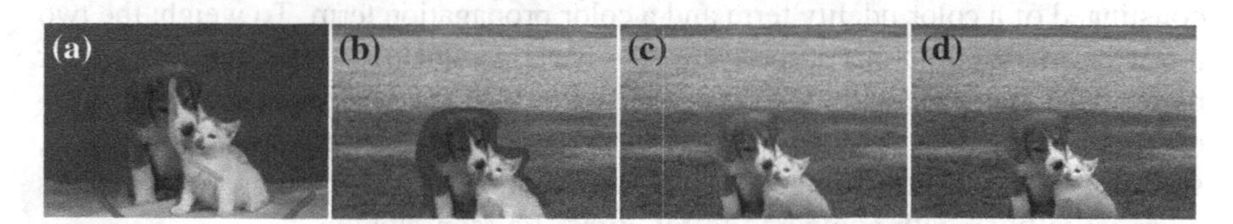

Fig. 32.4 Example of color preserved composting, our method only need additional four strokes (shown in **a**) besides specifying the ROI in this example

Fig. 32.5 More examples, the transparencies such as the windows shows appropriate color in **d**

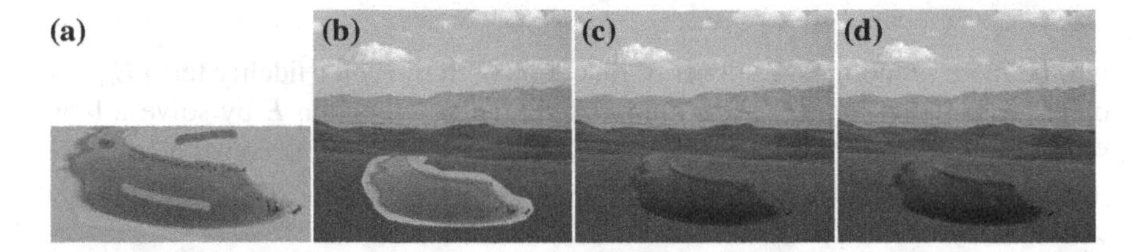

Fig. 32.6 More examples, besides obtaining the appropriate color of lake, our method make the luminance of lake be more consistent with that of target image

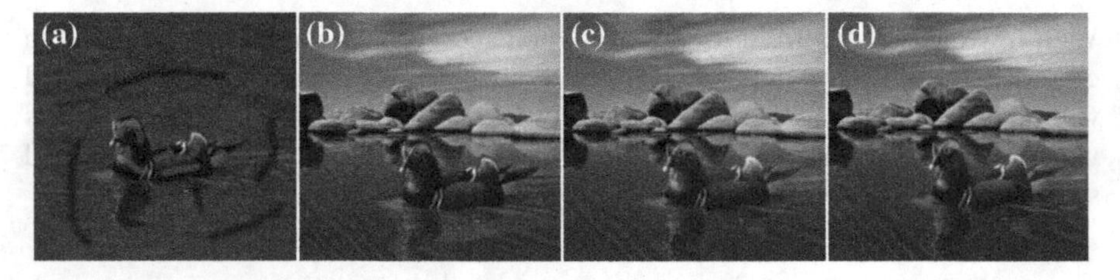

Fig. 32.7 More examples, the color of foreground object was preserved faithfully in our method

32.6 Conclusion

In this paper, we proposed a method to clone a ROI of source image to target image seamlessly while preserve the original color of foreground object contained in ROI. Instead of operating on all channels of pixels homogeneously, our method deals with the luminance and color components differently. We proposed a method to correct the color distortion by minimizing a closed-form energy function that is constituted of a color fidelity term and a color propagation term. To weight the two terms of energy function properly, we derive a weight function referred as color belief from the general geodesic distance transform. Experimental results demonstrate our method is suffered from little color distortion of foreground object, while keeps a seamless boundary and the transition near the foreground object is more visually smooth compared to the direct Poisson cloning.

References

1. Li Y, Sun J, Tang C, Shum H (2004) Lazy snapping. In: Proceedings of ACM SIGGRAPH 2004. New York, USA, pp 303–308
2. Rother C, Kolmogorov V, Blake A (2004) Grabcut-interactive foreground extraction using iterated graph cut. In: Proceedings of ACM SIGGRAPH 2004, pp 309–314
3. Chuang Y, Curless B, Salesin D et al (2001) A bayesian approach to digital matting. In: Proceedings of CVPR 2001. IEEE Computer Society, pp 264–271
4. Wang J, Cohen M (2007) Simultaneous matting and compositing. In: Proceedings of IEEE CVPR 2007. Washington DC, USA: IEEE Computer Society, pp 1–8
5. Wang J, Cohen M (2007) Image and video matting: a survey. Found Trends Comput Graph Vis 3(2):1–78
6. Perez P, Gangnet M, Blake A (2003) Poisson image editing. In: Proceedings of ACM SIGGRAPH 2003, ACM, New York, USA, pp 313–318
7. Jiaya J, Sun J, Tang C et al (2006) Drag-and-drop pasting. In: Proceedings of ACM SIGGRAPH 2006. ACM, New York, USA, pp 631–637
8. Sunkavalli K, Johnson M, Matusik W et al (2010) Multi-scale image harmonization. In: Proceedings of ACM SIGGRAPH 2010, Article no 125, pp 1–10
9. Farbman Z, Hoffer G, Lipman Y et al (2009) Coordinates for instant image cloning. In: Proceedings of ACM SIGGRAPH 2009. ACM, New York, USA, Article no, vol 67, pp 1–9
10. Agarwala A, Dontcheva M, Agrawala M et al (2004) Interactive digital photomontage. In: Proceedings of ACM SIGGRAPH 2004. ACM, New York, USA, pp 294–302
11. Boykov Y, Jolly MP (2001) Interactive graph cuts for optimal boundary & region segmentation of objects in n-d images. In: Proceedings of IEEE ICCV 2001. IEEE Computer Society, Washington DC, USA, vol 1, pp 105–112
12. Guo D, Sim T (2009) Color me right–seamless image compositing. In: Proceedings of the international conference on computer analysis of images and patterns 2009. Springer-Verlag, New York, USA, pp 444–451
13. Criminisi A, Sharp T, Rother C, P'erez P (2010) Geodesic image and video editing. ACM Trans Graph (TOG) 29(5):1–15

Chapter 33
A Novel Detection Algorithm of Double MP3 Compression

Pengfei Ma, Rangding Wang, Diqun Yan and Chao Jin

Abstract MP3 is the most widely used audio format nowadays in our daily life, while MP3 audio often be forged by an audio forger for their own benefits in some significant events, which will cause double MP3 compression. In this paper, it is the first time to calculate the statistical frequency of Huffman code table index, and a support vector machine is applied for classification to detect double MP3 compression. Experimental results demonstrate that the proposed method has low complexity and high accuracy, also the blank with the high bitrate in double MP3 compression detection is made up.

Keywords Huffman table index · Double MP3 compression · High bitrate

33.1 Introduction

With the development of the multimedia information technology, audio, video, images, and other multimedia information industry penetrate into every corner of our society, and bring us great convenience. However, everything have two sides, some nonprofessionals can modify the multimedia information without leaving obvious traces by using the cheap or free multimedia editing software. Although most people manipulate the multimedia information in order to make the multimedia become more perfect, if some criminals maliciously modified the multimedia information for their own interests to harm the harmonious society, the consequences would be disastrous [1]. For example, the tampered audio can be taken as evidence in court, which will result in the opposite outcome of the trial; false-quality audios spreading widely on the market may infringe the copyright of formal

P. Ma · R. Wang (✉) · D. Yan · C. Jin
College of Information Science and Engineering, Ningbo University, Ningbo, China
e-mail: wangrangding@nbu.edu.cn

A. A. Farag et al. (eds.), *Proceedings of the 3rd International Conference on Multimedia Technology (ICMT 2013)*, Lecture Notes in Electrical Engineering 278,
DOI: 10.1007/978-3-642-41407-7_33,

business; a speech which conveys orders in the military being tampered, can cause military conflict, and so on. Therefore, more and more people begin to pay attention to the security issues of multimedia information. The process of compressed audio tampering must take double audio compression [2–4]. As the current most popular audio format, MP3 audio can be easily manipulated by some audio editing software, such as Cool Edit, Gold Wave, and so on. Thus the double MP3 compression detection for audio forensics becomes more significant than before.

While image and video double compression detection have attracted many scholars' attention, research on double MP3 compression detection is rare, especially for the detection of the same bitrate. Yang et al. [2, 3] considered that the number of the MDCT coefficients between -1 and 1 in double compressed MP3 audio is less than that in single compressed MP3 audio. The method can detect the double compressed MP3 audio transcoded from lower bitrate to high bitrate, but the results show the low accuracy rate on the other opposite situation. Liu and Qiao et al. [4, 5] observed the statistical characteristics of the MDCT coefficients which exceed a certain threshold, a Support vector machine (SVM) is applied for classification, experiment results demonstrate that the algorithm can correctly detect whether the double compressed MP3 audio is transcoded from high bitrate to lower bitrate or from low bitrate to higher bitrate.

In summary, although many researchers have been devoted to the double MP3 compression under different bitrates, there is still a blank in the double compressed audio detection with the high bitrate, which means both the first and the second compressed bitrate are higher than that of normal. In this case, the encoder has enough bits to store the audio information, and the loss of the audio information is tiny after the first and the second compression, which result in the slight change of the effective features between them [6]. All the traditional algorithms of the detection are based on the statistical features of quantized MDCT coefficients. This paper presents a novel algorithm which utilizes the statistical probability of the Huffman table index as the characteristics to detect double MP3 compression.

The rest of this paper is organized as follows. In Sect. 33.2, the procedures of the Huffman coding in MP3 encoding are reviewed briefly. The statistical features of the Huffman code table index are observed between single and double compressed MP3 audios in Sect. 33.3. Experiments of detecting double MP3 compression are implemented based on the extracted features in Sect. 33.4. Finally, the conclusions and future work are summarized in Sect. 33.5.

33.2 Application of Huffman Code Table

As a lossy encoder in audio compression, MP3 encoder can maximize the filtered information that human auditory system is unable to perceive, without affecting the audio quality of the original hearing. Not only that, the encoder make use of the lossless Huffman coding, which compress the audio volume as tiny as possible so as to facilitate the network transmission.

MP3 encoding is in units of frames. Each frame contains two granules, and each granule contains two channels in stereo music. Each channel includes 576 quantized MDCT coefficients for Huffman coding at the end of the MP3 encoding [7]. As a kind of lossless encoding, Huffman coding is mainly used to improve the compression efficiency of MP3 encoding, so the selection of Huffman code table is related with the quantized MDCT coefficients closely. Quantized MDCT coefficients are in accordance with the order from low frequency to high frequency, therefore the energy value of the quantized MDCT coefficients comply with the contrary law, but not absolutely. According to these rules, the quantized MDCT coefficients in each channel are divided into three areas: the large value region (Big_value), the small value region (Count1) and the all-zero region (Rzero), as shown in Fig. 33.1.

Quantized MDCT coefficients are different in each channel, so the selection of the area size changes according to the energy values of the coefficients. As one of the MPEG encoding standard, MP3 encoder contains 34 Huffman code tables. The Rzero region is not necessary to be coded because all the values in this region are zeros. Tables 32 and 33 are only used for the Count1 region, in which the values are only 0 or 1, and each four coefficients are grouped for encoding. The Big_value region contains 0–31 tables, and every two coefficients are treated as a set for encoding. This region is further divided into three subregions: region0, region1, and region2, and the MP3 encoder selects the most appropriate Huffman code table for each subregion according to the maximum value in the region to ensure the lossless characteristic of Huffman encoding. So the selection of the Huffman code table in big value can better reflect the change details of the quantized MDCT coefficients between the single and double compressed MP3 audios.

33.3 Feature Extraction

So far, researchers [2–5] used to devote more efforts to the statistical characteristics of the features on MDCT coefficients and quantized MDCT coefficients. While these features are difficult to overcome the large amount of calculation and

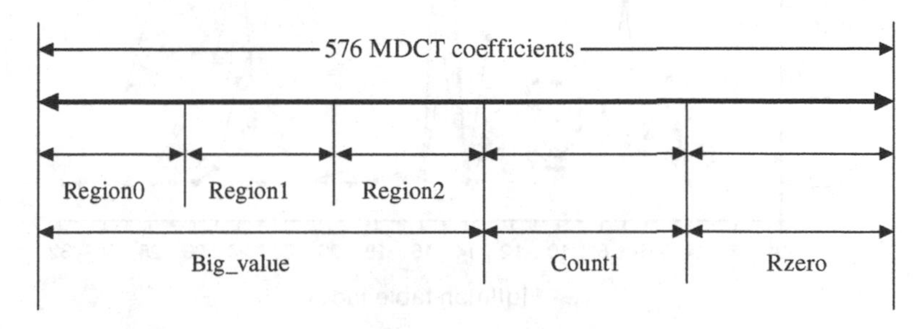

Fig. 33.1 The partition of quantized QMDCT coefficients

the shortcomings of the slight change between the single and the double compressed MP3 audios, especially when the audio samples are coded with the high bitrate, in which case the subtle changes on the MDCT coefficients are indistinguishable. MP3 audio is the bitstream coded form the quantized MDCT coefficients, so the selection on Huffman code table can reflect the fluctuation of the quantized MDCT coefficients in the process of MP3 encoding. In this paper, the selection changes of the Huffman code table between the single and double compressed MP3 audios are observed to forecast whether the audio has been suffered double compression.

33.3.1 Distribution of Huffman Code Table

In one same bitrate, the Huffman code table selection in different audio samples is similar, while the difference is obvious under different bitrates on the same audio sample [8], as shown in Figs. 33.2 and 33.3. The probability of the usage of small Huffman code tables is relatively large than that of normal when the low bitrate is used for encoding, on the contrary, the utilization of big Huffman code table will increase under the high bitrate encoding. The reasons can be explained as follows.

MP3 encoding is in units of frames. Each frame contains two granules, and each granule contains one channel in mono audio or two channels in stereo audio. The 576 quantized MDCT coefficients are formed in every channels, the frame size can

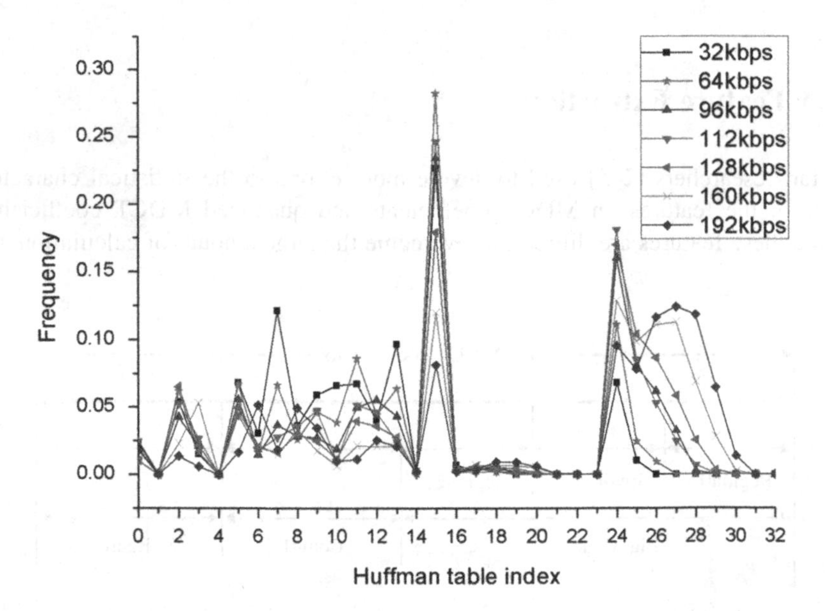

Fig. 33.2 The distribution of the Huffman table in single compressed MP3

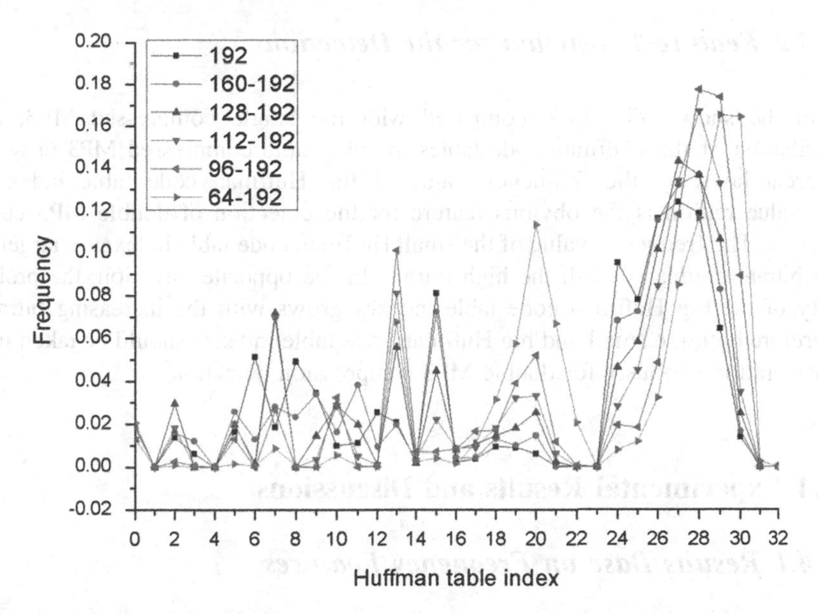

Fig. 33.3 The distribution of the Huffman table in double compressed MP3

be calculated according to the Formula 33.1. Where the Bitrate means the bitrate parameter set up at the beginning of the MP3 encoding. The Sampling rate represents the sampling rate when the audio sample is recorded. The Padding bit expresses the filling bit which is used to adjust the frame size and the value is 0 or 1. The frame size of MP3 and the bitrate hold the positive relationship approximately as shown in the Formula 33.1, so less audio information will be contained in each frame when the low bitrate is used for encoding, which result in the high probability of the low Huffman code tables selection.

$$\text{Frame}_{\text{size}} = 144 * \text{Bitrate}/_{\text{Sampling rate}} + \text{Padding bit} \quad (33.1)$$

When a MP3 audio is treated as the evidence in court or other forensics, the most important two cases on double MP3 compression detection are as follows: (1) MP3 audio transcoded form low bitrate to high bitrate; (2) MP3 audio transcoded with the same bitrate. However, MP3 audio transcoded from high bitrate to low bitrate has no practical significance, because every MP3 encoding is a kind of lossy compression, even though under the high bitrate [9]. The auditory quality of the audio will decline rapidly and the suspicion is caused easily when the MP3 audio is encoded with the high bitrate after decoding the low bitrate MP3. In this paper, the proposed method focuses the research on the MP3 audios transcoded from low bitrate to high bitrate and transcoded with the same bitrate.

33.3.2 Feature Extraction for the Detection

From the shown Fig. 33.3, compared with the single compressed MP3, the distribution of the Huffman code tables in the double compressed MP3 presents different laws, so the frequency value of the Huffman code table index in Big_value region is the obvious feature for the detection of double MP3 compression. The frequency value of the small Huffman code table indexes is larger in low bitrate compared with the high bitrate. In the opposite direction, the probability of the big Huffman code table indexes grows with the increasing bitrate. Therefore both the small and big Huffman code table indexes should be taken into consideration seriously for double MP3 compression detection.

33.4 Experimental Results and Discussions

33.4.1 Results Base on Frequency Features

The 687 single compressed MP3 audios are obtained with different bitrates of 56, 64, 80, 96, 112, 128, 160, and 192 kbps, respectively by MP3 encoder lame 3.99.5 which is the latest and most popular MP3 encoder nowadays [10]. Then each of the MP3 audio is decompressed and recompressed with the bitrate equal to or bigger than the bitrate in single compressed MP3 audio. These double compressed MP3 audios are served as negative samples, while the single compressed MP3 audios mentioned above are treated as positive samples.

We extract the positive frequency features of the Huffman code table index in the decoding process of the single compressed MP3 audios, and the negative frequency features in the procedure of decoding the double MP3 decompressed audios. Taking the distribution of the Huffman code table index in low bitrate and high bitrate into account, the frequencies of the table index of 0, 2, 7, 9, 12, 24, 25 26, $\leq$ 15 and >15 are selected as the positive and the negative frequency features. Support vector machine is used to distinguish the double compressed MP3 audios from the single compressed MP3 audios. The 70 % features are randomly chosen for training and the rest of them are used for testing in each of the experiments, finally the average testing accuracy is computed. Table 2 shows the average accuracy of the detection based on the frequency of the Huffman code table index.

In Table 33.1, the testing result 98.35 %, corresponding to that the value of 'B1' and 'B2' are 96 and 112, respectively, which means the detection accuracy between the 112 kbps-bitrate MP3 audios that are transcoded from the 96 kbps-bitrate MP3 audios and the single compressed MP3 audios with the 112 kbps bitrate. And so on.

Table 33.1 Test results based on the frequency of the Huffman table index (%)

B1	B2 32	64	96	112	128	160	192
32	86.15	99.89	99.77	99.99	99.93	99.91	99.90
64		86.56	99.13	99.71	99.91	99.78	99.87
96			84.69	98.35	99.90	99.72	99.64
112				82.19	97.82	98.32	98.76
128					79.17	96.47	97.73
160						80.16	88.52
192							81.25

33.4.2 Discussion

As the Table 33.1 show, the frequency features of Huffman code table index have the excellent detection results for the MP3 audios transcoded from low bitrate to high bitrate. As can be seen from the tables of testing results, in the case of one fixed bitrate in second compression, the lower bitrate in the first compression, the more serious distortion of audio quality and the more obvious deviation of the Huffman code table index distribution between the first and the second compression, which leads to high detection accuracy. With the bitrate increasing in first compression and approaching the bitrate in second compression little by little, the effective features become similar gradually and the detection accuracy gets worse and worse, especially for the high bitrate in the twice compression. Moreover the MP3 audio transcoded with the same bitrate is the extreme condition, and the detection accuracy decreased rapidly.

33.5 Conclusions and Future Work

In this paper, the theory of the Huffman coding in MP3 encoding is reviewed, and the statistical characteristics of the Huffman code table index are proposed as the features to detect double MP3 compression. Experiment results show that proposed method performs well for the double MP3 compression detection. Compared with the previous algorithm based on quantized MDCT coefficients, our method have high accuracy and low computational complexity, because the features can be obtained directly from the MP3 bitstream without fully decoding.

Markov model [11–14] plays an irreplaceable role for correlation detection in digital signal processing. The 576 coefficients in each channel have a certain correlation, so as to the arrangement of the Huffman code table indexes in the Big_vale region.

In the future, more efforts will be committed to apply the Markov model on double MP3 compression detection and researches on detecting MP3 audios which experienced twice or more MP3 compression from normal MP3 audios.

Acknowledgments This work is supported by the National Science Foundation of China (61170137), Doctoral Fund of Ministry of Education of China (20103305110002), Zhejiang Province Key teaching material Construction project (ZJB2009074), Scientific Research Fund of Zhejiang Provincial Education Department (Y201119434), Zhejiang Scientific and Technical Key Innovation Team of New Generation Mobile Internet Client Software (2010R50009), Open Fund of Zhejiang Provincial Top Key Discipline of Information and Communication Engineering (XKXL1310).

References

1. Gupta S, Cho S, Jay Kuo C-C (2012) Current developments and future trends in audio authentication. IEEE Multimedia 19(1):50–59
2. Yang R, Shi Y, Huang J (2010) Detecting double compression of audio signal. Media forensics and seeurity 11, vol 7541. San Jose, p 1–10
3. Yang R, Shi Y, Huang J (2009) Defeating fake-quality MP3. In: Proceedings of the 11th ACM workshop on multimedia and security, p 117–124
4. Liu Q, Sung AH, Qiao M (2010) Detection of double MP3 compression. Cogn Comput 2(4):291–296
5. Qiao M, Sung AH, Liu Q (2010) Revealing real quality of double compressed MP3 audio. ACM Multimedia p 1011–1014
6. D'Alessandro B, Shi Y (2009) MP3 bit rate quality detection through frequency spectrum analysis. In: Proceedings of the 11th ACM workshop on multimedia and security, p 57–61
7. Zhu J, Wang R, Li J, Yan D (2011) The filterbank in MP3 and AAC Encoders: a comparative analysis. In: 2011 international conference on electronics, communications and control, p 1110–1113
8. Tan J, Wang R (2008) The analysis of VLC-coded efficiency based on MPEG-I layer III audio. In: International conference on cyber worlds, p 136–141
9. Vercellesi G, Vitali A, Zerbini M (2007) MP3 audio quality for single and multiple enciding. In: IEEE International conference on multimedia and expo, p 1279–1282
10. LAME 3.99.5 (2013) MP3 encoder. http://lame.sourceforge.net/. Accessed 12 April 2013
11. Shi YQ, Chen C, Chen W (2006) A markov process based approach to effective attacking jpeg steganography. Information hiding p 249–264
12. Chen C, Shi YQ (2008) JPEG image steganalysis utilizing both intrablock and interblock correlations. Institute of software Chinese academy of sciences (ISCAS), Beijing, p 3029–3032
13. Chen C, Shi YQ, Su W (2008) A machine learning based scheme for double JPEG compression detection. In: 19th international conference on pattern recognition (ICPR), p 8–11, 1–4
14. Qiao M, Sung AH, Liu Q (2009) Steganalysis of MP3Stego. In: International joint conference on neural networks (IJCNN), p 2566–2571

Chapter 34
A Genre-Independent Chord Transcription System from Audio Using GMM-Based HMMs

Hao Wu, Dan Su, Yifang Wang and Xihong Wu

Abstract Chord transcription is an important task in music processing. In most current implementations, the chord transcription suffers greatly from the high acoustic variance caused by different music styles (such as genre, musician). In this paper, we describe a new style-independent acoustic chord transcription system, which can perform chord transcription on different styles of music directly and gives good frame-level recognition results. In this implementation, all music files are first used without genre clustering to train universal acoustic chord Hidden Markov models, whose state is modeled by single or multiple Gasussian mixture model. Then we extended such model with a probabilistic latent semantic analysis (PLSA) based approach to deal with the acoustic variations. Experimental results show that by the proposed PLSA-based approach, our genre-independent chord transcription system, although recognizing without any genre-specific information of testing data, has outperformed a genre-dependent system. Further analysis is also made to find the most important factors our PLSA-based models have captured.

Keywords Acoustic chord transcription · Gaussian mixture models (GMMs) · Genre-independent · Hidden Markov models (HMMs) · Probabilistic latent semantic analysis (PLSA)

H. Wu (✉) · D. Su · Y. Wang · X. Wu
Speech and Hearing Research Center, Peking University, Peking, China
e-mail: wuhao@cis.pku.edu.cn

D. Su
e-mail: sudan@cis.pku.edu.cn

Y. Wang
e-mail: yfwang@cis.pku.edu.cn

X. Wu
e-mail: wxh@cis.pku.edu.cn

A. A. Farag et al. (eds.), *Proceedings of the 3rd International Conference on Multimedia Technology (ICMT 2013)*, Lecture Notes in Electrical Engineering 278,
DOI: 10.1007/978-3-642-41407-7_34,

34.1 Introduction

A musical chord is a collection of many notes that are being played together. As a robust mid-level representation, chord sequences over time can be used to describe musical harmonic content and have been used in many applications such as musical information retrieval, music filtering, audio classification, etc., [1, 2].

Some progress has been made recently in chord transcription. By using symbolic data to train acoustic model, an extremely laborious task of human annotation of chord names and boundaries can be avoided. The acoustic feature vectors, information of chord names, and boundaries can be easily extracted from symbolic music files, thus make it possible to collect a large set of labeled training set with minimal amount of human labor.

Hidden Markov models (HMMs) have become a dominant methodology for speech recognition. But for chord recognition it was first introduced in [3, 4] and has been paid more attention since then. In those models, feature vectors extracted from musical audio were used as the feature observation and each chord is represented by a hidden state. The parameters for observation probability distribution and state transition probabilities are estimated in training stage, and in recognition stage, Viterbi search algorithm is used to decode an optimal chord sequence path for test musical data.

Many works [1, 5] have suggested the performance of HMMs have a strong dependency on training data styles and the amount of the training set. The model could perform well when the test data belongs to the same style with the training data and the training data set is large enough, otherwise, performance will degrade heavily. That means if we have no prior information about the style of the test data, we cannot choose its corresponding style-specific model to perform transcription; or if data in test set belong to different music styles from training set, we have to collect enough training data to train model for each style. However, collecting training data itself is a difficult task and it is impossible to some extent.

Until now, most works have treated "music style" here as "music genre," but "genre" is only a "coarse" categorization, even music with the same genre, they may belong to different "sub-genre," acoustic and chord transition characteristics may vary dramatically when they are composed by different artists and played with different instruments. Factors which lead to a certain genre or subgenre cannot be strictly defined, may involve emotion, timbre, etc. Also, as mentioned above, people cannot always get a large amount of training data for every genre of music data. Based on the above analysis, a genre-independent chord transcription system maybe more favorable and applicable, which take use of all training data of different genres, furthermore, unnecessary to get the genre-specific information of testing data in testing stage.

The work in this paper is motivated on resolving this problem. A genre-independent automatic chord transcription system is built, taking all styles of music data as a whole training set. Taking the variability of observation features into consideration, Gaussian mixture model (GMM) is used to model each state in HMMs. Then we adopt the method proposed in [6], the variations caused by timbre for chord transcription task is considered to be analogous to speaker variation in speech recognition tasks. The "trajectory folding phenomenon" caused by independence assumption of HMMs also exists and will lead to heavy performance degradation. Thus, a PLSA-based approach is used to deal with this problem for the general-dependent acoustic chord model. Besides, we also investigated what factor is most related with the latent topic produced by Probabilistic latent semantic analysis (PLSA) model.

The rest of the paper is organized as follows: Sect. 34.2 gives detailed about the feature used in our system. Section 34.3 describes the GMM-based HMMs and Sect. 34.4 explains how we adopt PLSA-based approach in HMMs. The experiments and results are presented in Sect. 34.5. Finally conclusion is given in the last section.

34.2 Feature Vectors

The most commonly used features for acoustic chord modeling are based on quantifying spectral components of 12 pitch classes. Its basic idea is to map the spectral components of a music signal into a chromatic scale, such as 12-dimensional pitch class profiles (PCP) introduced by [7], which has been widely used in most chord recognition systems. Christopher et al. [5] introduced a pitch space model, mapped the 12-D PCP to the interior space of a 6-D tonal centroid vector by projecting PCP on close harmonic relations, such as fifths and thirds. Since this paper studies identifying musical chords, tonal centroid is adopted as feature, which has been demonstrated more robust for identifying musical chords. Feature extractor and calculation could be divided into three steps: PCP calculation, PCP tuning, tonal centroid calculation.

PCP calculation. The first stage is the Constant Q transform, a frequency analysis similar to DFT, described in [8]. Different from DFT, the frequency domain channels are not linearly spaced but logarithmically spaced, corresponding to the frequency resolution of human ear.

Assuming $f_{\min}$ is the start point of the analysis, B is the number of bins per octave. The kth spectral component is defined as:

$$f_k = \left(2^{1/B}\right)^k f_{\min} \tag{34.1}$$

The spectral analysis is DFT-based, the constant Q transform X_{cq} of a temporal signal $x(m)$ is calculated as:

$$X_{cq}(k) = \sum_{n=0}^{N(k)-1} w(n,k)x(n)e^{-j2\pi f_k n} \tag{34.2}$$

where $w(k)$ is the analysis window, $N(k)$ is the window's length, both are functions of the bin position k. Once X_{cq} is computed, the PCP for a given frame can then be calculated as:

$$\text{Chroma}(b) = \sum_{m=0}^{M} |X_{cq}(b + m\beta)| \tag{34.3}$$

where $b \in [1, \beta]$ is the PCP bin number, and M is the total number of octaves in the constant Q spectrum. In this way, X_{cq} belong to the same pitch classes are summarized, present as the value of the pitch class in the PCP. In this paper, the raw audio is down sampled to 11.025 Hz, analysis is performed between $f_{\min} = 13.75$ Hz and $f_{\max} = 3.520$ Hz, so M is 8. The frame size is 8.192 samples and the hop size is 2.048 samples. Although there are 12 pitch classes, considering in real-world recordings are often not perfectly tuned, we set $\beta = 36$, do pitch class Profiles tuning in the following.

PCP tuning. In theory, concert pitch A4 is 440 Hz. But in real world, some instrument are not tuned perfectly, the tuning pitch could be different away from concert pitch. This could influent the estimation of chords, so PCP should be tuned. In this paper, the tuning algorithm presented by [5] is adopted. With the pitch class Profiles vector *Chroma*, a peak picking algorithm is first applied to obtain peaks of the pitch classes. Then, a quadratic interpolation is performed on these peak bins, obtain the position and magnitude of the peaks. A histogram is calculated over the length of the audio file and the maximum value in that histogram is taken as the center tuning value. With the position of the center tuning value, the ranges of all pitch class and the 12-d tuned pitch class Profiles TD_{Chroma} can be calculated.

Tonal centroid calculation. Chords have specific interval relations, details could find in [9]. Tonnetz is a harmonic network, a well-known planar representation of pitch relations. On the plane, the smaller the distances between chords are the closer harmonic relations they have. Based on the harmonic network, tonal centroid feature is designed to detect harmonic changes. With the Chroma vector *Chroma* the L1 norm is first done on the 12-D feature. Multiply TD_{Chroma} Vector by a transformation matrix Φ, get 6-D tonal centroid vector *Tonal*.

$$\text{Tonal}_n(i) = \frac{1}{\|\text{TD\_Chroma}\|} \sum_{j=0}^{11} \Phi(i,j)\text{TD\_Chroma}_n(j) \tag{34.4}$$

where

$$0 \le i \le 5,$$
$$0 \le j \le 11,$$
$$\Phi = [\varphi_0, \varphi_1, \ldots, \varphi_{11}],$$
$$\varphi_j = \begin{bmatrix} \Phi(0,i) \\ \Phi(1,i) \\ \Phi(2,i) \\ \Phi(3,i) \\ \Phi(4,i) \\ \Phi(5,i) \end{bmatrix} = \begin{bmatrix} r_1 \sin l \frac{7\pi}{6} \\ r_1 \cos l \frac{7\pi}{6} \\ r_2 \sin l \frac{3\pi}{2} \\ r_2 \cos l \frac{3\pi}{2} \\ r_3 \sin l \frac{2\pi}{3} \\ r_3 \cos l \frac{2\pi}{3} \end{bmatrix}$$

The values of r_1, r_2 and r_3 are 1, 1 and 0.5, corresponding to the radii of the fifths, minor thirds and major thirds circles.

Finally, we get the tonal centroid vector. With features for all frames and all musical audio, we could calculate the observation distribution for HMM.

34.3 GMM-Based HMMs for Chord Transcription

In the chord HMMs, we use 36 states HMMs, each state represents a single chord. Since augmented chords rarely appear in Western tonal music, we simply consider only major, minor and diminished chords for 12 pitch classes. Taking the variability of observation feature into consideration, single Gaussian for state output distribution is not adequate. Here, we model the observation distribution with GMMs. The mean vectors and covariance vectors of Gaussian mixtures are estimated from the features of the training data set. The transition probabilities between states correspond to transition probabilities from chord to chord can also be learned from the training data set.

The observation distribution. Traditionally, the observation distribution is modeled by a single multivariate Gaussian, defined by mean vector and covariance matrix of features. Due to different characters, the observation distributions belong to different styles show variances with each other.

In the chord HMMs, we use 36 states HMM, each state represent a single chords. Since augmented chords rarely appear in Western tonal music, we simply consider only major, minor, and diminished chords for 12 pitch classes. Considering the variability of observation feature, single Gaussian which previous work did is not adequate, we model the observation distribution by GMMs. The mean vectors and covariance vectors of Gaussian mixtures are estimated from the features of the training data set. The transition probabilities between states correspond to transition probabilities from chords to chords, can also be learned from the training data set.

Previous works have presented that music pieces belong to different genres show great variation in observation space, furthermore, the figures show that music pieces belong to the same genre still show obvious variation in observation features.

The transition probabilities. Since only 36 chords are considered and each chord is modeled by a state, in the ergodic HMMs, the transition probabilities matrix is 36 × 36 matrix $\Sigma_{36\times36}$, each element in this matrix $P(ij)$ represents the transition probability from the ith chord to the jth chord. Matrices of different genres show difference in the right and left third of matrices, especially the left. This is because in whatever genre, major and min chords are all transited to diminished chords with little probabilities. It shows that the matrix is strongly diagonal, that means chords have high probability of transiting to themselves. That because the frame in the experiment is longer than the chord's duration, so the chord and corresponding state does not change for several frames.

34.4 The PLSA-Based Approach

When we pool all the music data of different genres and train a general acoustic chord HMMs, great confusion is introduced. For two distinct music genres, the instrument classes and the chord transition characteristic are very different. So the confusability mainly lies in two aspects, i.e., the observation distribution and the transition probabilities. The proposed approach mainly deals with the former one. When we use GMMs as the state output distribution, HMMs describe trajectories inadequately due to the independence assumption. In speech recognition, this is called "trajectory folding phenomenon."

Figure 34.1 illustrates the so-called Trajectory folding phenomenon. Each chord name is modeled with a state and a state is modeled with two Gaussian pdfs (each circle represents a Gaussian). The HMMs has been trained with observations from two distinct music files, in which the instruments are different. It is reasonable to assume that, in each state, one Gaussian component corresponding to the observation distribution in one music file. The dashed lines represent the true trajectories for the two training music file. However, when presented with previously unseen input, the model may yield a high probability for path, which has never been observed in the training set. As a consequence, this trajectory folding phenomenon will result in performance degradation.

To solve this problem, we adopt the idea described in [6], apply it in chord transcription tasks, the acoustic variation caused by different instruments or instrument combination is analogs to speaker variations in speech recognition. In one music file, this c characteristic should be relatively stable. Thus, the PLSA-based approach which capture the latent factor information and enrich the trajectory modeling in HMMs will hopefully alleviate the confusion problem.

The PLSA-approach can be divided into training phase and decoding phase, in the training phase, the goal is to perform co-occurrence analysis between "dominant component" and music files and get the probabilistic model. In the decoding

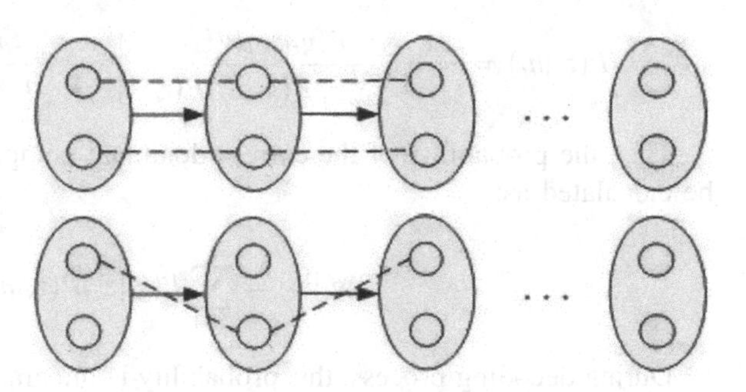

Fig. 34.1 An example to illustrate the trajectory folding phenomenon in acoustic chord HMMs

phase, the probabilistic model is integrated into one-pass decoding procedure. The detail is described as follows.

Training phase. The term "dominant component" is defined as in [6], given a frame of feature vector and its labeled state, the index k of the dominant component can be obtained as:

$$\arg\max_{k} w_k N(x;\ \mu_k,\ \Sigma_k) \tag{34.5}$$

In another word, the dominant component is the most likely Gaussian to generate the feature vector. Since the chord boundary information is in perfect alignment, the dominant component can be directly obtained from the training music data by calculating the probability of each frame with its corresponding state.

A co-occurrence matrix between Gaussian components and training music files can be built in which each row represents a Gaussian and each column represents a particular training music file. The matrix is a $M \times N$ matrix with $M = i \times j$ in which i is the total number of states in HMMs and j is the number of Gaussians per state. Each element in this matrix is the frequency of the Gaussian shows as dominant component in that particular training music file.

After PLSA performed on this co-occurrence matrix, the probabilities such as $P(z)$, $P(g|z)$, and $P(f|z)$ can be obtained. Here, f represents a training music file, g represents a Gaussian component, z can be considered as a topic or factor which lead to some specific characteristics in music.

Decoding phase. The probabilities obtained can be integrated into one-pass decoding procedure. It provides a measure of how the latent factors of the searching path consistent within a music file. The acoustic confusion can be effectively alleviated. The details are described as follows.

During decoding process, for the first dominant component to be evaluated, the distribution of the factor defaults to the distribution in the training data

$$P(z_k|h_1) = P(z_k) \tag{34.6}$$

Given the dominant component history h_i and current dominant component g_i, the distribution of latent speaker cluster can be adapted with

$$P(z_k|h_t) = \frac{1}{t+1} \frac{P(g_t|z_k)P(z_k|h_{t-1})}{\sum_{q=1}^{K} P(g_t|z_q)P(z_q|h_{t-1})} + \frac{t}{t+1} P(z_k|h_{t-1}) \tag{34.7}$$

Then the probability of the current dominant component given the history can be calculated as:

$$P(g_t|h_t) = \sum_{k=1}^{K} P(g_t|z_k)P(z_k|h_t) \tag{34.8}$$

During decoding process, this probability is integrated with the standard output probability density using a linear interpolation method as in equation (34.9) to constrain the searching path, making the searching path consistent with a latent speaker cluster, thereby the trajectory folding phenomenon could be alleviate.

$$\log p = (1 - \lambda) \log p(x|s_t) + \lambda P(g_t|h_t) \tag{34.9}$$

where $p(x|s_t)$ is the standard output probability density of current state t given the feature vector x.

34.5 Experimental Results and Analysis

Training and testing data. In order to train chord recognition models, large number of audio sets and corresponding chord labels are needed. Also, in order to validate the train data dependency phenomenon and prove the PLSA-based approach works, data sets should belong to different genres. In our experiments, MIDI files which belong to four different genres are used, details are show in Table 34.1.

Rock and Pop music MIDI files are collected from [10]. Rock music are all of the Beatles, Pop music belong to more than 200 singers. Classical music files are collected from [11], belongs to Bach, Beethoven, Handel, Haydn and so on.

Harmony analysis is done on MIDI files by the Melisma Music Analysis, available on [12]. It is a musical analyze system, could extract information like chord, key from MIDI files. Configure the system, make it generate chord labels every beat as ground truth. In the meantime, WAVE format audio which are down sampled to 11.025 Hz are synthesized from MIDI files by Timidity ++. Then tonal centroid features are extracted from audio. The windows length and hop size are 8.192 and 2.048 samples, respectively. With chord labels and features, transition

Table 34.1 The statistics of the training and test data (s)

Genre	Training data (s)	Test data (s)
Classical	99376.6	3056.5
Pop	60384.1	2654.1
Rock	27994.6	1194.4

Table 34.2 Chord recognition accuracy on different single Gaussian HMMs

Test data	Training data			
	Classical	Pop	Rock	All
Classical	61.92	56.33	59.46	60.78
Pop	64.44	56.33	62.54	63.17
Rock	65.84	44.65	68.27	62.12

probability matrix and parameters for Gaussian mixtures are estimated. Initial state probability for each state is 1/36, no prior to anyone.

Evaluation of general model. In order to examine the model's dependency on the training data, we train three genre-dependent models with classical, pop, and rock music files, respectively. A general model is also trained with all the training data. For all the models, single Gaussian output distribution is used. Cross-recognition experiments were performed, the frame-level recognition accuracies are show in Table 34.2.

From the results, we can observe that models perform different. For Rock and Classical's test data, best performance can be obtained when genre-dependent models are used. For the Pop's test data, the highest accuracy appear when recognized with classical. The result is consistent with the conclusion in [1] that the model trained on classical data is more robust to the change in musical genre of the test input. That is, the classical model performs equally well on all the test sets. This is partially because the model trained on the classical music have largest amount of training data and better to generalize. However, although there is also a large training set for test data for pop music, it gets worst performance when the pop music model parameter is used. That indicates genre is not a robust factor to arrange music files for chord recognition, even belong to a same genre, model parameters can also vary dramatically. In addition, even with largest amount of training data, the general model performs worse on all different test set than genre-dependent models. That is because great confusion is introduced when all the training data sets are used.

Evaluation of GMMs. It is a natural extension from single Gaussian to Gaussian mixtures as the form of output distribution when all the training data is used. We examined the performance of GMM-based HMMs for general model. Different numbers of Gaussian are evaluated. Table 34.3 shows the results on frame-level accuracy.

Table 34.3 Chord recognition accuracy for test data on different Gaussian HMMs

Test data	Number of Gaussions			
	1	4	8	16
Classical	60.78	65.03	62.56	59.29
Pop	63.17	65.85	62.74	59.11
Rock	62.12	65.87	61.17	56.96

When Gaussian number is 4, the general model performs best for different test data, so for the experiments below, number of Gaussian is set to be 4.

When Gaussian number further increases, as can be seen from Table 34.3, for the 8 and 16 Gaussian mixtures, performances degrades fast. This is partially because the amount of the training data is not enough, while another important reason is that "trajectory folding phenomenon" become serious due to the poor trajectory modeling capability of standard HMMs. This is also the reason why the 4 Gaussian mixture general model still can not outperform the genre-dependent models as shown in Tables 34.2 and 34.3. Thus, the proposed PLSA-based approach can be applied to solve this problem and improve the performance, implementing a more applicable system for genre-independent chord transcription task with the general model.

Evaluation of the number of topics. We evaluated the PLSA-based approach. The Gaussian number is fixed as 4, the probabilities such as $P(z)$, $P(f|z)$, and $P(f|z)$ were obtained and integrated into decoding process as in Eqs. (34.6–34.9). Detail is described as in Sect. 34.4. First, we perform experiments to choose an appropriate number of latent topics. When the number of topic is 1, it degenerates to standard decoding procedure. Here, the interpolation weight is set to be 5. According to the results shown in Table 34.4, we set the topic number to be 5 in all the following experiments.

Evaluation of interpolation weight. The fourth experiment is to choose the interpolation weight λ in Eq. (34.9). The experimental results are given in Fig. 34.2.

As the figure shows, when the interpolation weight λ is small, recognition accuracy improves for all the three test set as λ increases, the best performance is obtained when λ is between 0.3 and 0.5. Based on the results, We fixed the interpolation weight to be 0.3.

Comparison with genre-dependent models. With the topic number fixed to be 5 and the interpolation weight fixed to be 3. We compared our proposed genre-independent system (Ind*) with the genre-dependent system (Dep). The results when performing a standard decoding with a general model without the PLSA-based approach are also given (Indep). Recognition results are given in Table 34.5. Here, detailed accuracy for each music file is given.

We can observe that our proposed genre-independent system, with a general model and proposed PLSA-approach performs better than a simple use of general model, most of the recognition accuracy even higher than the genre-dependent

Table 34.4 Chord recognition accuracy for test data with different topic number on universal HMM

Test data	Number of topics			
	1	3	5	7
Classical	62.88	66.63	66.73	66.57
Pop	63.81	66.00	66.57	65.99
Rock	64.74	68.81	70.18	70.52

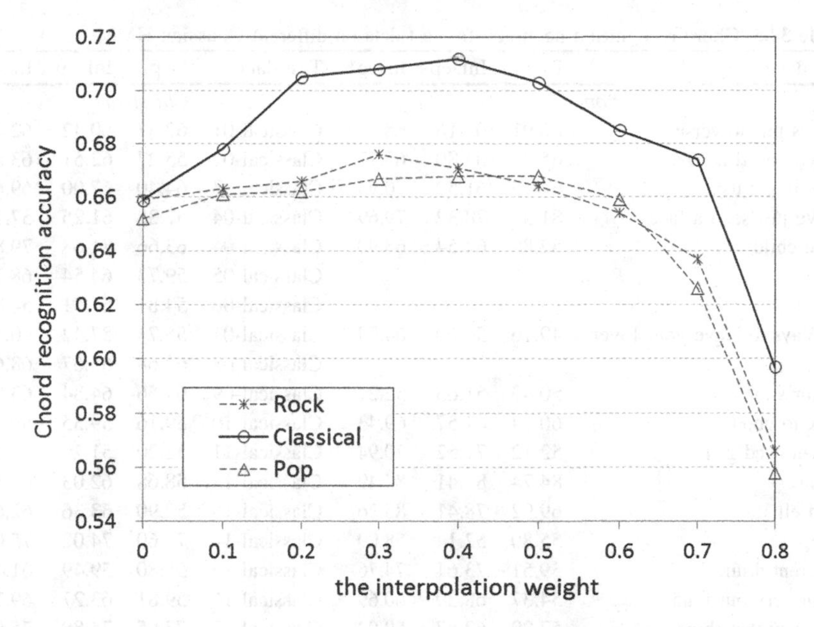

Fig. 34.2 Chord recognition accuracy on the three test set with different interpolation weight

system. The PLSA-based approach has brought a significant improvement as we expected, alleviate the trajectory folding phenomenon in chord HMMs and genre dependency problem.

34.6 Analysis of the Topics in PLSA-Based Approach

It is interesting to investigate what the true factor is best relevant with the topic obtained in the proposed PLSA-based approach. As we have assumed, the acoustic variation caused by different instrument is analogs to speaker variation in speech recognition task. So we do an analysis about the relation between instrument and topics.

The joint probabilities between each two pieces of music in the training set can be calculated with the probabilities as below,

$$P(f_1,f_2) = \sum_z P(f_1|z)P(f_2|z)P(z) \tag{34.10}$$

In musical MIDI files, most of them have several program names, such as piano, Celesta, or violin, then corresponding audio files are played by several instruments. The similarity in timbre between two music files can be measured by the number of the same instruments. If instrument is the crucial factor that affects

Table 34.5 Chord recognition accuracy for test data on different Gaussian HMMs

Test data	Dep	InDep	Indep*	Test data	Dep	InDep	Indep*
Pop				*Classical*			
Across the universe	66.91	63.16	68.11	Classical-01	62.37	60.42	62.87
All you need is love	65.77	63.80	65.77	Classical-02	55.27	62.51	63.83
Back in the ussr	64.45	61.52	70.95	Classical-03	64.40	67.00	69.60
I have just seen a face	81.32	78.32	79.69	Classical-04	57.38	61.25	57.19
Little child	62.88	62.54	68.90	Classical-04	63.66	81.68	79.87
Rock				Classical-05	59.73	64.54	68.13
				Classical-06	54.81	54.51	54.81
50 Ways to leave your lover	49.16	54.60	64.09	Classical-07	55.74	57.42	60.77
				Classical-08	62.68	65.67	68.65
Arthur's theme	50.49	51.63	52.21	Classical-09	64.56	64.84	63.06
Back to good	60.04	64.57	69.43	Classical-10	59.16	59.35	59.45
Brown eyed girl	52.12	71.52	70.94	Classical-11	52.20	51.29	55.45
China	84.74	83.41	84.49	Classical-12	58.68	62.03	64.12
Cord ell	69.92	78.41	83.26	Classical-13	52.90	63.36	62.67
Creep	55.80	52.14	58.09	Classical-14	71.60	74.02	75.00
Different drum	39.51	73.61	74.76	Classical-15	55.80	59.49	61.47
Do not cry out loud	54.37	68.39	60.69	Classical-16	59.81	63.27	69.27
Doo wop that thing	57.29	62.87	59.02	Classical-17	73.95	74.88	75.69
Do you know the way to san	48.54	52.04	52.33	Classical-18	64.48	67.46	68.91
				Classical-19	53.26	65.85	67.31
Morning glory	33.22	50.74	58.65	Classical-20	82.29	80.20	82.29
No frontiers	77.12	92.08	90.77	Classical-21	59.78	61.30	65.08
				Classical-22	79.76	73.31	79.47

the topics in our PLSA-based approach. Two music files which have more same instruments should result in higher joint probability.

We extract instrument information from training MIDI files and statistic the number of same instruments each two audio files have. The joint probability of

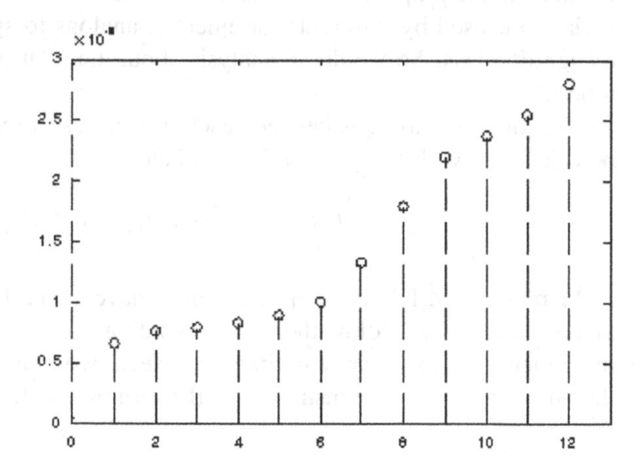

Fig. 34.3 The average joint probability of each two music files which have certain number of same instruments

each two audio files were calculated, the average joint probability of each pair which have certain number of the same instruments is presented in Fig. 34.3.

As can be seen in Fig. 34.3, the more same instruments two music files have, the higher average joint probability is obtained. This result indicates that the topics in our PLSA-approach do capture the timbre characteristics in music

34.7 Conclusion

In this paper, we describe a new method to build a genre-independent acoustic chord transcription system. Symbolic data have been used to produce the training audio wave files and chord transcriptions in order to avoid human labors. Music files of three genres are collected, which are Classical, Pop, and Rock. First, a general model is trained. Ergodic HMMs are adopted and we experimented the extension from single Gaussian model to Gaussian mixture models for modeling each state in HMMs. Then we propose a PLSA (probabilistic latent semantic analysis) based approach to deal with the acoustic variations. Experimental results show that by the proposed PLSA-based approach, our genre-independent chord transcription system, although recognizing without any genre-specific information of testing data, has obvious improvement compared with a genre-dependent system.

Further analysis is also made to verify our assumption that the topics in our PLSA-based approach is most related with music timbre caused by different instruments.

Acknowledgments The work was supported in part by the National Natural Science Foundation of China (60435010; 60535030; 60605016), the National High Technology Research and Development Program of China (2006AA01Z196; 2006AA010103)

References

1. Lee K, Slaney M (2008) Acoustic chord transcription and key extraction from audio using key-dependent HMMs trained on synthesized audio. IEEE Trans Audio Speech Lang Process 16(2):291–301
2. Orio N (2006) Music retrieval: a tutorial and review. In: Oard DW (ed) Foundations and trends in information retrieval, vol 1. Now Publishers Inc, vancouver, pp 1–90
3. Raphael C (2002) Automatic transcription of piano music. In: Proceedings of ISMIR, pp 15–19
4. Sheh A, Ellis DPW (2003) Chord segmentation and recognition using em-trained hidden markov models. In: Proceedings of ISMIR 2003, pp 185–191
5. Harte C, Sandler M, Gasser M (2006) Detecting harmonic change in musical audio. In: Proceedings of the 1st ACM workshop on audio and music computing multimedia. ACM, Santa Barbara, pp 21–26

6. Su D, Wu X, Chi H (2007) Probabilistic latent speaker analysis for large vocabulary speech recognition. In: Proceedings of interspeech, Antweerp, Belgium, pp 27–31
7. Fujishima T (1999) Realtime chord recognition of musical sound: a system using common lisp music. In: Proceedings of the international computer music conference, pp 464–467
8. Brown JC (1991) Calculation of a constant q spectral transform. J Acoust Soc Am 89:425–434
9. Cohn R (1998), Introduction to neo-riemannian theory: a survey and a historical perspective. J Music Theory 42(2):167–180
10. Midi DB: free MIDI tracks and karaoke song files. http://www.mididb.com. Accessed 7 May 2013
11. Muse data: an electronic library of classical music scores. http://www.musedata.org/. Accessed 6 June 2013
12. The melisma music analyzer. http://www.link.cs.cmu.edu/music-analysis. Accessed 6 June 2013

Chapter 35
Multiclass Color-Texture Image Segmentation Based on Random Walks Framework Integrating Compact Texture Information

Chanchan Qin, Guoping Zhang, Guoqing Li, Liu Chen and Jing Ge

Abstract In this paper, we propose an interactive multiclass color-texture image segmentation method. A new feature descriptor is designed by using the covariance matrices of coordinates, color with compact texture information and then integrated into random walks method to obtain the segmentation result. In this paper, we use multiscale nonlinear structure tensor (MSNST) to describe the texture feature of an image. Since the MSNST matrices set have different feature structures from color and coordinate vector, they cannot be used to construct covariance matrices directly. To address this problem and obtain the compact texture information simultaneously, we use the Isometric Mapping (Isomap) dimensionality reduction techniques for each scale of MSNST in tensor space. Experiments using synthesis texture images and real natural scene images demonstrate the superior performance of our proposed method.

Keywords Interactive image segmentation · Random walks · Texture information · Dimensionality reduction

35.1 Introduction

How to partition an image into a set of nonoverlapping regions under the complex background environment effectively and accurately has become an important and difficult problem in the field of computer vision and image analysis. It has found a

C. Qin (✉) · G. Zhang · L. Chen · J. Ge
The College of Physical Science and Technology, Central China Normal University, Wuhan 430079, Hubei, China
e-mail: qinchanchan@163.com

G. Li
The College of Mathematics and Computer Science, Bijie University, Bijie 551700 Guizhou, China

A. A. Farag et al. (eds.), *Proceedings of the 3rd International Conference on Multimedia Technology (ICMT 2013)*, Lecture Notes in Electrical Engineering 278, DOI: 10.1007/978-3-642-41407-7_35,

wide range of applications, such as target detection [1], medical image processing [2], image retrieval [3], pattern recognition [4], etc. In the past few decades, many methods have been proposed to segment the color image. There are two main segmentation categories: automatic methods and interactive methods.

Automatic approaches [5–7] provide segmentation results without any prior knowledge about the image and do not require any user interaction. Due to the amount of information contained in images and their unpredictable complexity, automatically partitioning a natural image into significative regions to represent distinct scenes is an open problem and until now still is not effectively solved.

Correspondingly, interactive image segmentation method, which allows the users integrate their empirical knowledge and subjective requirements to obtain segmentation results through simple and intuitive user interaction, has been developed in recent years very quickly. This technique can extract the target object more accurately and seems more flexible and practical in the applications of natural scene images. In the last few decades, a lot of useful interactive image segmentation methods have been proposed. Such as intelligent scissors [8], snakes [9], Graph cuts [10], Lazy Snapping [11], GrabCut [12]. These approaches have attracted great attention due to their high efficiency and accuracy, but they all for binary segmentation of image and can only interactively distinguish the object from the background.

Grady et al. [13] have proposed a multiclass interactive image segmentation method named random walks. This method considers the image as a weighted graph. Nodes of this graph correspond to pixels in the image and edges are placed between nearby pixels. The edge weights are treated as probabilities of a particle at one node traveling to a neighboring node. Given a small number of pixels with user-defined labels, one may compare the probability that a particle at any unlabeled pixel travels first to the foreground or background seeds and then assign the pixel the same label. It shows that the random walks algorithm finds an exact global energy minimum and produces high-quality, visually pleasing results, especially in the presence of weak boundaries and noise [13, 14].

In this paper, we investigate an interactive multiclass color-texture segmentation method based on random walks method. Since color-texture image contains rich color information and complex texture patterns, with the traditional multiclass interactive random walks image segmentation methods only rely on color information is usually difficult to obtain satisfactory segmentation results. Texture information is often a more appropriate discriminating feature and commonly used in the field of image analysis. It describes various physical properties of the surface area, including the perceived brightness, uniformity, density, roughness, etc. It reflects not only the spatial distribution of grayscale images, but also contains information about the image surface with the surrounding environment. Han et al. [15] proposed to exploit the multiscale nonlinear structure tensor (MSNST) to describe the texture feature of images. Although this method has both the omni-directional compression description ability of structure tensor and the powerful description ability of Gabor wavelet transform in scale space, the segmentation process of [15] treats the color and texture features separately, thus, it may ignore the intrinsic relationship between different features. To address this problem, we design a new strong feature descriptor for local pixel neighborhoods by using covariance matrices of low-level features as proposed by [16, 17]. Additionally, to construct covariance matrices of different feature structures directly and obtain the compact texture information simultaneously, we use the dimensionality reduction techniques for each scale of MSNST in tensor space.

The rest of the paper is organized as follows: Sect. 35.2 introduces our interactive multiclass color-texture image segmentation method in detail. Section 35.3 presents a number of experimental results. Finally some conclusion remarks are drawn in Sect. 35.4.

35.2 Interactive Multiclass Color-Texture Image Segmentation

In this section, we first briefly review the random walks image segmentation algorithm [13] (Sect. 35.2.1). Following, the new strong color-texture feature descriptor is introduced in detail (Sect. 35.2.2).

35.2.1 Random Walks Segmentation

For the Random walks image segmentation [13], an image is represented as an undirected graph, where each vertex corresponds to a pixel of an image and each edge interconnect neighboring pixels. The weight of an edge e_{ij} is denoted as w_{ij}.

It is common to use the edge weighting function

$$w_{ij} = \exp\left(-\beta\left(g_i - g_j\right)^2\right) \tag{35.1}$$

Where g_i indicates the image feature at pixel i. The value of β represents the only free parameter in this algorithm.

Random walks algorithm was motivated by placing random walkers at pixels and noting which they first arrive at. The problem is exactly has the solution as the Dirichlet problem [18].

This method can interactive partition an image into multiclass different regions and has been successfully applied in the field of image segmentation. However, computing the color values simply may not be enough to obtain satisfaction segmentation results. Image segmentation with various features can greatly improve the segmentation performance. To enhance the feature description ability, a new strong color-texture feature descriptor for local pixel neighborhoods was introduced by fusing coordinate, color, and compact texture information through covariance matrices in the following section.

35.2.2 Construction of Feature Descriptor

In [15], Han et al. proposed to exploit the (MSNST) to describe the texture feature of images. Compared with the traditional Gabor wavelet [19], (MSNST) simplifies the descriptive ability in scale space. Additionally, this method takes into account the omni-directional texture of structure tensor at the same time.

Given the image I, multiscale structure tensor (MSST) can be obtained by using the nonorthogonal (redundant) discrete wavelet frameworks [20]. Following, the technology of nonlinear diffusion filtering has been adopted, which can smoothing the noises and enhance the edges simultaneously, to obtain MSNST. For most of the natural image, the texture information is mainly focused on a few scales and the overlarge S will include too much redundant and meaningless information. In this paper, we only use the first two scales, which $s = 0$ and $s = 1$, in feature description.

Since the MSNST matrices set have different feature structures from color and axis vector, they cannot be used to construct covariance matrices directly. To address this problem and obtain the compact texture information simultaneously, we use the dimensionality reduction techniques for each scale of MSNST in tensor space. Considering that MSNST matrices do not lie in a Euclidean space, traditional linear techniques, such as Principal Component Analysis (PCA) [21] is difficult to capture the intrinsic geometry. Therefore, we map the MSNST matrices from the tensor space to a low-dimensional space by adopting the Isometric Mapping (Isomap) algorithm [22]. This nonlinear method estimated the geodesic distances $d_M(i, j)$ between all pairs pixel i, j from N_p MSNST matrices on the manifold M by computing their shortest path distances $d_G(i, j)$ in the graph G over each scale of MSNST matrices, and then applied classical MDS [23] to the matrix of graph distances $D_G = \{d_G(i, j)\}$ to obtain the d-dimensional coordinate as M_s. Here, we set $d = 2$. Therefore, to represent the multiscale texture information with vector representation, we can connect the M_s at two scales together to obtain a 4-dimensional vector as $x = (M_0, M_1)$.

Following, we use a 9-dimensional feature vector f by integrating the coordinates, compact texture information, and the Lab color (for Lab, it was shown to be approximately perceptually uniform) for constructing covariance matrices which is defined as:

$$f = [x\,y\,L\,a\,b\,\mathrm{M}_0(1)\mathrm{M}_1(1)\,\mathrm{M}_1(2)] \tag{35.2}$$

Where x and y are the normalized pixel coordinates, L, a and b are the pixel values of the Lab color space and $M_0(1)\,M_0(2)\,M_1(1)\,M_1(2)$ are the corresponding first and second scale compact texture information.

Then, we define a fixed neighborhood size $N \times N$ for every pixel and calculate the symmetric 9×9 covariance matrix as in [17]. Since covariance matrices are symmetric, we get a 36-dimensional feature vector per pixel containing color and compact texture information of local neighborhood. Note that such local covariance matrices of pixel features can be calculated very efficiently by integral images

[16]. Taking into account the Riemannian structure of the covariance matrices, the distances of two covariance matrices can be measured in manifold space as in manifold space as in [24].

35.3 Experiments

In our experiments, we provide a substantial of synthesis color-texture images and real natural scene images for comparison, and these images are all from the natural color-texture database MIT VisTex [25], Berkeley segmentation database BSD300 [26].

Our segmentation method was implemented in MATLAB. Notice that the sources codes of the popular random walks implementations were downloaded from the author's websites [27].

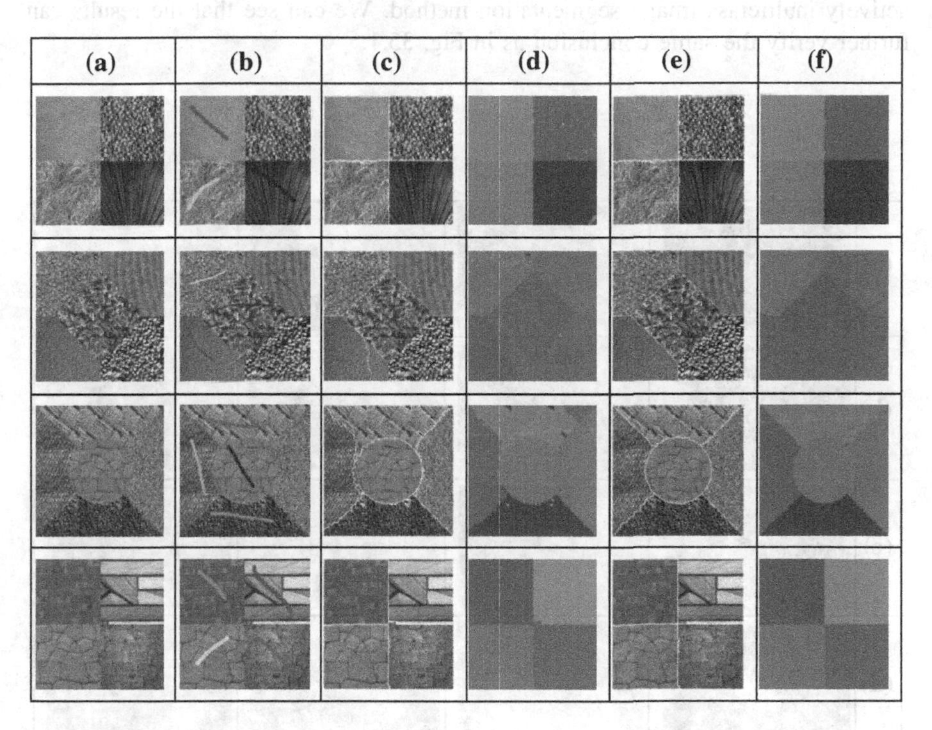

Fig. 35.1 Comparison of segmentation results for the synthesis color-texture images based on the MIT VisTex database. **a** Original image. **b** The original image and the scribbles. **c** The original image and the boundaries (use color feature only). **d** the result denoted by the average color of the regions (use color feature only). **e** the original image and the boundaries (use our proposed feature descriptor). **f** The result denoted by the average color of the regions (use our proposed feature descriptor)

There are a large number of parameters that must be appropriately predefined for the implementation of the proposed method. Parts of the default values for these parameters have been given when we described the corresponding algorithms. For the point of clarity and integrity, we give the description of the parameters setting again. The implementation of the multiscale structure tenor is the same as in [15]. Consider that large-scale does not work well in feature description. We set the scale number S of extracted multiscale structure tensor in Formula (35.1) as 2. The fixed neighborhood size N is fixed as 15.

In Fig. 35.1, we test some synthesis color-texture images based on MIT VisTex [25] database. The input by user includes multiple groups of scribbles which are distinguished by their colors. One group of scribbles with the same color represent one image region, and the number of the different colors of the scribbles means how many classes the user is going to partition the images into. From the comparisons of segmentation results in Fig. 35.1, we can see that our approach can obtain excellent results.

In Fig. 35.2, we select some testing natural images from the Berkeley segmentation database BSD300 [26] to test the performance of the proposed interactively multiclass image segmentation method. We can see that the results can further verify the same conclusion as in Fig. 35.1.

Fig. 35.2 Segmentation results for natural images on the Berkeley segmentation database BSD300 [26]. **a** Original image. **b** The original image and the scribbles. **c** The original image and the boundaries (use our proposed feature descriptor). **d** The result denoted by the average color of the regions (use our proposed feature descriptor)

35.4 Conclusion

An interactive multiclass color-texture image segmentation method integrating a new feature descriptor based on the random walks framework is presented to achieve an improved segmentation performance. We propose to use the covariance matrices of coordinates, color, and compact texture information to describe the feature of images. Consider that the MSNST matrices set have different feature structures from color and coordinate vector, we use the Isometric Mapping (Isomap) dimensionality reduction techniques for each scale of MSNST in tensor space. The comparison experiments demonstrate that the proposed method can achieve more superior interactive multiclass image segmentation results according to the simply input scribbles with different colors by user.

References

1. De Grandi GD, Lee J-S, Schuler DL (2007) Target detection and texture segmentation in polarimetric SAR images using a wavelet frame: theoretical aspects. IEEE Trans Geosci Remote Sens 45(11):3437–3453
2. Tu Z, Bai X (2010) Auto-context and its application to high-level vision tasks and 3D brain image segmentation. IEEE Trans Pattern Anal Mach Intell 32(10):1744–1757
3. Akakin HC, Gurcan MN (2012) Content-based microscopic image retrieval system for multi-image queries. IEEE Trans Inf Technol Biomed 16(4):758–769
4. Yue Y, Shi Q, Hu G, Wang Ja (2011) A composed statistical pattern recognition and geosciences analysis approach for segmentation-based remotely sensed imagery classification. International conference on geoinformatics
5. Shi J, Malik J (2000) Normalized cuts and image segmentation. IEEE Trans Pattern Anal Mach Intell 22(8):888–905
6. Paragios N, Osher S (2003) Geometric level set methods in imaging, vision, and graphics. Springer, New York
7. Bresson X, Esedoḡlu S, Vandergheynst P, Thiran J-P, Osher S (2007) Fast global minimization of the active contour/snake model. J Math Imaging vis 28(2):151–167
8. Mortensen EN, Barrett WA (1998) Interactive segmentation with intelligent scissors. Graph Models Image Process 60(5):349–384
9. Kass M, Witkin A, Terzopoulos D (1988) Snakes: active contour models. Int J Comput Vision 1(4):321–331
10. Boykov YY, Jolly M-P (2001) Interactive graph cuts for optimal boundary and region segmentation of objects in N-D images. Proceedings of international on conference on computer vision
11. Li Y, Sun J, Tang C-K, Shum H-Y (2004) Lazy snapping. ACM Trans Graph 23(3):303–308
12. Rother C, Kolmogorov V, Blake A (2004) Grabcut: interactive foreground extraction using iterated graph cuts. ACM Trans Graph 23:309–314
13. Grady L (2006) Random walks for image segmentation. IEEE Trans Pattern Anal Mach Intell 28(11):1768–1783
14. Grady L, Funka-Lea G (2004) Multi-label image segmentation for medical applications based on graph-theoretic electrical potentials. Computer vision and mathematical methods in medical and biomedical image analysis. Springer, New York, pp 230–245

15. Han S, Tao W, Wang D, Tai X, Wu X (2009) Image segmentation based on grab cut framework integrating multiscale nonlinear structure tensor. IEEE Trans Image Process 18(10):2289–2302
16. Porikli F, Tuzel O, Meer P (2006) Covariance tracking using model update based on lie algebra. IEEE Computer society conference on computer vision and pattern recognition
17. Donoser M, Urschler M, Hirzer M, Bischof H (2009) Saliency driven total variation segmentation. IEEE 12th international conference on computer vision
18. Kakutani S (1945) Markov processes and the Dirichlet problem. Proc Jap Acad 21:227–233
19. Manjunath BS, Ma W (1996) Texture features for browsing and retrieval of image data. IEEE Trans Pattern Anal Mach Intell 18(8):837–842
20. Mallat S (1999) A wavelet tour of signal processing. Academic press, San Diego
21. Jolliffe IT (1986) Principal component analysis. Springer, New York
22. Tenenbaum JB, De Silva V, Langford JC (2000) A global geometric framework for nonlinear dimensionality reduction. Science 290(5500):2319–23
23. Borg I, Groenen PJ (2005) Modern multidimensional scaling: theory and applications. Springer, New York
24. Förstner W, Moonen B (1999) A metric for covariance matrices. Quo vadis geodesia. 113–128
25. http://vismod.media.mit.edu/vismod/imagery/VisionTexture/vistex.html. Accessed 10 Jul 2013
26. Martin D, Fowlkes C, Tal D, Malik J (2001) A database of human segmented natural images and its application to evaluating segmentation algorithms and measuring ecological statistics. IEEE international conference on computer vision
27. http://cns.bu.edu/~lgrady/random_walker_matlab_code.zip. Accessed 10 Jul 2013

Chapter 36
Statistical Modeling of Speech Spectra in the Fan-Chirp Transform Domain

Sichen Zheng, Hongwei Wu, Qingyuan Xu and Yibiao Yu

Abstract The fan-chirp transform is a transform method that matches the characteristics of the speech signal. We use the curve fitting tool to study the probability distribution of speech spectra obtained by the fan-chirp transform in order to apply the results to the statistical model-based speech processing. The experimental results demonstrate that the clean speech spectra are best described with Gamma distribution for the real part, imaginary part, and amplitude. For the white noise, the real part and imaginary part of speech spectra are best described with the Laplacian model while the amplitude is best modeled with the Gamma distribution. In other noisy cases, the real part, imaginary part, and amplitude of spectra are all best described with Gamma distribution. The phase spectrum is a nonuniform distribution for the clean speech while it is uniform for the noisy speech.

Keywords Curve fitting · Fan-chirp transform · GOF (Goodness-of-Fit) test · Speech modeling

36.1 Introduction

Methods based on the statistical model of speech have been widely employed in speech processing applications, especially in speech enhancement, for several decades. The successful minimum mean square error (MMSE) method [1] assumes that the spectral components obey the Gaussian model thus the amplitude for each signal spectral component is Rayleigh distributed. And later it is extended to the Gamma [2], super Gaussian [3], and Laplacian [4] model. The corresponding Maximum A Posterior (MAP) estimators are obtained from super Gaussian [5] and

---

S. Zheng · H. Wu (✉) · Q. Xu · Y. Yu
School of Electronics and Information, Soochow University, Suzhou 215006, China
e-mail: wuhwei@suda.edu.cn

A. A. Farag et al. (eds.), *Proceedings of the 3rd International Conference on Multimedia Technology (ICMT 2013)*, Lecture Notes in Electrical Engineering 278,
DOI: 10.1007/978-3-642-41407-7_36,

generalized Gamma [6] model. These works are in the discrete Fourier transform (DFT) domain. Gazor et al. conducted some research on the speech modeling in the discrete cosine transform (DCT) and the Karhunen–loeve transform (KLT) domain[7, 8] and applied the results in speech enhancement [9, 10]. In any way, there is controversy about what the true distribution of speech spectral components is. However, the best fitting model will tend to gain the best results in the application such as speech enhancement.

The fan-chirp transform (FChT) [11, 12] is a new time frequency analysis method, which is more advantageous over the Fourier transform (FT) for harmonically related signals such as human speech. Based on our previous work on FChT [13], we are interested in the distribution model of speech spectra in the FChT domain. We collect data of the real part, imaginary part, spectral amplitude, and phase for clean and noisy speech spectra at several signal-to-noise ratios (SNRs). The curve fitting is carried out for three models: Gaussian, Laplacian, and Gamma. And several goodness-of-fit (GOF) tests are conducted: sum of squares due to error (SSE) test, Pearson's Chi squared test, and Kolmogorov–Smirnov (KS) test. The results show that the real part, imaginary part, and amplitude of clean speech spectra in the FChT domain can be best described with Gamma distribution while its phase is not the case of uniform distribution as postulated in existing researches in the field of speech enhancement. The GOF test results also show that in the white noise case, Laplacian is preferred for the real part and imaginary part and that Gamma is preferred for the amplitude. For the other noises, the spectral amplitude can be best described with the Gamma model and if neglecting the KS statistics the real and the imaginary part are also Gamma distributed. The noisy phase follows a uniform distribution.

The paper is organized as follows. The fan-chirp transform is briefly described in Sect. 36.2. The modeling and graphical comparison are presented in Sect. 36.3. Several GOF tests and their results are discussed in Sect. 36.4. Finally the conclusion and future work are presented in Sect. 36.5.

36.2 Fan-Chirp Transform

In this section, we briefly introduce the concept of FChT. Considering the fact that speech signal is processed frame wise with a finite duration, we use the following definition of FChT for a signal $x(t)$, centered at the origin, with duration T:

$$X(f, \alpha) = \int_{-\frac{T}{2}}^{\frac{T}{2}} x(t)\sqrt{|\phi_{\alpha}'(t)|}\mathrm{e}^{-j2\pi f \phi_{\alpha}(t)}\mathrm{d}t \qquad (36.1)$$

and the function $\phi_\alpha(t)$ is defined as:

$$\phi_\alpha(t) = \left(1 + \frac{1}{2}\alpha t\right)t \tag{36.2}$$

The signal $x(t)$ is non-zero only during the interval of $-T/2 \le t \le T/2$. In order to recover $x(t)$ from $X(f, \alpha)$, the chirp rate α must be constrained to $|\alpha| < 2/T$.

The fast computation of FChT can be implemented with the fast Fourier transform (FFT) by warping the time axis. We denote $\Psi_\alpha(\tau) = \phi_\alpha^{-1}(t)$ as the inverse of $\phi_\alpha(t)$. Since $\phi_\alpha(t)$ is quadratic, the solution of interest is:

$$\Psi_\alpha(\tau) = -\frac{1}{\alpha} + \frac{\sqrt{1 + 2\alpha\tau}}{\alpha}. \tag{36.3}$$

With this variable substitution, the FChT becomes:

$$X(f, \alpha) = \int_{\phi_\alpha(\frac{-T}{2})}^{\phi_\alpha(\frac{T}{2})} \tilde{x}(\tau)\tilde{\rho}(\tau)e^{-j2\pi f\tau}d\tau, \tag{36.4}$$

where $\tilde{x}(\tau) = x(\Psi_\alpha(\tau))$ is a time-warped version of $x(t)$ and $\tilde{\rho}(\tau) = 1/\sqrt[4]{1 + 2\alpha\tau}$ is a scaling function on the time-warped axis. The samples of $\tilde{x}(\tau)$ are readily interpolated from the available sampling points of $x(t)$ on the original time axis t. Thus the formula (36.4) can be implemented using the FFT in the discrete time.

The outstanding characteristic of FChT is that it can project the signal along many related directions in the time–frequency plane for a harmonic signal. The spectrum width under FChT is independent of the frequency variation rate in that it projects the signal along the directions of frequency variation while FT projects the signal onto the frequency axis. Therefore its advantage over FT becomes more obvious for rapidly time-varying spectral components. The FChT spectrum is more energy compact than the FT spectrum. For space limitation, the illustration of spectrum comparison is omitted.

36.3 Modeling and Graphical Comparison

Three models are used for speech modeling in the FChT domain, which are the most-common models: Gaussian, Laplacian, and Gamma. Their pdfs are as follows:

- Gaussian: $f_X(x) = \left(1 \middle/ \sqrt{2\pi b^2}\right) \exp((-x^2/2b^2)$,
- Laplacian: $f_X(x) = (1/2b) \exp(-|x|/b)$,
- Gamma: $f_X(x) = |x|^{a-1} \exp(-|x|/b)/(\Gamma(a)b^a)$

where a, b are constants, $\Gamma(\bullet)$ denotes the Gamma function.

Speech material is taken from www.dailywav.com and sampled at 8 kHz, with the length of 20 s equally taken from four female and four male speakers. Six noises from the NOISEX-92 database [14] are white, babble, f16 fighter, factory I, factory II, and Volvo car. The noisy speeches are made by mixing the clean speech and noises at various SNRs. Each signal frame is 512 points with a shift of 64 points and is hamming windowed. The chirp rate for FChT is estimated from the clean speech using the mean shift algorithm [15]. In the FChT domain, the real part, imaginary part, amplitude and phase of speech spectra are separately modeled for the clean and noisy cases at several SNRs. Their statistical intervals in the histogram are 0.01. Optimization of model parameters is obtained by the curve fitting toolbox in MATLAB, whose algorithm is based on the maximum likelihood criterion. The outputs of curve fitting toolbox are the fitting curves, model parameters and SSE, etc.

Figure 36.1 shows the fitting curves. (a) is for the real part of FChT spectra of the clean speech, and (b) is its zoom into the interval 0.0–0.3. It can be clearly seen that the Gamma model is preferred to the other two models. The imaginary part has a similar result and its illustration is omitted. As for the white noise case, the real part of noisy spectra is best modeled with the Laplacian distribution and the illustration is also omitted. The other two graphs in Fig. 36.1 are about amplitude spectra. (c) is the zoom into the amplitude spectra during the interval 0–1.0 for the clean speech, and (d) is for the noisy signal at 10 dB SNR with the white noise. Obviously, the Gamma model is the best choice for amplitude spectra of both the clean and noisy speech. It should be noted that the center is shifted to the peak of distribution and a multiplying factor is used for the other two models when fitting amplitude spectra.

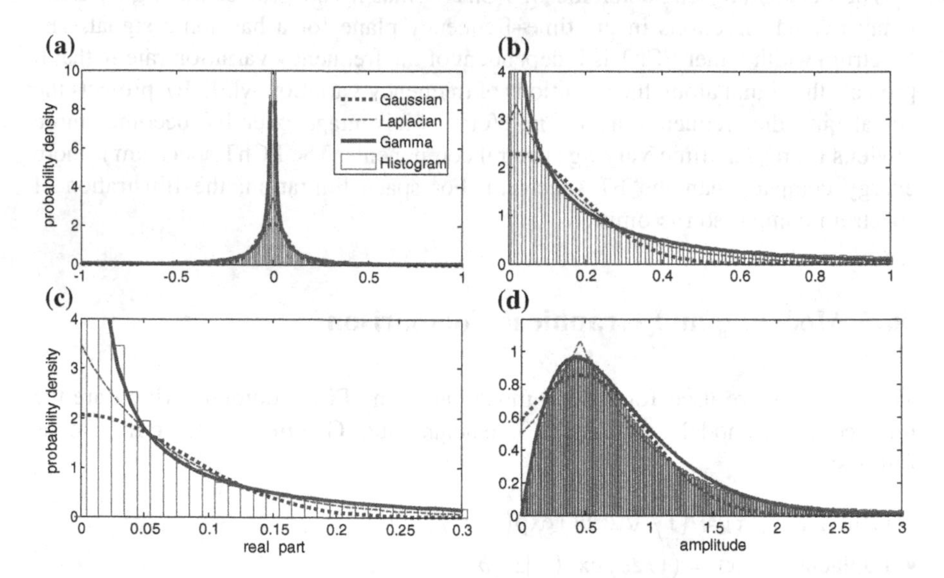

Fig. 36.1 Histograms of FChT spectra and three fitting curves. **a** whole look of the real part fitting, **b** zoom into the interval 0–0.3 of (a), **c** amplitude of clean speech, **d** amplitude of noisy speech at 10 dB SNR, white noise

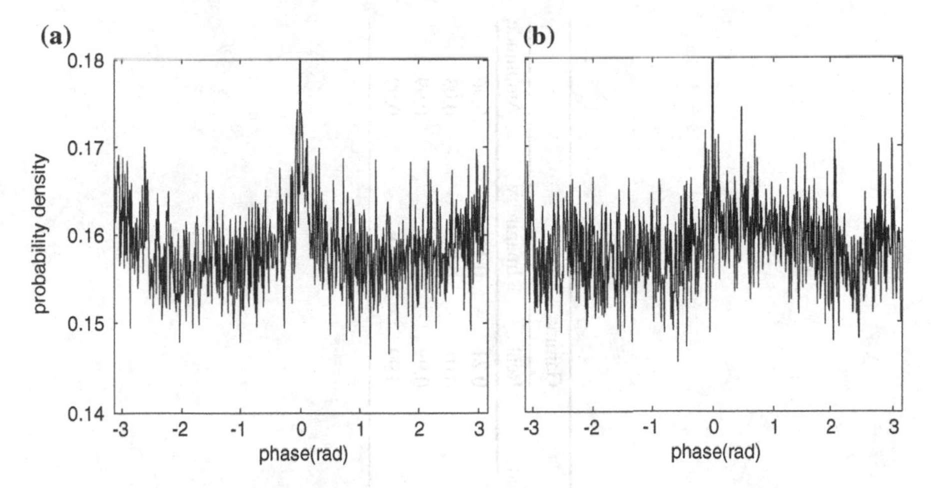

Fig. 36.2 Phase distribution of FChT spectra, **a** clean, **b** white noise at 10 dB SNR

As for the phase of speech spectra, it is always assumed to be uniformly distributed for simplicity whether the speech is clean or noisy, which is not true in reality. Figure. 36.2 shows that the frequency of phase in the neighborhoods of 0 and $\pm\pi$ is much higher than those in the other phase region. It also shows that the frequency of phase in neighborhoods of $\pm\pi/2$ is the least. The uniform distribution scenario of phase of noisy speech is illustrated in the right graph of Fig. 36.2.

36.4 Modeling and Graphical Comparison

The following GOF tests are conducted: SSE test, Pearson's Chi squared test, and KS test.

36.4.1 SSE

The Curve Fitting Toolbox has four GOF statistics for parametric models: sum of squared error (SSE), root mean squared error (RMSE), R^2, and adjusted-R-square. A SSE value closer to 0 indicates a better fit, so does RMSE. R^2 and adjusted R^2 can take on any value ≥ 1, with a value closer to 1 indicating a better fit. From Table 36.1 we can see that, Gamma is the best model, in that its SSE and RMSE are the lowest and its R^2 and adjusted R^2 are the highest among three models no matter for the real part, imaginary part, or amplitude of the clean speech.

The above four statistics are consistent even in the noisy case, thus only SSE measure is used in the next. The imaginary part is not illustrated because it is quite

Table 36.1 Four GOF statistics of real part, imaginary part, and amplitude of the clean speech

GOF	Gaussian			Laplacian			Gamma		
	Real	Imaginary	Amplitude	Real	Imaginary	Amplitude	Real	Imaginary	Amplitude
SSE	2.82	2.68	7.80	1.19	1.05	3.33	0.21	0.11	0.96
RMSE	0.04	0.04	0.09	0.02	0.02	0.06	0.01	0.01	0.03
Rsquare	0.92	0.93	0.88	0.97	0.97	0.95	0.99	1.00	0.99
Adj-rsquare	0.92	0.93	0.88	0.97	0.97	0.95	0.99	1.00	0.99

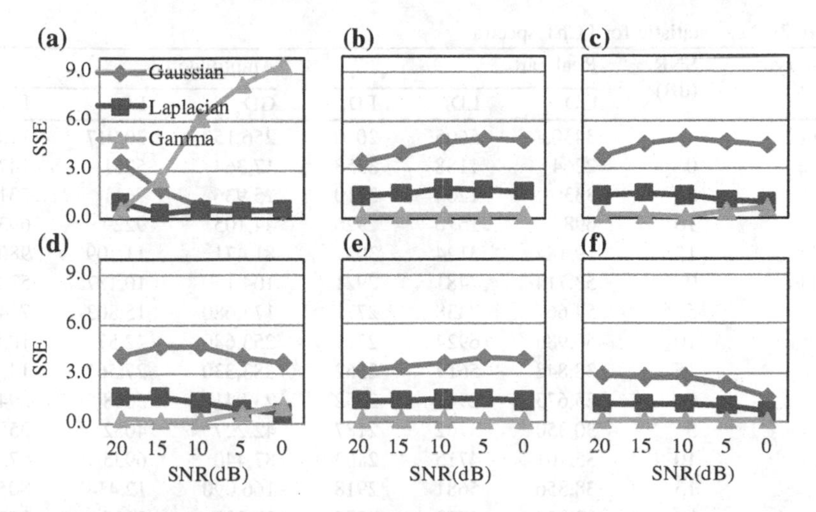

Fig. 36.3 SSE statistics of the real part of FChT spectra with different SNRs. **a** white, **b** babble, **c** f16 fighter, **d** factory I, **e** factory II, **f** Volvo car

similar to the real part. The SSE statistics of the real part in noisy cases are illustrated in Fig. 36.3. We can see that the Laplacian model is favored for the white Gaussian noise and that the Gamma model is favored for the other noises. The SSE statistics of the amplitude are illustrated in Fig. 36.4. We can see that the Gamma model is favored for most cases except that the Laplacian model is favored for white Gaussian noise with SNR > 10 dB and for f16 fighter and

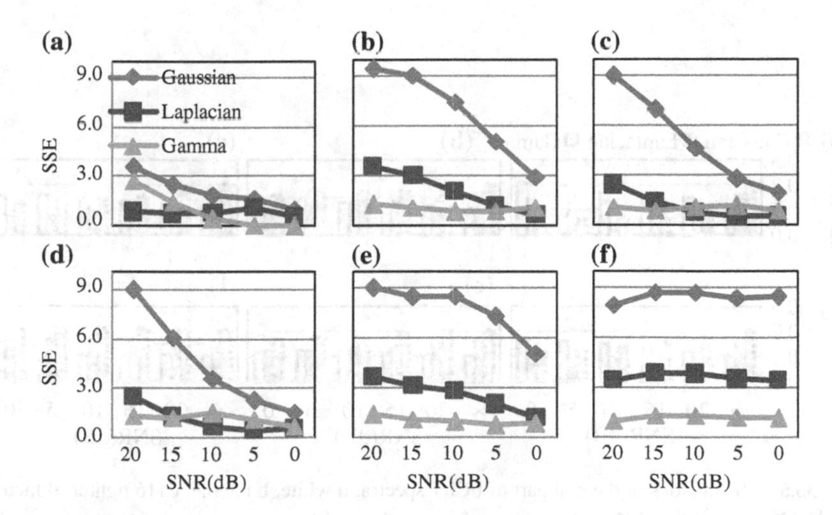

Fig. 36.4 SSEs statistics of the amplitude of FChT spectra with different SNRs. **a** white, **b** babble, **c** f16 fighter, **d** factory I, **e** factory II, **f** Volvo car

Table 36.2 χ^2 statistic for FChT spectra

Noise type	SNR (dB)	Real part			Amplitude		
		GD	LD	ΓD	GD	LD	ΓD
Clean		34505	5605	2070	256,150	30,917	11,047
White	0	2264	1188	3358	17,364	5381	1475
	5	3330	1202	3210	25,936	7611	2311
	10	6984	1570	2926	44,105	9229	6931
	15	19,184	3124	3031	81,471	11,109	9802
Babble	0	53,714	7481	2921	104,150	10,137	5337
	5	54,667	7438	2771	173,680	15,803	7242
	10	50,986	6924	2711	253,680	22,577	10,846
	15	38,848	5617	2362	283,330	27,561	14,136
F16	0	25,675	2943	2052	23,041	3048	2941
	5	30,350	3762	2187	42,227	4052	3557
	10	35,701	4715	2463	87,440	6935	4728
	15	38,356	5681	2918	166,090	12,414	8050
FactoryI	0	23,934	2872	2275	26,339	3617	3723
	5	30,106	3841	2418	48,705	4805	4687
	10	41,256	5735	2967	101,970	7908	6276
	15	48,332	7087	3334	202,910	15,515	10,598
FactoryII	0	38,788	5598	2630	141,700	11,929	5492
	5	45,774	6435	2689	258,000	22,270	10,013
	10	39,258	6194	2733	290,050	29,437	13,173
	15	38,041	6164	2271	261,800	29,953	11,834
Volvo	0	24,949	4386	1954	262,600	32,774	12,545
	5	34,190	5766	2274	255,110	33,217	14,215
	10	34,916	6280	2142	270,140	35,626	15,163
	15	34,597	5832	1990	285,620	33,969	13,078

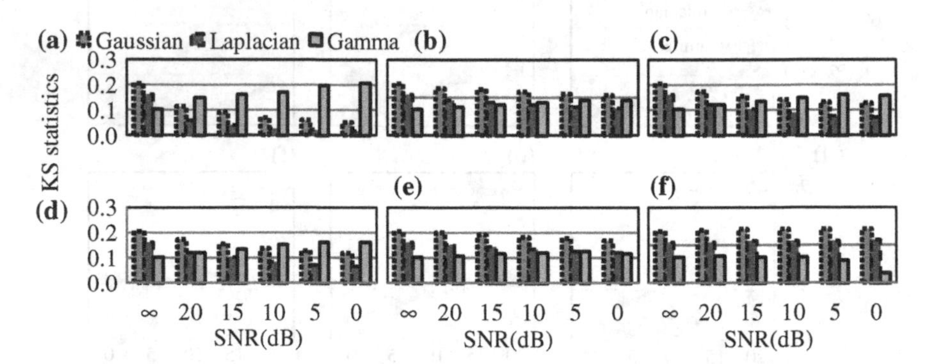

Fig. 36.5 KS statistics of the real part of FChT spectra. **a** white, **b** babble, **c** f16 fighter, **d** factory I, **e** factory II, **f** Volvo car

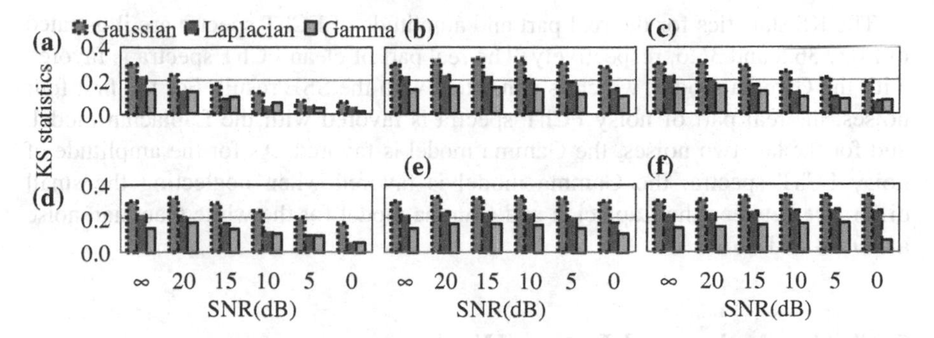

Fig. 36.6 KS statistics of the amplitude of FChT spectra. **a** white, **b** babble, **c** f16 fighter, **d** factory I, **e** factory II, **f** Volvo car

factory I noise with SNR < 10 dB. Since the SSE difference between Laplacian and Gamma is very small for f16 fighter and factory I noise, the Gamma model is reasonably chosen at low SNR.

36.4.2 Pearson's Chi squared Test

The χ^2 test compares experimental data with some given pdfs and measures the distortion between them by $\chi^2 = \sum_{i=1}^{n} (O_i - E_i)^2 \Big/ E_i$, where the sample space is partitioned into n intervals; O_i is the value of observed frequency in the ith interval; E_i is an expected (theoretical) frequency in the ith interval, asserted by the null hypothesis [8]. It is calculated for each 1 s time frame with the frame shift of 250 ms. In Table 36.2 the results of the real part and amplitude are summarized. The smaller the χ^2 test value, the better the fit. It can be clearly seen that the Gamma distribution is more suitable for both the real part and amplitude whether in the clean or noisy case except that the Laplacian distribution is favored for the real part in the case of white noise. The shadow items are neglected because they are very close to their row neighbors with a square frame.

36.4.3 KS Test

The KS statistic for a given cumulative distribution function $F(x)$ is $D_n = \sup_x |F_n(x) - F(x)|$, where $\sup_x$ is the supremum of the set of distances. The empirical distribution function F_n for nth observations X_i is defined as $F_n(x) = \frac{1}{n}\sum_{i=1}^{n} I_{X_i \leq x}$, where I_{X_i} is the indicator function, which is 1 if $X_i \leq x$ and 0 otherwise [10, 16]. The smaller the KS statistic, the better the hypothesized model fits with the empirical distribution.

The KS statistics for the real part and amplitude of FChT spectra are illustrated in Figs. 36.5 and 36.6, respectively. The real part of clean FChT spectra is favored with the Gamma model, which is consistent with the SSE result. For the first four noises, the real part of noisy FChT spectra is favored with the Laplacian model, and for the last two noises, the Gamma model is favored. As for the amplitude of noisy FChT spectra, the Gamma model is favored when neglecting the small difference between the Laplacian and Gamma model for the white Gaussian noise at 10 dB and 15 dB.

36.5 Conclusion and Future Work

To the best of our knowledge, this is the first study on the probability distribution of FChT spectra. We studied the distribution model for the real part, imaginary part, amplitude and phase of the speech spectra in the FChT domain. We conducted the graphical evaluations and the numerical evaluations; the former refers to the fitting curves and the latter refers to several GOF tests.

The major conclusions drawn from this work are as follows:

- The amplitude spectra is Gamma distributed in that Gamma is a flexible distribution form with a shape parameter and a scale parameter while Gaussian and Laplacian are symmetric forms with a position parameter and a scale parameter. For the noisy speech, the distribution of amplitude is both sided but not symmetric, which can be fitted only by Gamma model. For the clean speech, Gaussian and Laplacian model cannot fit both the distribution peak and tail simultaneously even with a multiplying factor. Thus, the Gamma model is the best choice.
- The real part of clean speech is Gamma distributed. There are many components from silent intervals, and they are around zero. Among the three models, only Gamma can have a high peak and a long tail simultaneously. Thus, the Gamma model most fits the distribution of the real part spectra.
- The real part of speech with the white noise is Laplacian distributed. Because the white noise has a flat spectrum, so that the components around zero is not as many as the clean case, i.e., the peak is lowered. But there is still a long tail. Thus the Laplacian model is favored since it has a longer tail than the Gaussian model. However their difference is not much.
- For the other noises, the real part of noisy speech is Gamma distributed. Based on the above explanation, the main reason is that noises are not stationary and they contribute much more to the number of components around zero than the white noise, i.e., the histogram is high peaked.
- The phase of clean speech is not uniformly distributed. If only the chirp rate is accurate for vowels, the components in the FChT domain are real and least likely to be imaginary ones. Since vowels constitute most of the speech, thus the neighborhoods of 0 and $\pm\pi$ have a higher density and those of $\pm\pi/2$ have a least density.

It is worthwhile to conduct speech enhancement based on the statistical modeling of speech in a transform domain. As the main future work, we plan to enhance the noisy speech based on the modeling results and characteristics of speech in the FChT domain.

References

1. Ephraim Y, Malah D (1984) Speech enhancement using a minimum mean-square error short-time spectral amplitude estimator. IEEE Trans Acoust Speech Signal Process 32(6):1109–1121
2. Martin R (2002) Speech enhancement using MMSE short time spectral estimation with gamma distributed speech priors. In: Proceedings of the IEEE International Conference Acoustics, Speech, Signal Process, vol. 1, Orlando, pp I253–I256
3. Martin R (2005) Speech enhancement based on minimum mean-square error estimation and supergaussian priors. IEEE Trans Speech Audio Process 13(5):845–856
4. Chen B, Loizou PC (2007) A Laplacian-based MMSE estimator for speech enhancement. Speech Commun 49(2):134–143
5. Lotter T, Vary P (2005) Speech enhancement by MAP spectral amplitude estimation using a super-gaussian speech model. EURASIP J Appl Signal Process 7:1110–1126
6. Tran HD, Takeda K, Itakura F (2005) Generalized Gamma modeling of speech and its online estimation for speech enhancement. ICASSP, pp 181–184
7. Zhang W, Gazor S (2002) Statistical modeling of speech signals. ICSP pp 480–483
8. Gazor S, Zhang W (2003) Speech probability distribution. IEEE Signal Process Lett 10(7):204–207
9. Gazor S, Zhang W (2005) Speech enhancement employing Laplacian-Gaussian mixture. IEEE Trans Speech Audio Process 13(5):896–904
10. Chang JH, Gazor S, Kim NS, Mitra SK (2007) Multiple statistical models for soft decision in noisy speech enhancement. Pattern Recogn 40(3):1123–1134
11. Kepesi M, Weruaga L (2006) Adaptive chirp-based time-frequency analysis of speech signals. Speech Commun 48(5):474–492
12. Weruaga L, Kepesi M (2007) The fan-chirp transform for non-stationary harmonic signals. Signal Process 87:1504–1522
13. Wu HW (2011) On spectral aliasing of the fan-chirp transform. In: International conference on graphic and image processing (ICGIP), pp 82855R-1–82855R-9
14. Varga AP, Steeneken HJM, Tomlinson M et al (1992) The Noisex-92 study on the effect of additive noise on automatic speech recognition [Report]. DRA Speech Research Unit, Malvern
15. Wu HW, Wu ZY, Zhao L (2007) Improved mean-shift-based pitch determination. J SE Univ (Engl Ed) 23(4):494–499
16. Glen AG, Leemis LM, Barr DR (2001) Order statistics in goodness-of-fit testing. IEEE Trans Reliab 50(2):209–213

In [illegible] work, [illegible] conduct speech enhancement based on the statistical modeling of speech in a [illegible] domain. [illegible] future work, we plan to enhance the noisy speech based on the [illegible] results and characteristics of speech in the [illegible].

References

1. Ephraim Y, Malah D (1984) Speech enhancement using a minimum mean-square error short-time spectral amplitude estimator. IEEE Trans Acoust Speech Signal Process [illegible]
2. Martin R (200[illegible]) Speech enhancement using MMSE short time spectral estimation with gamma distributed speech priors. In: Proceedings of the IEEE International Conference on Acoustics, Speech and Signal Processing, Orlando, [illegible]
3. Martin R (2005) Speech enhancement based on minimum mean-square error estimation and supergaussian priors. IEEE Trans Speech Audio Process [illegible]
4. Chen B, Loizou PC (2007) A Laplacian-based MMSE estimator for speech enhancement. Speech Commun 49(2):134–143
5. Lotter T, Vary P (2005) Speech enhancement by MAP spectral amplitude estimation using a super-gaussian speech model. EURASIP J Appl Signal Process 7:1110–1126
6. [illegible] (2005) Generalized Gamma modeling of speech and its online estimation for speech enhancement. ICASSP, [illegible]
7. Zhang W, Gazor S (200[illegible]) Statistical modeling of speech signals. ICSP, pp 480–483
8. Gazor S, Zhang W (2003) Speech probability distribution. IEEE Signal Process Lett [illegible]
9. Gazor S, Zhang W (2005) Speech enhancement employing Laplacian-Gaussian mixture. IEEE Trans Speech Audio Process [illegible]:896–904
10. Chang JH, Gazor S, Kim NS, Mitra SK (2007) Multiple statistical models for soft decision in noisy speech enhancement. Pattern Recogn 40(3):1123–1134
11. Kepesi M, Weruaga L (2006) Adaptive chirp-based time-frequency analysis of speech signals. Speech Commun 48(5):474–492
12. Weruaga L, Kepesi M (2007) The fan-chirp transform for non-stationary harmonic signals. Signal Process [illegible]:1504–1522
13. [illegible] (201[illegible]) On statistical modeling of fan chirp transform. In: International conference [illegible] graphic and image processing (ICGIP), [illegible]
14. Varga AP, Steeneken HJM, Tomlinson M, Jones D (1992) The NOISEX-92 study on the effect of additive noise on automatic speech recognition. Technical Report, DRA Speech Research Unit, [illegible]
15. [illegible]
16. [illegible]

Chapter 37
Multiphase Image Segmentation from a Statistical Framework

Jiangxiong Fang, Huaxiang Liu, Juzhi Deng, Yulin Gong, Haning Xu and Jun Liu

Abstract The study is to investigate a new representation of a partition of an image domain into a number of regions using level set method derived from a statistical framework. The proposed model is composed of evolving simple closed planar curves by a region-based force determined by maximizing the posterior image densities over all possible partitions of the image plane containing two terms: a Bayesian term based on the prior probability, a regularity term adopted to avoid the generation of excessively irregular and small segmented regions. This formulation leads to a system of coupled curve evolution equations, which is easily amenable to a level set implementation, and an unambiguous segmentation because the evolving regions form a partition of the image domain at all time during curve evolution. Given these advantages, the proposed method can get good performance and experiments show promising segmentation results.

Keywords Multiphase image segmentation · Level set · Statistical approach

37.1 Introduction

Image segmentation is a fundamental problem in image processing and computer vision. Its goal is to partition a given image into several parts in each of which the intensity is homogeneous. It plays an important role in numerous useful applications, e.g., SAR image processing, [1] biomedical image processing, [2] scene

J. Fang (✉) · H. Liu · J. Deng · Y. Gong · H. Xu · J. Liu
Department of Nuclear Engineering and Physical Geography,
East China Institute of Technology,
Nanchang 330013, China
e-mail: fangchj2002@163.com

J. Fang
Jiangxi Province Key Lab for Digital Land, Fuzhou 344000, China

A. A. Farag et al. (eds.), *Proceedings of the 3rd International Conference on Multimedia Technology (ICMT 2013)*, Lecture Notes in Electrical Engineering 278,
DOI: 10.1007/978-3-642-41407-7_37,

interpretation, and [3] video image analysis; [4] since it facilitates the extraction of information and interpretation of image contents. Over these decades, many approaches have been developed to solve the image segmentation problem. Researchers have also done great efforts to improve the performance of the image segmentation algorithms. However, it is still a difficult problem to solve for complicated images.

In recent years, level set method [5] is the most important and successful method for image segmentation. Chan-Vese (C–V) model [6] is one of the most popular active contour models based on Mumford-Shah segmentation formulas. With no reliance on the gradient to stop the propagation process, the model becomes an energy minimizing segmentation which can be seen as a particular case of the minimal partition problem. Later, Vese, and Chan proposed a multiphase level set framework [8] represented by multiple level set functions. But the interiors of two or more curves may overlap, leading to ambiguous segmentation.

For any statistical approach to image segmentation, the problem estimating the probability density remains a challenging issue. Zhu and Yuille [9] established relations between statistical methods and the cartoon limit of the Mumford-Shah functional. Cremers et al. [10] surveyed several classes of region-based level set segmentation methods and described how they may derive from a common statistical framework. As opposed to other statistical approaches using particular distribution models, the pixel intensities in each region are obtained by nonparametric density estimation. Kim et al. [11] proposed an information theoretic approach that maximizes the mutual information between the region labels and the pixel intensities, subject to curve length constraints, to formulate the segmentation. Chen [12] proposes an alternative criterion derived from the Bayesian risk classification error, which can avoid the generation of excessively irregular and small segmented regions.

The classical multiphase level set segmentation on statistical model is difficult because two or more closed curves unambiguously partitioning the image domain into disjoint regions may overlap, which leads to ambiguous segmentation. To solve these problems, we propose a new representation which divides a partition of an image domain into a number of regions derived from a statistical framework. The proposed model is composed of evolving simple closed planar curves by a region-based force determined by maximizing the posterior image densities over all possible partitions of the image plane containing three terms: a Bayesian term based on the prior probability, a regularity term adopted to avoid the generation of excessively irregular, and small segmented regions. The formulation relies on the optimum decision of pixel classification and the estimates of prior probabilities, which make the segmentation results more reliable in theory and practice.

The remaining of this paper is organized as follows. In Sect. 37.2, multiphase segmentation framework, including image segmentation as Bayesian inference, energy functional for multiphase image segmentation, level set implement, and description of our proposed method is presented. Experimental results illustrating the performances of segmentation are discussed in Sect. 37.3. Finally concluding remarks and future works are given in Sect. 37.4.

37.2 Multiphase Segmentation Framework

37.2.1 Image Segmentation as Bayesian Inference

Let $\Omega \subset R^d$ be a given vector valued image, and $I : \Omega \rightarrow R^n$ be the image domain, and d is the dimension of the vector $\boldsymbol{I}$. The image being partitioned into N regions is to find a partition $\{R_i\}_{i=1}^N$ from the image domain Ω so that each region is homogeneous. In this case, it is convenient to cast segmentation in a Bayesian framework [10] by maximizing a posteriori (MAP) estimation. The problem would consist of finding an optimal partition $\{R_i\}_{i=1}^N$ which maximizes the a posteriori probability $p(\{R_i\}_{i=1}^N|I)$ over all possible N-region partitions of the image plane Ω.

$$\{R_i\}_{i=1}^N = \arg\max_{R_i \subset \Omega} p(\{R_i\}_{i=1}^N|I) = \arg\max_{R_i \subset \Omega} p(I|\{R_i\}_{i=1}^N)p(\{R_i\}_{i=1}^N) \tag{37.1}$$

Maximization of the a posteriori probability (37.1) is equivalent to minimizing its negative logarithm. Here, we assume that $I(x)$ is independent of $I(y)$ for $x \neq y$, the function $E[\{R_i\}_{i=1}^N]$ is defined as follows:

$$E[\{R_i\}_{i=1}^N] = -\sum_{i=1}^N \int_{x \in R_i} \log p(I(x)|\{R_i\}_{i=1}^N)\mathrm{d}x - \log p(\{R_i\}_{i=1}^N) \tag{37.2}$$

where $p(I(x)|R_i)$ denotes the probability of observing an image $\boldsymbol{I}$. The first term, also called the Bayesian term, computers the conformity of image data within each region $R_i, i = 1, \ldots, N$, to a parametric distribution $p(I(x)|R_i)$. The Gaussian distribution has been considered in most studies because it can reduce the computational cost and simply be used. In this case, the Bayesian term are expressed as follows:

$$\begin{aligned} E^B[\{R_i\}_{i=1}^N] &= -\sum_{i=1}^N \int_{x \in R_i} \log p(I(x)|\{R_i\}_{i=1}^N)\mathrm{d}x \\ &= \sum_{i=1}^N \left(\frac{(I(x) - u_i)^2}{2\sigma_i^2} + \log\left(\sqrt{2\pi\sigma_i^2}\right)\right) \end{aligned} \tag{37.3}$$

where u_i and σ_i denote the mean value and variance within the region R_i.

The second term in (37.2), also called regularization term, is commonly used for smooth segmentation boundaries.

$$E^R[\{R_i\}_{i=1}^N] = -\text{In } (p(\{R_i\}_{i=1}^N)) = \lambda \sum_{i=1}^{N-1} \int_{\gamma_i} \mathrm{d}s \tag{37.4}$$

where ∂R_i is the boundary of the region R_i and λ is a positive factor.

37.2.2 Energy Functional For Multiphase Image Segmentation

$$
\begin{aligned}
R_1 &= R_{\vec{\gamma}_1} \\
R_2 &= R^c_{\vec{\gamma}_1} \cap R_{\vec{\gamma}_2} \\
R_k &= R^c_{\vec{\gamma}_1} \cap R^c_{\vec{\gamma}_2} \cap \cdots \cap R_{\vec{\gamma}_k} \\
&\cdots \\
R_N &= R^c_{\vec{\gamma}_1} \cap R^c_{\vec{\gamma}_2} \cap \cdots \cap R_{\vec{\gamma}_{N-1}}
\end{aligned}
\quad (37.5)
$$

Let a family $\{\vec{\gamma}_i\}_{i=1}^N$: $[0,1] \to \Omega$ of plane curves parametrized by the arc parameter $s \in [0,1]$. To guarantee an unambiguous segmentation, we use a representation of a partition of the image domain by the following explicit correspondence between the family of regions $\{R_{\vec{\gamma}_i}\}$ enclosed by the curves $\{\vec{\gamma}_i\}_{i=1}^{N-1}$. The regions of partition $\{R_i\}_{i=1}^N$ of the image domain Ω are showed in (37.6) and the partition representation is shown in Fig. 37.1 with five regions divided by four curves. With representation of a partition of the image domain into ***N*** regions, *N*-1contours $\{\vec{\gamma}_i\}_{i=1}^{N-1}$ are represented by the zero level set of *N*-1 Lipschitz function. The energy functional is defined as:

$$
\begin{aligned}
E[\{\vec{\gamma}_i\}_{i=1}^{N-1}] &= \int_{R_1} \omega_1(x)\mathrm{d}x + \cdots + \int_{R_N} \omega_N(x)\mathrm{d}x + \lambda \sum_{i=1}^{N-1} \int_{\gamma_i} \mathrm{d}s \\
&= \sum_{i=1}^{N} \int_{R_i} \omega_i(x)\mathrm{d}x + \lambda \sum_{i=1}^{N-1} \int_{\gamma_i} \mathrm{d}s
\end{aligned}
\quad (37.6)
$$

where $\omega_i(x)$ is the intensities of the regions drawn from a Gaussian distribution. And extending the C–V model [10], one can implement the functional (8) by:

$$
E[\{\vec{\gamma}_i\}_{i=1}^{N-1}] = \sum_{i=1}^{N} \int_{R_i} \omega_i(x)\mathrm{d}x + \lambda \sum_{i=1}^{N-1} \int_{\gamma_i} \mathrm{d}s = \sum_{i=1}^{N} \chi_{R_i} \int_{\Omega} \omega_i(x)\mathrm{d}x + \lambda \sum_{i=1}^{N-1} \int_{\vec{\gamma}_i} \mathrm{d}s \quad (37.7)
$$

where the parameter χ_{R_i}, $i = 1, 2, \cdots N$denotes the characteristic function of *i-th* region and makes $\sum \chi_{R_i} = 1$, i.e., $\chi_{R_i} = 1$ if $x \in R_i$ and $\chi_{R_i} = 0$ if $x \in R_i^c$; c_i is the

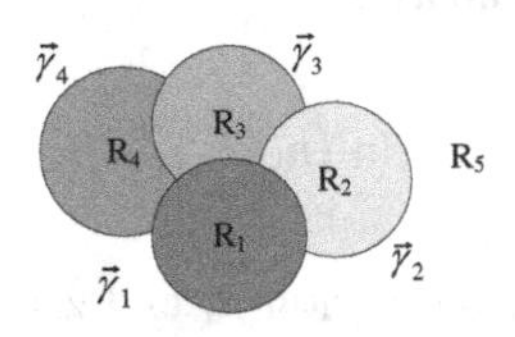

Fig. 37.1 Representation of a partition of the image domain by explicit correspondence between regions of segmentation (illustration for five regions)

i-th region constants that approximate the *i-th* region intensity in the image $\boldsymbol{I}$. Let Hbe the Heaviside function,χ_{R_i} is represented as follows:

$$\begin{cases} \chi_{R_i}(\vec{\gamma}_i) = \chi_{R_{\vec{\gamma}_1}^c}(\vec{\gamma}_k)\chi_{R_{\vec{\gamma}_2}^c}(\vec{\gamma}_k)\cdots\chi_{R_{\vec{\gamma}_{i-1}}^c}(\vec{\gamma}_i)\chi_{R_{\vec{\gamma}_i}}(\vec{\gamma}_i) \quad = \prod_{i=1}^{k-1}[1 - H(\vec{\gamma}_i)]\cdot H(\vec{\gamma}_i) \quad i = 1,\ldots,N-2 \\ \chi_{R_{N-1}}(\vec{\gamma}_{N-1}) = \chi_{R_{\vec{\gamma}_1}^c}(\vec{\gamma}_{N-1})\chi_{R_{\vec{\gamma}_2}^c}(\vec{\gamma}_{N-1})\cdots\chi_{R_{\vec{\gamma}_{N-2}}^c}(\vec{\gamma}_{N-1})\chi_{R_{\vec{\gamma}_{N-1}}^c}(\vec{\gamma}_{N-1}) = \prod_{i=1}^{N-1}[1 - H(\vec{\gamma}_i)] \quad k = N-1 \end{cases} \tag{37.8}$$

where $\boldsymbol{H}$ denotes the heaviside step function. In practice, the Heaviside function $\boldsymbol{H}$ and the derivative of H_ε are approximated by a smooth function $H_\varepsilon$defined as:

$$H_\varepsilon(x) = \frac{1}{2}\left[1 + \frac{2}{\pi}\arctan\left(\frac{x}{\varepsilon}\right)\right] \quad \text{And} \quad \delta_\varepsilon(x) = H_\varepsilon'(x) = \frac{1}{\pi}\frac{\varepsilon}{\varepsilon^2 + x^2} \tag{37.9}$$

37.2.3 Level Set Implementation

To minimize the functional E in Eq. (37.7) with respect to the curves $\{\vec{\gamma}_i\}_{i=1}^{N-1}$, we considered it is performed by embedding by the family $\{\vec{\gamma}_i : [0,1] \to \Omega\}_{i=1}^{N-1}$ of plane curves parameterized by arc parameter $s \in [0,1]$. The Euler–Lagrange descent equation corresponding to $\vec{\gamma}_i$ is obtained by embedding the curve into a family of one-parameter curves $\vec{\gamma}_i : [0,1] \times R^+ \to \Omega, i = 1,2,\ldots,N-1$ by solving the evolution equations:

$$\frac{d\vec{\gamma}_i}{dt} = -\frac{\delta E}{\delta\vec{\gamma}_i}, \quad i = 1,2,\cdots,N-1 \tag{37.10}$$

The functional derivatives $\delta E/\delta\vec{\gamma}_i$ can be easily computed by suitably rewriting the area integrals appearing in the energy functional. By compute the functional derivatives $\delta E/\delta\vec{\gamma}_i$ for all i, the minimization of the energy functional is obtained and given the following evolution equations:

$$\begin{cases} \frac{d\vec{\gamma}_1}{dt} = -\left[(\omega_1(\vec{\gamma}_1) - \Phi_1(\vec{\gamma}_1)) + \lambda k_1\right]\vec{n}_1 \\ \vdots \\ \frac{d\vec{\gamma}_i}{dt} = -\left[\chi_{R_{\vec{\gamma}_1}^c}(\vec{\gamma}_i)\cdots\chi_{R_{\vec{\gamma}_{i-1}}^c}(\vec{\gamma}_i)(\omega_i(\vec{\gamma}_i) - \Phi_i(\vec{\gamma}_i)) + \lambda k_i\right]\vec{n}_i \\ \vdots \\ \frac{d\vec{\gamma}_{N-1}}{dt} = -\left[\chi_{R_{\vec{\gamma}_1}^c}(\vec{\gamma}_{N-1})\cdots\chi_{R_{\vec{\gamma}_{N-1}}^c}(\vec{\gamma}_{N-1})(\omega_{N-1}(\vec{\gamma}_{N-1}) - \Phi_{N-1}(\vec{\gamma}_{N-1})) + \lambda k_{N-1}\right]\vec{n}_{N-1} \end{cases} \tag{37.11}$$

where k_i is the mean curvature function of $\vec{\gamma}_i$ and $\Phi_i(\vec{\gamma}_i)$ is defined as follows:

$$\phi_i(x) = \omega_{i+1}(x)\chi_{R_{\vec{\gamma}_{i+1}}}(x) + \omega_{i+2}(x)\chi_{R_{\vec{\gamma}_{i+1}}^c}(x)\chi_{R_{\vec{\gamma}_{i+2}}}(x) + \cdots + \omega_{N-1}(x)\chi_{R_{\vec{\gamma}_{i+1}}^c}(x) \cdots \chi_{R_{\vec{\gamma}_{N-2}}^c}(x)\chi_{R_{\vec{\gamma}_{N-1}}}(x) + \omega_N(x)\chi_{R_{\vec{\gamma}_{i+1}}^c}(x) \cdots \chi_{R_{\vec{\gamma}_{N-2}}^c}(x)\chi_{R_{\vec{\gamma}_{N-1}}^c}(x) \tag{37.12}$$

To implement the curves evolution equations in (37.11), a better alternative is to represent the curve $\vec{\gamma}_i$ implicitly by discretizing the interval [0, 1] on which the curves $\{\vec{\gamma}_i\}_{i=1}^{N-1}$ are defined using the zero level set function. With level sets, the curve $\vec{\gamma}_i$ is represented implicitly by the zero level set function $u_i : R^2 \to R$, i.e., we define $\vec{\gamma}_i$ as the set $u_i(x,t) = 0$. This implies that:

$$\frac{\mathrm{d}u_i(x,t)}{\mathrm{d}t} = \frac{\partial u_i}{\partial t}(x,t) + \vec{\nabla} u_i \cdot \frac{\mathrm{d}\vec{\gamma}_i}{\mathrm{d}t} = 0 \tag{37.13}$$

The level set evolution equation minimizing the functional (37.11) is, therefore, given by the following system of coupled partial deferential equations:

$$\begin{cases} \frac{\mathrm{d}u_1(x,t)}{\mathrm{d}_t} = -\left[\omega_1(x) - \Phi_1(x) + \lambda k_{u_1}\right]\left\|\vec{\nabla} u_1(x,t)\right\| \\ \vdots \\ \frac{\mathrm{d}u_i(x,t)}{\mathrm{d}_t} = -[\chi_{u_1(x,t)<0}(x)\cdots\chi_{u_{i-1}(x,t)<0}(x)(\omega_i(x) - \Phi_i(x)) + \lambda k_{u_i}]\|\nabla u_i(x,t)\| \\ \vdots \\ \frac{\mathrm{d}u_{N-1}(x,t)}{\mathrm{d}_t} = -[\chi_{u_1(x,t)<0}(x)\cdots\chi_{u_{N-1}(x,t)<0}(x)(\omega_{N-1}(x) - \Phi_{N-1}(x)) + \lambda k_{u_{N-1}}]\|\nabla u_{N-1}(x,t)\| \end{cases} \tag{37.14}$$

where $\Phi_k(\vec{\gamma}_k)$ are defined as follows, respectively:

$$\Phi_i(x) = \omega_{i+1}(x)\chi_{u_{i+1}(x,t)>0} + \cdots + \omega_{N-1}(x)\chi_{u_{i+1}(x,t)<0}\cdots\chi_{u_{N-2}(x,t)<0}\chi_{u_{N-1}(x,t)>0} + \omega_N(x)\chi_{u_{i+1}(x,t)<0}\cdots\chi_{u_{N-2}(x,t)<0}\chi_{u_{N-1}(x,t)<0} \tag{37.15}$$

With $\chi_{u_i(x,t)>0} = H(\vec{\gamma}_i)$ if $u_i(x,t) > 0$and $\chi_{u_i(x,t)\le 0} = 1 - H(\vec{\gamma}_i)$ if $u_i(x,t) \le 0$; k_{u_i} is the curvature of the level set of u_i and is given as a function as follows:

$$k = \vec{\nabla} \cdot \frac{\vec{\nabla} u_i}{\left\|\vec{\nabla} u_i\right\|} = \frac{u_{xx}u_y^2 - 2u_x u_y u_{xy} + u_{yy}u_x^2}{(u_x^2 + u_y^2)^{3/2}} \tag{37.16}$$

where u_x, u_y, u_{xx}, u_{yy} and u_{xy} are computed as follows:

$$u_x = \frac{1}{2h}(u_{i+1,j} - u_{i-1,j}), u_y = \frac{1}{2h}(u_{i,j+1} - u_{i,j-1}),$$
$$u_{xx} = \frac{1}{h^2}(u_{i+1,j} + u_{i-1,j} - 2u_{i,j}), u_{yy} = \frac{1}{h^2}(u_{i,j+1} + u_{i,j-1} - 2u_{i,j}), \quad (37.17)$$
$$u_{xy} = \frac{1}{h^2}(u_{i+1,j+1} - u_{i-1,j+1} - u_{i+1,j-1} + u_{i-1,j-1})$$

where h is the grid spacing. And the optimal estimates for the mean u_i and the variance σ_i can be computed:

$$u_i = \frac{\int \chi_{R_i}(\vec{\gamma}_i)I(x)\mathrm{d}x}{\int \chi_{R_i}(\vec{\gamma}_i)\mathrm{d}x}, \qquad \sigma_i^2 = \frac{\int \chi_{R_i}(\vec{\gamma}_i)(I(x) - u_i)^2\mathrm{d}x}{\int \chi_{R_i}(\vec{\gamma}_i)\mathrm{d}x} \quad (37.18)$$

Thus, the segmentation procedure is summarized as follows:

Step 1: Set initial parameters: set a given fixed region number N, the number of iterations *Max_iter*, and the weighted factor α

Step 2: Initialize the curves $\{\vec{\gamma}_i\}_{i=1}^m$, with $\{\vec{\gamma}_i^k\}_{i=1}^{N-1}$ defined as the distance function from initial curves

Step 3: For each $\vec{\gamma}_i$, compute the corresponding u_i and $\sigma_i, i = 1, \cdots N$ as the averages for each region

Step 4: For each $\vec{\gamma}_i$, compute $\vec{\gamma}_i^{k+1}$ by solving the following:

$$\vec{\gamma}_i^{k+1} = \vec{\gamma}_i^k - \Delta t \frac{\partial E}{\partial t}(\{\vec{\gamma}_i\}_{i=1}^{N-1}) \quad (37.19)$$

where Δt is the time step

Step 5: Reinitialize each curve $\vec{\gamma}_i^k$ locally from the queue $\{\vec{\gamma}_i\}_{i=1}^m$ to the signed distance function to the curve

Step 6: Run the process until the curves $\{\vec{\gamma}_i\}_{i=1}^m$ reaches convergence

37.3 Experiments and Results

To demonstrate the effectiveness of our proposed method, a large number of tests with different image types have been tested. We do a large number of experiments on synthetic images, medical image, SAR image, and remote sensing image The proposed method is implemented using the Matlab and C programming language and experimented on a Pentium IV 2 × 2.2 GHz computer with Memory 3G. In our experiments, we fixed to the parameter of the model $\lambda = 0.1$ and the time step $\Delta t = 0.2$.

To further validate the effectiveness of our proposed method, two synthetic images with noise model are tested. Figure 37.2 depicts the segmentation results. The objects can be partitioned into two regions and the noise in the image disappeared. The final position of the curves at convergence is shown in the second column; the third column shows the segmentation results. From the results, our method for the image segmentation with noise is more effective.

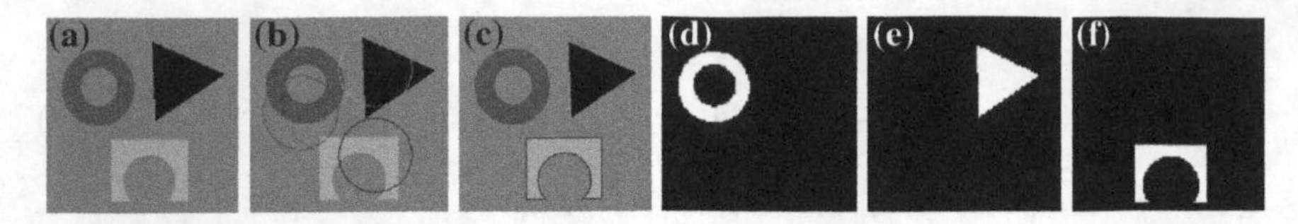

Fig. 37.2 The synthetic image segmentation using our proposed level set method. **a** The input image; **b** The initial *curves* in the input image; **c** shows the final segmentation result of contour images; **d–f** Final segmentation result of the three regions

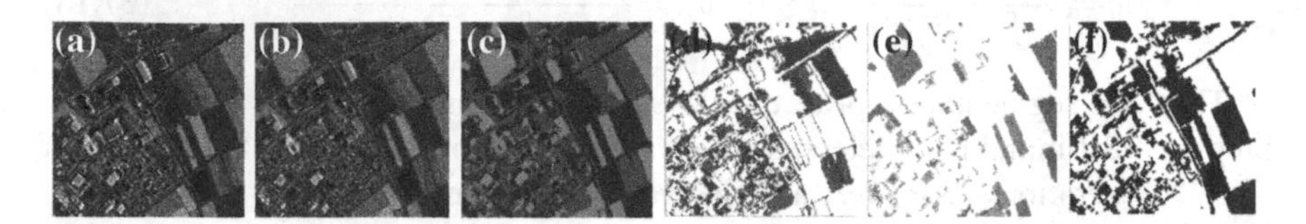

Fig. 37.3 Segmentation results for the SAR image using our proposed level set method: **a** the initial location of the curves; **b** the final location of the curves; **c** the final segmentation result; **d–e** segmentation regions corresponding to the objects; **f** segmentation region corresponding to the background

Figure 37.3 shows the segmentation results for SAR image and remote sensing image, respectively. With the initialization (*N* regions, *N*-1 curves corresponding to *N*-1 circles), these images have been segmented correctly. We can notice that some curves disappear at convergence shown in Fig. 8, leading to the same correct segmentation into three regions.

37.4 Conclusion

We proposed an efficient level set method on statistical approach with multiple regions, which is guaranteed to be a partition of the image domain. This ensures that no ambiguities arise when assigning points to the various segmented regions. The experimental results show that the final segmentation remains a partition of the image domain and the method is efficient. The future work will consider how to build more effective schemes to distinguish foreground from background during segmentation.

Acknowledgments: The research described in this paper was funded by National international technology cooperation plan (2007DFA20790), Jiangxi province scientific and technological achievements promotion plan(GanCaiJiao[2011]243), Jiangxi Province Key Lab for Digital Land (DLLJ201301), Science & technology Project of Jiangxi Province Education Department (GJJ13446).

References

1. Ayed B, Mitchie A, Belhadj Z (2005) Multiregion level-set partitioning of synthetic aperture radar images. IEEE Trans Pattern Anal Mach Intell 7:793–800
2. Pham DL, Xu C, Prince J (2000) Current methods in medical image segmentation. Ann Rev Biomed Eng 2:315–338
3. Cai Q, Aggarwal J (1999) Human motion analysis: a review. Comput Vis Image Underst 73:428–440
4. Zhao T, Nevatia R, Wu B (2008) Segmentation and tracking of multiple humans in crowded environments. IEEE Trans Pattern Anal Mach Intell 30:1198–1211
5. Osher S, Sethian JA (1988) Fronts propagating with curvature-dependent speed: algorithms based on Hamilton-Jacobi formulations. J Comput Phys 79(1):12–49
6. Chan T, Vese L (2001) Active contours without edges. IEEE Trans Image Process 10:266–277
7. Mumford D, Shah J (1989) Optimal approximation by piecewise smooth functions and associated variational problems. Commun Pure Appl Math 42:577–685
8. Vese L, Chan T (2002) A multiphase level set framework for image segmentation using the Mumford and Shah model. Int J Comput Vision 50:271–293
9. Zhu SC, Yuille A (1996) Region competition: Unifying snakes, region growing, and Bayes/MDL for multiband image segmentation. IEEE Trans Pattern Anal Mach Intell 18(9):884–900
10. Cremers D, Rousson M, Deriche R (2007) A review of statistical approaches to level set segmentation: Integrating color, texture, motion and shape. Int J Comput Vision 72:195–215
11. Kim J, Fisher JW, Yezzi A et al (2002) Nonparametric methods for image segmentation using information theory and curve evolution. Int Conf Image Proc 3:797–800
12. Chen YT (2010) A level set method based on the Bayesian risk for medical image segmentation. Pattern Recognit 43:3699–3711
13. Mansouri A-R, Mitiche A, Vazquez C (2006) Multiregion competition: a level set extension of region competition to multiple region image partitioning. Comput Vis Image Underst 101:137–150
14. Brox T, Weickert J (2006) Level set segmentation with multiple regions. IEEE Trans Image Process 17:3213–3218

Chapter 38
A Hierarchical Feature Extraction Scheme with Special Vocabulary Generation for Natural Scene Classification

Tian Luo, Zhuo Su and Xiaonan Luo

Abstract To automatically classify natural scenes instead of manual ways, this paper proposes a novel approach to recognize scene categories. First, we extract appearance features from an image similar to a pyramid. Then, the visual words are generated from different classes separately based on Bag of Words (BOW) model. At last, Spatial Pyramid Matching (SPM) algorithm is used to obtain histogram of visual words and Support Vector Machine (SVM) is applied to classification. There are two contributions in this paper: one is that we partition an image into patches at different resolution levels and use multiple descriptors to obtain some omissive image information; the other is that visual words are formed by performing K-means clustering from each category and concatenated to form a dictionary distinguish to traditional BOW. We present satisfactory performances on a large scale of 13 categories dataset.

Keywords Scene classification · Bag of words · Spatial pyramid matching · Support vector machine

38.1 Introduction

Scene classification creates a foundation for a further recognition of objects, so it is a meaningful task to identify the semantic category an image belongs to. They can be divided into two categories in previous researches. The first one is based on

T. Luo · Z. Su (✉) · X. Luo (✉)
National Engineering Research Center of Digital Life, State-Province Joint Laboratory of Digital Home Interactive Applications, School of Information Science and Technology, Sun Yat-sen University, Guangzhou 510006, China
e-mail: suzhuoi@gmail.com

X. Luo
e-mail: lnslxn@mail.sysu.edu.cn

A. A. Farag et al. (eds.), *Proceedings of the 3rd International Conference on Multimedia Technology (ICMT 2013)*, Lecture Notes in Electrical Engineering 278,
DOI: 10.1007/978-3-642-41407-7_38,

global features: feature extraction based on it exploits low-level pixel information to depict visual contents of an image. At present, the best way about global features regards a spatial envelope model [1] as its feature descriptor. However, these methods have high computational consumption during the feature extraction and are sensitive to image scale or brightness. Classification accuracy is also low for complex scenes. The other one is based on the "Bag of Words" (BOW) [2], several researchers combined this model with other new models to perform image categorization. Li et al. [3] presented a Bayesian hierarchical model to learn and recognize natural scene categories. But they did not employ spatial information among local features. The pyramid match kernel proposed by Lazebnik et al. [4] was a superior way as for the performance of matching and classification. But sharp slowdowns in performance would appear as dimension increases. Allowing for spatial relationships between two images, Grauman et al. [5] learned from Grauman's idea and proposed a spatial pyramid match kernel.

Although BOW model has a large application in the image categorization, image retrieval and video retrieval fields, current approaches are limited. Traditional model partitions an image into several segments just at a certain resolution level, which is likely to ignore some details from an image and will influence on semantic representation, then all captured features are pooled into a single set, it will challenge clustering algorithms when clusters amounts of features. In addition, similar characteristics from different types of images can be confused, leading to lower discernibility among visual words. Recently, a series of approaches [6–10] were proposed to perfect original ways for classification and yield desirable results, i.e., a group coding strategy based on saliency coding (SaC) [7] had been proposed to improve coding process when using Bow model. [6] introduced a bag-of-multimedia-words model.

In this paper, we propose a modified Spatial Pyramid-BOW model to overcome traditional drawbacks of BOW and spatial pyramid model. This model can take more details of images in the spatial field into account through hierarchical feature extraction, and enhance accuracy of creating vocabularies. Novel aspects of this model include an efficient scheme for feature extraction, and a special way of constructing a vocabulary. A specific procedure is illustrated in Fig. 38.1.

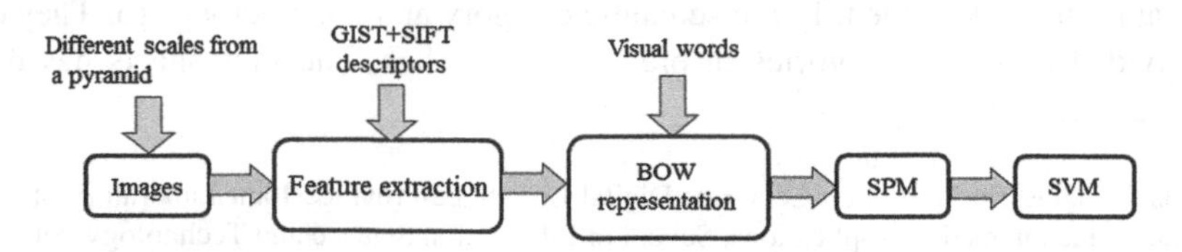

Fig. 38.1 The pipeline of our method. Given an image, we form a series of images in different scales, capture multiple descriptors in a distinctive way, and compute BOW histograms based on the BOW model. Then SPM model is combined to form a high dimension vector representation of the image, while SVM can be used to classify uncertain scenes

38.2 Features Extraction

There are various methods to obtain image regions, such as image block or image segmentation. We use image block to obtain local patches at first. Pyramid is a principle form of image multiscale representation. In this paper, we build a three-level pyramid constructed by images in different scales in order to obtain more discriminative images. Lots of work about feature extraction are based on single descriptors, LBP, HOG, SIFT, SURF, etc., which may fail to carry more useful information [11, 12]. Against its drawbacks, a new scheme is chosen to compensate. GIST is a rapid computational model of the whole features for real-world scenes, while SIFT is a local feature descriptor invariant to image scaling, translation, and rotation [13]. So we combine global GIST features and local SIFT features to form different descriptors, as illustrated in Fig. 38.2. Multiple feature descriptors can represent the image from macroscopic and microcosmic angles and provide more full information.

A dense feature representation has been used rather than sparsely detected keypoints when capture local features. The local image patches are extracted by regular grid division all over the image. Feature descriptors are computed on the 16×16 patches with 8-pixel spacing. At the first level of the pyramid, source image scale is retained and the local patch size is 16×16 pixels; at the second level, image scale has been narrowed and the patch size is 8×8 pixels; the third level, image scale is narrowed further and the patch size is 4×4 pixels. In the process of extraction, we first divide a patch into several 4×4 sub-regions and count gradient histograms from eight directions, so every sub-region is deemed as a seed point, then concatenate sixteen seed points into a 128-dimension feature vector. Three kinds of descriptors from three levels are saved finally.

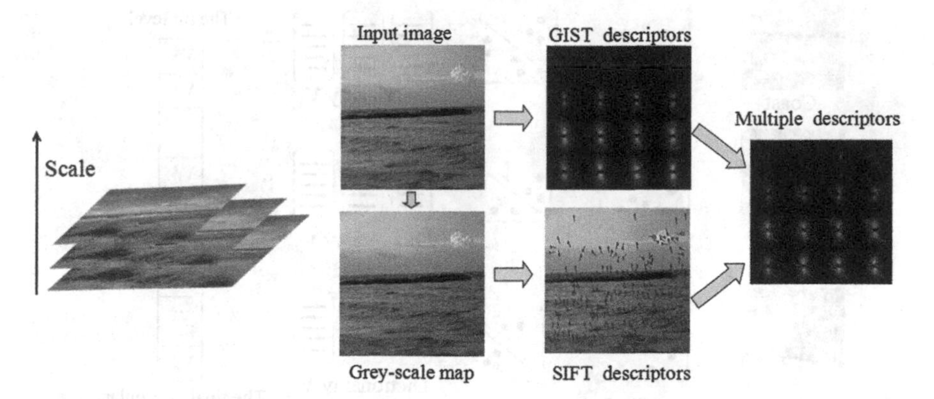

Fig. 38.2 Two types of descriptors are fused. We combine global GIST [1] and local SIFT [13] into a whole descriptors. On the left of it are a set of images in different scales from the coast scene

38.3 Discrete Categories Visual Words Generation

Having extracted all features of an original image, traditional way to generate visual words collects all of vectors into a holistic set of features $\mathbf{S} = \{\mathbf{S}_1, \mathbf{S}_2, \mathbf{S}_3, \ldots, \mathbf{S}_n\}$ and create a visual vocabulary (can be also called a codebook) using clustering algorithms, which incorporates k visual words, $\mathbf{V} = \{\mathbf{V}_1, \mathbf{V}_2, \mathbf{V}_3, \ldots, \mathbf{V}_k\}$. Each word in the codebook is a clustering center, representing similar patches. But this way is imperfect: first, an image is partitioned into numerous subregions, the number of our image patches reaches 961 at most in the experiment, if we resemble 13 types of scene and cluster all features, it will challenge clustering algorithms; second, visual words' ability to discriminate become weaker, some similar regions in different scenes can be mapped into the same word, such as sea in a coast image may be thought as lawn instead. Against deficits above, we create visual words separately for each scene at a specific level and create an individual codebook, then synthesize situations from all levels to form a final vocabulary. Specific implementations are as follows:

1. At the ith level, we extract keypoints features from each image in m categories, such as a set belongs to the jth class at the ith level, denoted as $\mathbf{S}_i^j = \{\mathbf{S}_i^1, \mathbf{S}_i^2, \mathbf{S}_i^3, \ldots, \mathbf{S}_i^n\}$, $j = 1,2,\ldots,n$;
2. For m categories $\mathbf{S}_1, \mathbf{S}_2 \ldots \mathbf{S}_m$ at the ith level, generate visual words using K-means to create respective vocabulary, denoted as $\mathbf{V}_c = \{\mathbf{V}_i^1, \mathbf{V}_i^2, \mathbf{V}_i^3, \ldots, \mathbf{V}_i^k\}$, $c = 1,2,3.,\ldots,m$;
3. Pool all kinds of vocabularies as a final vocabulary $\mathbf{V} = \{\mathbf{V}_1, \mathbf{V}_2, \mathbf{V}_{3,\ldots,} \mathbf{V}_m\}$.

   Generation process is demonstrated in Fig. 38.3:

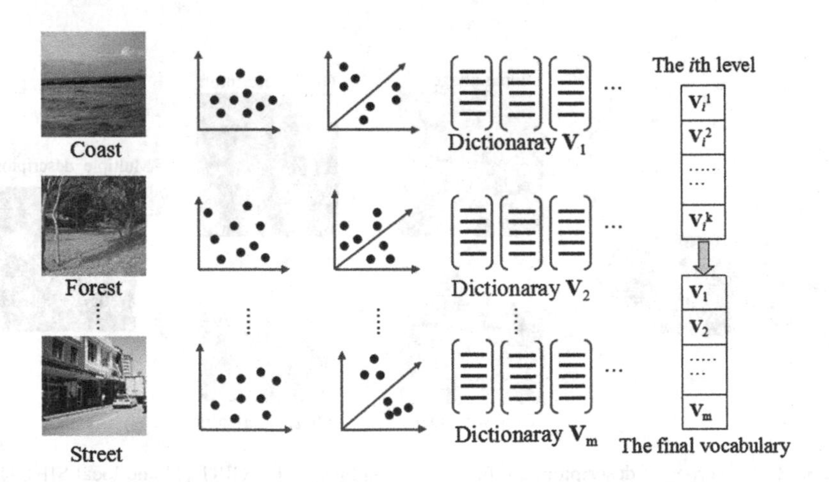

Fig. 38.3 Construction of a visual vocabulary

38.4 Spatial Pyramid Matching

A pyramid match kernel is designed to match two collections of features for orderless images as introduced in [4, 5]. Considering spatial information, spatial pyramid matching scheme is proposed in [5]. We first describe the fundamentals of pyramid matching: it works by dividing feature space into a sequence of increasingly coarser grids like a pyramid and taking a weighted sum of the number of matches at different resolution levels, in order to find similarity between two sets of features. At any specified level, two points are said to match if they fall into the same cell of the grid. Let X and Y be two sets of features in d-dimensional feature space, specifically, we construct a sequence of grids at resolutions from 0 to L for each dimension, at level L, the grid has 2^l cells, for a total of $D = 2^{dl}$ cells.

Let H_X^l and H_Y^l be the histograms of X and Y at level l, so $H_X^l(i)$ and $H_Y^l(i)$ denote the numbers of points from X and Y that fall into the ith cell of the grid, then the histogram intersection kernel function [4] is applied to describe the number of matches at level l, as shown in (38.1), $I(H_X^l, H_Y^l)$ is abbreviated as I^l.

$$I(H_X^l, H_Y^l) = \sum_{i=1}^{D} \min(H_X^l(i), H_Y^l(i)). \tag{38.1}$$

Note that the finer level $l + 1$ incorporates the number of matches at level l, therefore, the number of new matches found at level l is given by $I^l - I^{l+1}$, for $l = 0, \ldots, L$. The similarity between features at finer resolution is higher than that at coarser one, so the weight at level l is set to $\frac{1}{2^{L-l}}$. Putting all levels together, we form the pyramid match kernel:

$$k^L(X, Y) = I^L + \sum_{l=0}^{L-1} \frac{1}{2^{L-l}}\left(I^l - I^{l+1}\right) = \frac{1}{2^L} I^0 + \sum_{l=1}^{L} \frac{1}{2^{L-l+1}} I^l. \tag{38.2}$$

Spatial pyramid matching works for two-dimensional image space with pyramid match approach, and use traditional visual words description in feature space. In a spatial pyramid model, two points are considered to match when they belong to the same visual word. All feature vectors are quantized into N visual words. Two sets of 2-dimensional vectors are given for each visual word, X_n and Y_n, representing the coordinates of features that correspond to the visual word n in the respective images.

A discrete example [5] is shown in Fig. 38.4, the image is quantified into three types of visual words, the top displays three different levels of resolution for this image, then we count the features that fall in each spatial bin for any word at any level and weight each spatial histogram. The final spatial pyramid matching kernel of L levels is denoted as (38.3):

$$K^L(X, Y) = \sum_{n=1}^{N} K^L(X_n, Y_n). \tag{38.3}$$

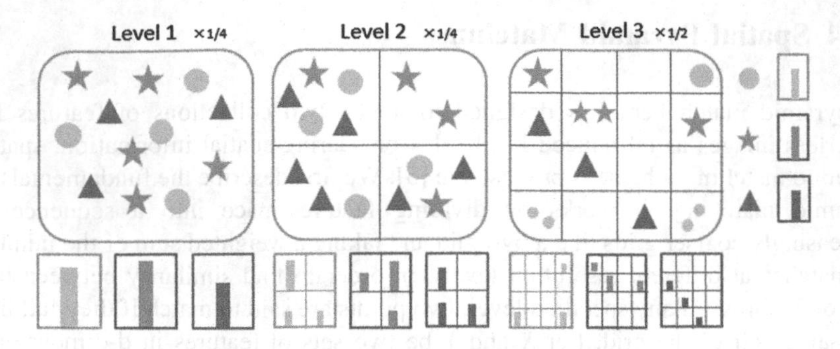

Fig. 38.4 A three-level spatial pyramid model: the image has three feature types, indicated by circles, stars, and triangles. An image is subdivided into three resolution levels

38.5 Experiments and Evaluation

Our set of experiments is on a dataset composed of 13 scene categories, which were provided by Li and Perona [3]. This dataset includes 13 categories: bedroom, coast, forest, highway, inside city, kitchen, mountain, living room, office, open country, street, suburb, tall building, Fig. 38.5 has displayed four categories of them from the training and testing set, respectively. Each category has 200–400 images, and we resize nonuniform images into 256 × 256 pixels.

For each category, 100 images are selected randomly as a training set, 50 images as a testing set. When multiple descriptors are input into K-means clustering algorithm, the vocabulary size are set as $N = 300$ or $N = 400$. The resolution level is set to three in the process of pyramid building. At the stage of categorization, SVM with different kernels are contrasted in Spatial Pyramid-BOW model and the modified model. All experiments are repeated five times. Table 38.1 lists two model's accuracies with different kernel functions, discriminative features are

Fig. 38.5 Example images from the 13 scene categories database: the four images on the left are from training set, the right four images are from the testing set

Table 38.1 Classification accuracy rate (%) on two models with different kernels

	Spatial pyramid-bow model				Modified spatial pyramid-bow model			
	RBF kernel		Histogram intersection		RBF kernel		Histogram intersection	
L	$N = 300$	$N = 400$	$N = 300$	$N = 400$	$N = 300$	$N = 400$	$N = 300$	$N = 400$
(0)	69.2	69.2	70.4	70.4	65.0	67.5	71.9	**71.3**
(0,1)	65.0	66.3	72.7	73.8	70.0	70.0	75.8	75.8
(0,1,2)	63.5	63.5	76.6	76.5	**77.7**	68.1	**80.0**	75.4

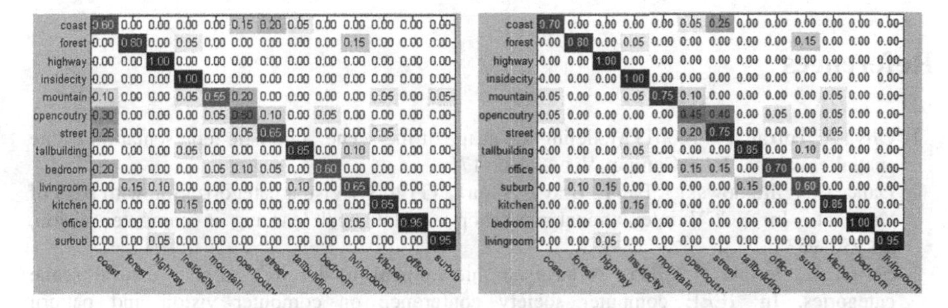

Fig. 38.6 Classification rates confusion table: the left one shows classification rates before taking our scheme, the one on the right shows classification rates after improvement

conducive to our performance when introducing a adapted model. We have used two kernels in SVM, one of them is Radial Basis Function (RBF), another is Histogram Intersection Kernel, as for performance, the latter is better than the former after validated. At $L = (0, 1, 2)$, $N = 300$, accuracy increases from 76.54 to 80 % when using the adapted model, and the bold data show the highest average classification rates of two models. And they are displayed in a confusion table, respectively. The diagonal shows classification rates for individual catcgorics, and the entry in the mth row and nth column is the percentage of images from class m that were misidentified as class n, illustrated by Fig. 38.6.

38.6 Conclusions

In this paper, we present a hierarchical feature extraction scheme with special codebook generation for natural scene classification. We build a pyramid to get local patches from an image in various scales and combine two types of descriptors. For each scene class, multiple descriptors are mapped into respective vocabulary and connected into a final one. Each patch of an image in the training and testing sets can be represented by visual words in the codebook and BOW histograms are computed. Then we make use of a spatial pyramid model to form pyramid histograms and concatenate all of them to form a high dimension vector representation of the image. Finally, SVM with a histogram intersection kernel is applied to realize classification of uncertain scenes. In the future work, we can deploy this approach with PLSA model to further recognize objects in an image, or apply it to other aspects such as clothing scenes more than natural scenes.

Acknowledgments This research is supported by NSFC-Guangdong Joint Fund (U0935004, U1135003), and the National Key Technology R&D Program (2011BAH27B01, 2011BHA16B08).

References

1. Oliva A, Torralba A (2001) Modeling the shape of the scene: a holistic representation of the spatial envelope. Int J Comput Vis 42:145–175
2. Dance C, Willamowski J, Fan L, Bray C, Csurka G (2004) Visual categorization with bags of keypoints. In: ECCV international workshop on statistical learning in computer vision, pp 950–953
3. Fei-fei L, Perona P (2005) A bayesian hierarchical model for learning natural scene categories. In: IEEE computer society conference on computer vision and pattern recognition, 524–531
4. Grauman K, Darrell T (2005) The pyramid match kernels: discriminative classification with sets of image features. ICCV 2:1458–1465
5. Lazebnik S, Schmid C, Ponce J (2006) Beyond bags of features: spatial pyramid matching for recognizing natural scene categories. Proc IEEE CVPR 2:2169–2178
6. Znaidia A, Shabou A, Le Borgne H, Hudelot C (2012) Bag-of-multimedia-words for image classification. In: 21st international conference on pattern recognition, ICPR, 1509–1512
7. Wu Z, Huang Y, Wang L, Tan T (2012) Group encoding of local features in image classification. In: 21st international conference on pattern recognition, 1505–1508
8. Ergul E, Arica N (2012) Scene classification using spatial pyramid of latent topics. In: 20th international conference on pattern recognition, 3603–3606
9. Li B, Xiong W, Hu W, Ding X (2012) Context-aware affective images classification based on bilayer sparse representation. In 20th ACM international conference on multimedia, 721–724
10. Yang J, Yu K, Gong Y, Huang T (2009) Linear spatial pyramid matching using sparse coding for image classification. In: IEEE computer society conference on computer vision and pattern recognition workshops, CVPR Workshops, pp. 1794–1801
11. Prabhakar CJ, Kumar PU (2012) LBP-SURF descriptor with color invariant and texture based features for underwater images. In: Proceedings of the eighth Indian conference on computer vision, graphics and image processing, no. 23
12. Xie L, Tian Q, Zhang B (2012) Spatial pooling of heterogeneous features for image applications. Association for Computing Machinery, Nara, pp 539
13. Lowe D (2004) Distinctive image features from scale invariant keypoints. Int J Comput Vis 60(2):91–110

Chapter 39
Intuitive Game Design for Early Learning in Music

Szu-Ming Chung and Chun-Tsai Wu

Abstract This research developed an intuitive interactive game to prepare subjects for early stage learning in music. The design content is based on Bruner's knowledge map and applied Orff and Kodály's philosophy and teaching to inspire music intelligence in children and prepare them for music learning later in life. Experimental research is also established in this study. Via experimental research, the authors test the records, analyze, and discuss the influential factors on music learning and the development of musical intelligence. This paper presents the first half of experimental research design, including exploring the design principles, developing musical games in an intuitive and interactive manner, and constructing a device on the android system with Flash.

Keywords Intuitive game design · Interactive game design · Music learning game · Android system · Multi-touch screens

39.1 Introduction

This research developed an intuitive interactive game to prepare subjects for early stage learning in music. For the following reasons, the android system, Flash, and tablets (multi-touch screens) are chosen to be the research tools: (1) the motions of multi-touch gestures provide a platform for intuitive and interactive design, which is advantageous for inspiring early stage musical activities primed for later stage

S.-M. Chung (✉) · C.-T. Wu
Department of Digital Content Design, Ling Tung University, Taichung, Taiwan, Republic of China
e-mail: szuming@teamail.ltu.edu.tw; smc200312@hotmail.com

C.-T. Wu
e-mail: ltctht53@teamail.ltu.edu.tw

A. A. Farag et al. (eds.), *Proceedings of the 3rd International Conference on Multimedia Technology (ICMT 2013)*, Lecture Notes in Electrical Engineering 278,
DOI: 10.1007/978-3-642-41407-7_39,

instrumental learning; (2) intuitive animations can easily be designed and completed to motivate visual activities and physical moves and interactions; (3) when the interactive installment/platform becomes a virtual reality, the visual and touch layout increases a user or learner's sense of reality and challenge; (4) they are interesting, inspiring, and easy to operate and manipulate by instinct for young children; and (5) they can be easily combined with other technologies to record every gesture made by the player/user/learner.

A good interactive design not only lies in the visual arts simulating motive to use, but also launching a personal mental activity and relationship with this interactive design. By utilizing these technological devices, experimental research can also be established. The experiment results and teaching effects are not obtained from standard tests but from actual recordings of players interacting with a machine. Through experimental research, the authors observe and record to discover how kindergarten children start a personal mental activity, start being motivated to learn, start knowing and recognizing the musical elements, and then start using them to engage in creating music. Via experimental research, the authors test the records, analyze, and discuss the influential factors on music learning and the development of musical intelligence. This paper presents a research design, including exploring the design principles, developing musical games in an intuitive and interactive manner, and constructing a device to record the player/user/learner's gestures.

39.2 Musical Development

Almost all human beings have (96 %) innate musical ability unless they have genetic defects [1]. A child's development is affected by many factors, such as basic biological potential, maturity, experience, opportunities, interests, education, family, peers, and sociocultural connotations [2]. However, musical development is influenced by experience, opportunities, and motivation (Sloboda's definition of influencing factors) [3]. Other researchers emphasize acculturation (enculturation) and the impact of cultural experiences. Many research proved that musical development is closely related to the environment at home. What music or songs they hear at home is greatly different from in schools. Sometimes the interference of parents decides a child's development in music. Most people think that musical talent is very rare, such as that absolute pitch is god given. But scientific studies [4, 5] found the following phenomena: (1) hidden in their innocent appearance, all human babies show amazing music-related skills and talents; (2) these talents show a typical developmental process independent from training or education; (3) parents or caregivers can provide an environment with an accelerated development of musical skills; (4) musical talent does not have a large impact on later musical development which is influenced, rather by mundane reasons: such as parental support, involvement, and continued practice [1].

Inspired by Piaget's developmental psychology, many music studies found that there are stages in the musical development of children. Although the distinction is not clear, it can be drawn out as a spiral diagram of musical development [1]. From birth to 4-years old is the first stage. At this stage, children will explore a variety of sounds including musical sounds, instrumental sounds, and singing voices. Their explorations will focus on volume, dynamics, tone quality, and they will try to repeat the rhythmic and melodic patterns. Five- to nine-years old is the second stage. At this age, children will try to convey emotion through music and stories, especially through songs, and will also notice changes in the speed and intensity of sound. They can follow traditional music activities and demonstrate a sense of rhythm, tempo, and fixed music phrases.

39.3 Music Teaching Methods

Orff's concept of teaching music starts with early childhood education and rhythmic training. All development stages center on musical rhythm, beginning with the initial language, including poetry, fairy tales, drama, rhymes, proverbs, poetic prose, calls, folksongs, pentatonic, church modes, accompaniment, rhythmic instruments, improvised language performances, folk dance, dance, mime, and other percussion-related activities [6].

Kodály used so-fa syllables and hand signs combined with rhythmic syllables to work as the start of musical learning activities for young children. Hand sign syllables combined with body movements are especially suitable for children who cannot read. They are also good preparation for music reading.

The teaching methods of Orff and Kodály share great resemblance [7]. Both of them advocated the sequence of music learning and process of language learning (patterns, rhymes, and proverbs), singing (two to five tones pentatonic melodies), playing musical instruments and accompaniment (Orff used percussion instruments and Kodály used recorders), improvisation, and rhythmic activities (various sounds made by the body, corresponding to a variety of accompaniment, songs, games, and musical forms).

39.4 Interactive Design and Music Learning

Interactive design involves extensive research, a majority of which is borrowed from a great part of the research of psychology, memory, and cognition [8]. For children, learning is to experience "how it sounds?" and "How does it feel?" Between the sensation and perception they feel new experiences. The perceptual experience becomes a part of cognition through judging, classifying, and memorizing. On perception studies, Gibson suggested that the survival of humans depends not on changing their perception of the world, but perceiving the world around them. The external behavior of seeing, hearing, smelling, tasting, and touching is to discover the external stimulus and to receive the best information [9, 10]. When we listen to music, we connect what we hear to concert halls, and all the movements and behavior of performers. These links became an expected result of concerts. For example, if we hear Rimski-Korsakov's Bumblebee, we will naturally think of a bumblebee's flight. This is based on the fact that how an insect flaps its wings generates the indicator of rhythm, movement, and action [11].

Ecological research methods have great influence on rhythm perception and cognition. A sense of rhythmic timing is generated from a special attention to sources of sound, including its dynamics and varieties of levels. We constantly anticipate future events greatly based on the rhythmic timing of incidences and a deep understanding of associated events. To hear the difference between two tones requires a minimum of 20 ms inter-onset interval [12, 13] (ms IOI). Regardless of the length of time between two tones, these perceptual activities produce a different recognizable space due to different forms and complexity. The minimum spacing should be 100 ms for it to be recognizable. The more important implication is that the perception of the action and visual perception have an absolute relationship [14]. In a specific time frame, the continuous visual stimulus can lead to an illusion [15]. There is a remarkable parallel relationship between visual perception and actions.

Tapping to a metronome directly involves and develops tempo synchronization and a sense of timing. We can produce very consistent percussive action without listening to or following a metronome [16–19]. Following pitched notes to tap is more accurate than following a metronome, because the perceptual competence of auditory recognition allows us to easily identify melodic patterns [20]. Melody is composed of a string of different pitches. It creates the shape of patterns and direction. With such perceptual competence, we can distinguish between the theme (figure) and non-thematic elements or patterns (background). A string of notes in a different octave cannot easily be recognized; conversely, a melodic line within an octave whose proximity, similarity, good continuation, and coherence are easily recognized by auditory competence [21].

Children can recognize chords without formal music training because there is no difference in listening ability between 6 and 11-year-old children. The major factor is that the vast majority of music they have heard is accompanied by chords [22].

The formation of knowledge cannot be attained by intuition and become meaningful to individuals [23]. According to Croce [24], the hierarchy of knowledge formation is through sensory impressions, through intuition knowledge, and then to achieve logical knowledge. Intuitive knowledge links dynamic modes, graphics, and various representations. Therefore, intuitive knowledge is the core and the central conversion station of all the senses and meanings. Intuitive knowledge is a realization of daily activities. It is essentially an imaginative movement. In the process of movement, we create formative images. When we hear a melody, we do not hear different pitches, but organize it into a meaningful image or feeling [25], this argument proved that we cannot test musical ability by analytic method, because it will contradict the intuitive knowledge and experience.

39.4.1 Computer Learning and Visual Design Principle

In a classroom setting, the context or time to construct meaning on a personal level is unexpectedly limited [26]. Computer- assisted learning is regarded as competing with machines and technology to achieve a mastery of skills [27]. The traditional design theory of the computer-assisted learning focuses on results. Conversely, learning theory focuses on the process. Recent research and development of these theories seem to show that they complement each other. Both agree that the design process is to construct something meaningful to a learner. Design results require an objective learning environment that also cultivates meaning on a personal level. When learners produce something concrete and sharable, the process is the most meaningful.

Many researchers found that once the user focuses their attention on the tactile pattern, it will be difficult for them to transfer to the auditory or visual mode [28–30]. If visual stimulation can occur a little earlier than tactile stimulation, it will synchronize the inner occurred tactile stimulation and visual stimulation, and shift the attention to visual patterns [29].

39.4.2 Game Design Content

Bruner's knowledge map [31] constructs its symbolic system through sensory experience and imagery thinking. Based on Bruner's knowledge map, the researchers designed an intuitive way for players to interact with the game content. Such a game design offers an auditory-visual experience of rhythms, melodies, and chords. Through interacting with the game interface, the player establishes his/her knowledge map of the Western music system which can be used to express, communicate, and create music. This game's design refers to the perception and cognitive learning process mentioned in this paper. It also applies Orff and Kodály's philosophies and teachings to inspire music intelligence in children and prepare them for music learning later in life. The game content design progresses in a sequential two levels each with four stages arranged from easy to difficult. Each stage begins with an animation of the gameplay. The following section includes the music game design, flowchart, and visual design.

Level 1

1. Tapping game: this game establishes the feeling of steady beat and basic rhythmic patterns. By tapping, the player/learner has to match the beating sound (1 beat per a second) from an easy beginning of two beats, to a medium

difficulty of 4 and 8 beats, and then to a hard difficulty of 12 and 16 beats (2 or 4 beat patterns).

2. Ladder game: this game establishes listening skills of absolute pitches on a 7 step ladder (the order of a rainbow colors represents the order of an octave scale of do, re, mi, fa, so, la, and si/ti).
3. Marble game: this game establishes the feeling of up/down melodic lines. The player has to use a shaking gesture to match the up/down melodic lines. If replaying the stage, the player will hear similar up/down melodic lines with different instruments.
4. Ocean bubble game: this game offers a space for player to compose music freely by using knowledge learned in prior games. The player can tap single colored bubbles to form a piece of music.

Level 2

1. Hopscotch game: this game provides listening experience of intervals (steps)—dissonant seconds (steps of two), and consonant thirds (steps of three). By matching the boxes of hopscotch, the player can hear and know consonant/dissonant intervals.
2. Block stacking game: this game provides listening practice of consonant thirds and dissonant second (similar to tile matching puzzle game).
3. Necklace game: this game constructs the knowledge of musical form. The player can choose 4 note patterns (by different colored tones) to create a necklace (musical forms of AA, ABA, AABA, ABABA; Capital letters of A and B represent different patterns).
4. Ocean bubble game: this game appears again as in level 1. Instead of flicking single colored bubbles, the player can choose any string of four colored bubbles (4 note patterns) emerging from the ocean space. After finishing, the player can choose to listen, compose a new one or end the game.

Flowchart

The flowcharts for level 1 and level 2 are listed below. Each of them shows four stages in sequence. The player has to pass the prior level or stage to progress to the next stage (Figs. 39.1, 39.2, 39.3, 39.4).

Gameplay Interface

The following visual design displays a sample of the sequential steps which the player will experience in this game (Figs. 39.5, 39.6, 39.7, 39.8).

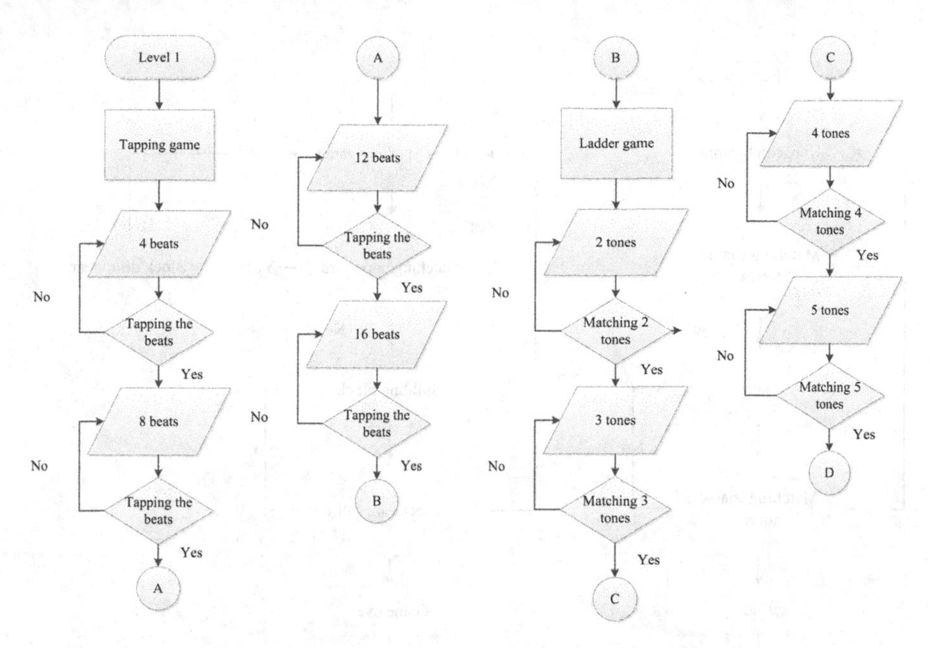

Fig. 39.1 Level 1-game stage 1 and 2

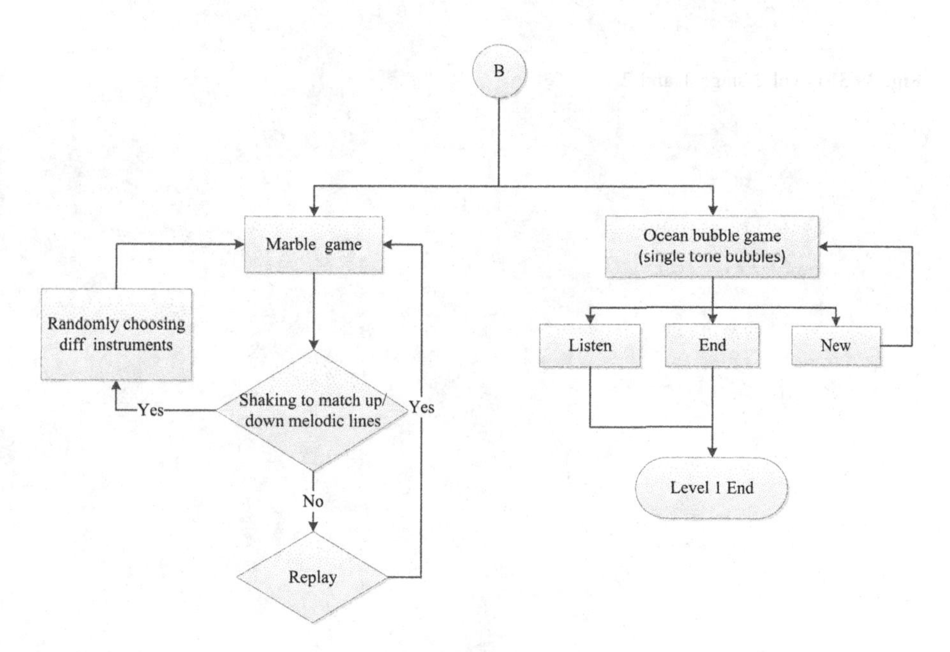

Fig. 39.2 Level 1-game stage 3 and 4

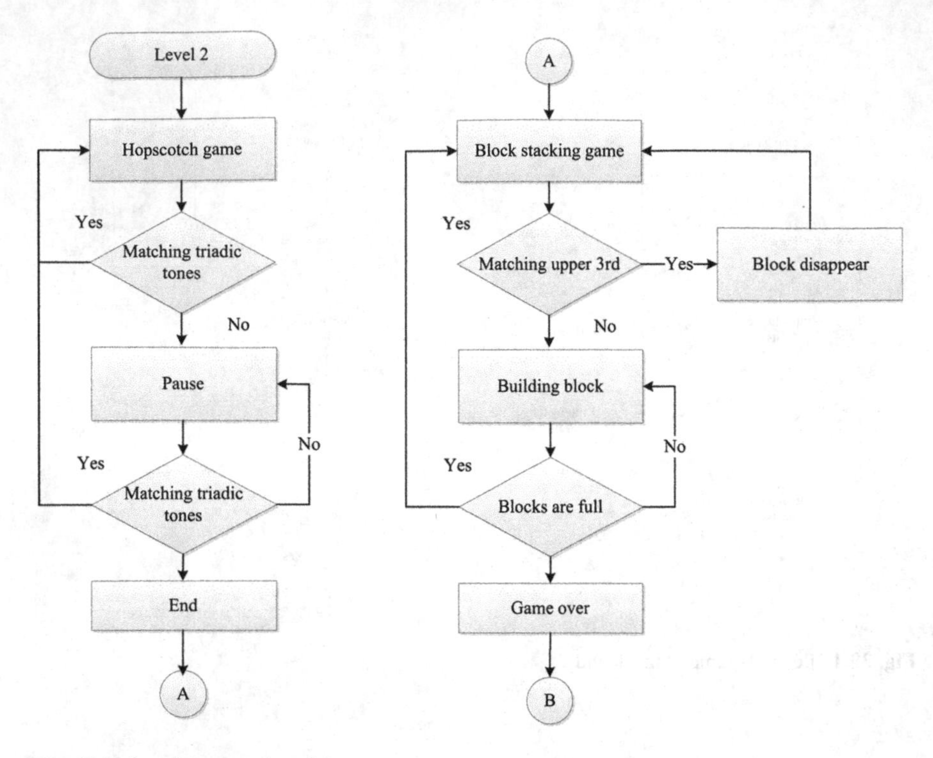

Fig. 39.3 Level 2-stage 1 and 2

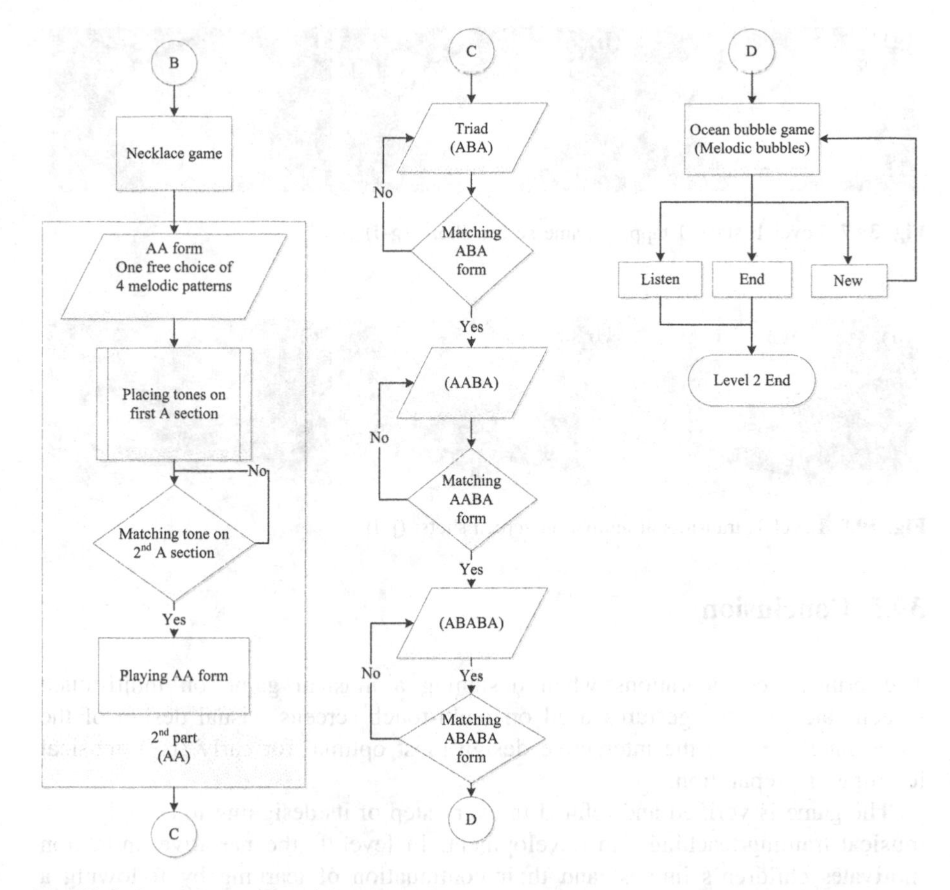

Fig. 39.4 Level 2-stage 3 and 4

Fig. 39.5 Beginning animation screen shots: **(a–c)**

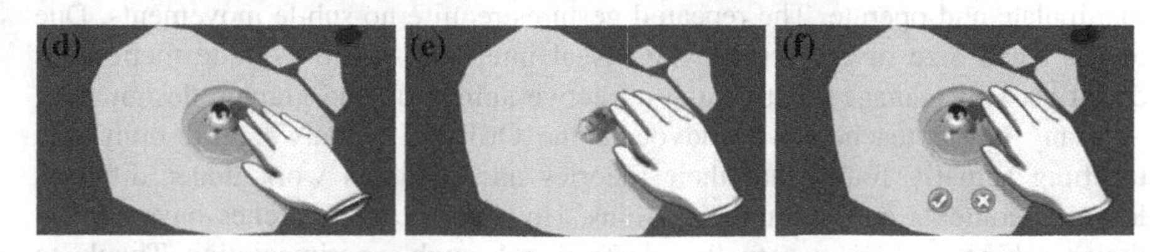

Fig. 39.6 Level 1, stage 1 gameplay animation screen shots: **(d–f)**

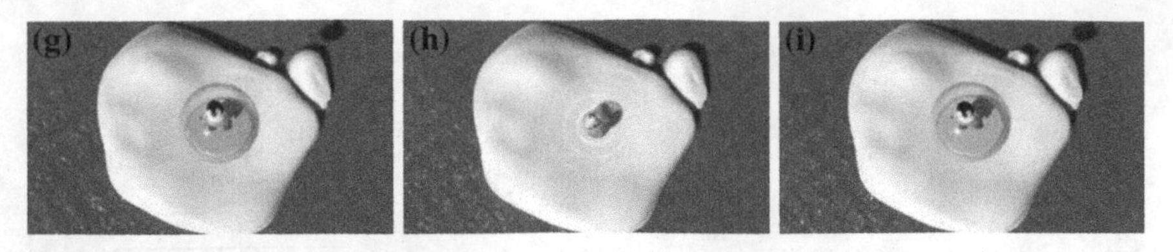

Fig. 39.7 Level 1, stage 1 tapping game screen shots: **(g–i)**

Fig. 39.8 Level 1, transitional animation screen shots: **(j–l)**

39.5 Conclusion

The primary considerations when designing a musical game on multi-touch screens includes the gestures used on multi-touch screens, visual design of the game interface, and the interactive design most optimal for early stage musical learning or preparation.

The game is verified and refined in every step of its designing to comply with musical training/teaching and development. In level 1, the narrative animation motivates children's interest and their continuation of learning by following a learning agent (a penguin). The sequential learning in level 1 progresses from the beat, the single pitches (a rainbow ladder implies diatonic and pentatonic scales), up/down scales, to free-writing of single tones. In level 2, it progresses from consonant/dissonant tones, triad and non-triad tones, musical forms, to the free writing of melodic patterns.

There are several disadvantages to using multi-touch screens. To protect the child's vision, a film screen protector is applied to the tablet. This results in a less sensitive touch screen. The multi-touch screens have limited gestures available for use. They are, however, enough for building the preparatory learning blocks for young children. The limited gestures are relatively easy for young children to manipulate and operate. The repeated gestures require no subtle movements. Due to the small size of the screen, the visual interface should exclude meticulous detail yet still manage creative and attractive animation and graphic design.

Many music teachers who advocate the Orff and Kodály's philosophy and teaching actually found that their theories and methods work under different learning contexts and cultural situations. However, few researches have proved their teaching methods to actually be effective through experimentation. Thanks to technology and methodology of experimental research, this music learning game can be designed and records every move of what the player/learner has acted and reacted to in the process of playing the game. The experiment and analysis of this experimental research will be continued in the following year.

Acknowledgment This research is funded by National Science Council in Taiwan, ROC (Case number: NSC 101-2410-H-275-009-).

References

1. Lehmann AC, Sloboda JA, Woody RH (2007) Psychology for musicians: understanding and acquiring the skills. Oxford University Press, New York
2. Welch G (1998) Early childhood musical development. Res Stud Music Educ 11:27–41
3. Mihill C (1993) The myth of the 'gifted' musician. The Guardian, 1 September
4. Runfola M, Swanwick K (2002) Developmental characteristics of music learners. In: Colwell R, Richardson C (eds) A project of music educators national conference. Oxford University Press, New York
5. Gembris H (2002) The development of musical abilities. In: Colwell R, Richardson C (eds) The new handbook of research on music teaching and learning. Oxford University Press, New York, pp 487–508
6. Warner B (1991) Orff-Schulwerk: application for the classroom. Prentice Hall, Englewood Cliiffs
7. Wheeler L, Raebeck L (1972) Orff and Kodály: adapted from elementary school. Wm. C. Brown Company Publishers, Dubuque
8. Kolko J (2007) Thoughts on interaction design. Brown Bear LLC, Savannah
9. London J (2004) Hearing in time: psychological aspects of musical meter. Oxford University Press, New York
10. Gibson JJ (1982) A note on problems of vision to be resolved. In: Reed E, Jones R (eds) Reasons for realism: selected essays of James J. Gibson. Erlbaum, Hillsdale, pp 391–396
11. Gjerdingen RO (1994) Apparent motion in music? Music Percept 11(4):335–370
12. Hirsh IJ (1959) Auditory perception of temporal order. J Acoust Soc Am 31(6):759–767
13. Hirsh IJ, Monohan CB et al (1990) Studies in auditory timing: 1. Simple Pattern Percept Psychophysics 47(3):215–226
14. Wertheimer M (1912/1961) Experimentelle Studien über das Sehen von Bewegung. In: Classics in psychology (Reprinted and trans: Thorne Shipley). Philosophical Library, New York, pp 1032–89
15. Bregman AS (1990) Auditory scene analysis: the perceptual organization of sound. MIT Press, Cambridge, pp 173–184
16. Bregman AS (2005) Auditory scene analysis and the role of phenomenology in experimental psychology. Can Psychol 46(1):32–40
17. Woodrow H (1932) The effect of rate of sequence upon the accuracy of synchronization. J Exp Psychol 15(4):357–379
18. Dunlap K (1910) Reaction to rhythmic stimuli with attempt to synchronize. Psychol Rev 17(6):399–416
19. Stevens LT (1886) On the time sense. Mind 11(43):393–404
20. Dowling WJ (1994) Melodic contour in hearing and remembering melodies. In: Aiello R, Sloboda JA (eds) Musical perception. Oxford University Press, New York, pp 173–190
21. Meyer LB (1956) Emotion and meaning in music. The University of Chicago Press, Chicago
22. Colwell R (ed) (2006) MENC handbook of musical cognition and development. Oxford University Press, New York

23. Polanyi M, Prosch H (1975) Meaning. University of Chicago Press, Chicago
24. Swanwick K (1994) Musical knowledge: intuition, analysis, and music education. Routledge, New York
25. Warnock M (1976) Imagination. Faber, London
26. Kafai Y, Resnick M (1996) Constructionism in practice: designing, thinking, and learning in a digital world. Lawrence Erlbaum Associates, New Jersey
27. Howard RW (1995) Learning and memory: major ideas, principles, issues and applications. Praeger Publishers, Westport
28. Spence C, Nicholls MER, Gillespie N, Driver J (1998) Cross-modal links in exogenous covert spatial orienting between touch, audition and vision. Percept Psychophys 60:544–557
29. Spence C, Shore DI, Klein RM (2001) Multimodal prior entry. J Exp Psychol Gen 130:799–832
30. Turatto M, Galfano G, Bridgeman B, Umiltà C (2004) Space-independent modality-driven attentional capture in auditory, tactile and visual systems. Exp Brain Res 155:301–310
31. Bruner JS (1971) The relevance of education. WW Norton & Company Inc., London

Chapter 40
Human Action Recognition with Block-Based Model and Flow Histograms

Jincai Song and Fuqiao Hu

Abstract We propose a compact representation for human action recognition by employing human block-based model and local optical flow features. The contour descriptor based on block-based model proposed in this paper can be used to represent the movement of human body exactly by dividing silhouette into different blocks. Meanwhile, we propose an accurate and stable optical flow descriptor for motion information. Finally, our action feature vector is constructed by using contour and optical flow descriptor together with global velocity information. In addition, we adopt the well-known Bag-of-words (BoW) model to obtain the final video level representations. Our approach is tested on two public datasets: Weizmann and KTH. Experimental results show that our approach achieves a 100 % test accuracy on Weizmann dataset and outperforms some state-of-the-art techniques on KTH dataset.

Keywords Action recognition · Block-based model · Optical flow · Bag of word

40.1 Introduction

Human action recognition is of great significance in various applications such as visual surveillance, human-computer interaction, and automatic video annotation. However, how to make the recognition faster and more accurate is still an open issue in the computer vision community.

Human action can be represented with a range of features. In the recent approaches, the silhouette, optical flow, spatial-temporal interest points, HOG3D, motion SIFT, and other features are widely applied. Folgado et al. [1] used the

J. Song (✉) · F. Hu
System Control and Information Processing Institute, Shanghai Jiao Tong University, 200240 Shanghai, China
e-mail: sjcsjtu@163.com

A. A. Farag et al. (eds.), *Proceedings of the 3rd International Conference on Multimedia Technology (ICMT 2013)*, Lecture Notes in Electrical Engineering 278,
DOI: 10.1007/978-3-642-41407-7_40,

spatial-temporal silhouette analysis for monitoring of human activity. Chaudhry et al. [2] proposed a motion descriptor based on histograms of oriented optical flow (HOOF) to describe human action. Laptev et al. [3] first introduced the notion of 'space-temporal interest points' to identify human actions. Blank et al. [4] defines actions as three-dimensional shapes induced by the silhouettes in the space-time volume [4]. Dollar et al. [5] propose a local descriptor based on histograms of oriented 3D spatial-temporal gradients. Those 3-D temporal and spatial features demonstrate the robustness to partial occlusions, nonrigid deformations, significant changes in scale and viewpoint, but the feature extraction is relatively complicated with high computational cost.

In this study, we demonstrate how we can utilize a novel contour descriptor and a dense representation of optical flow together with global velocity information for human action recognition. We use Bag-of-words (BoW) model [6] to obtain the video level representations, where each video sequence is represented as a bag of words histogram vectors. This paper uses histogram intersection kernel SVMs in the classification step [7]. The experiment shows that the proposed approach has an encouraging recognition performance with relatively low computational cost.

40.2 Our Human Action Recognition Approach

In this section, we first introduce a new contour descriptor based on the block-based model [1], which is described for monitoring of human activity. Then we also use a compact optical flow descriptor for motion information. Using contour and flow descriptor together with motion context information, we obtain the video level representations with BoW model.

40.2.1 Human Block-Based Model

The model presented in this work consists of six regions with the same height (see blocks $B_1,\ldots, B_6$ in Fig. 40.1a). Each of these regions can be delimited by a bounding box which is called as "block." In this case, the blocks of this division correspond to areas related to the physical position of certain parts of the body. For example, standing and in a position of repose, the correspondences are as follows: head is in $B_1,\ldots,$ feet are in B_6. Therefore, if we analyze gait recognition, we know that in a normal situation these are between blocks B_5 or B_6. This simplifies the problem and reduces it to a local analysis of these blocks. This model can also deal with lateral and frontal views. A human blob divided into blocks in lateral (a, b) and frontal (c, d) views as in Fig. 40.1a.

We use Fig. 40.1b to help us define the parameters and significant points used in the model. First, the upper and lower points (P_u and P_L) are defined, which limit the height of the set of blocks, H_T, enable us to divide into the different blocks,

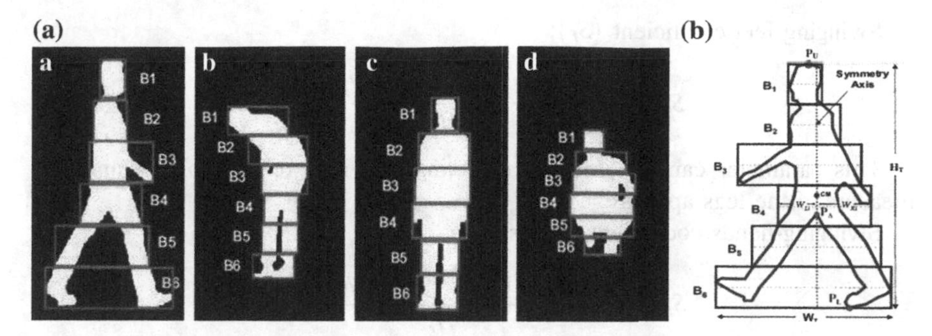

Fig. 40.1 **a** shows the human block-based model, **b** indicates primary points and parameters

B_i, $i = 1\ldots6$. All blocks have the same height ($H_{\mathrm{Bi}} = H_T/6$). Then we can define a new parameter: the width of each block, W_{Bi}.A symmetry axis is defined by the vertical line that passes through the centre of mass of the entire blob (CM). This axis allows us to measure the block width on each side (W_{Li}, W_{Ri}). We normalize the width parameters with respect to the distance to the camera.

From this initial points and measures, a set of secondary parameters are defined related to different situations:

The changed width vector (CW^T) is defined as follows:

$$CW_i^T = \frac{W_{\mathrm{Bi}}(t)}{W_{\mathrm{Bi}}(t-T)}, \quad i = 1,\ldots,6 \tag{40.1}$$

where each component contains the relation between the block width in a frame *t* and the preceding *t-T* frame for each block B_i.

The changed centre of mass vector (ΔCM^T):

$$\Delta CM^T = \left(CM_x(t) - CM_x(t-T),\ CM_y(t) - CM_y(t-T)\right) \tag{40.2}$$

where *t* represents the current frame instant, *t-T* represents the preceding *T* frame and CM_x and CM_y are the *x* and *y* coordinates of *CM*, respectively.

The symmetry vector (*S*) is defined by:

$$s_i = \frac{\min(W_{\mathrm{Li}}, W_{\mathrm{Ri}})}{\max(W_{\mathrm{Li}}, W_{\mathrm{Ri}})}, \quad i = 1,\ldots,6 \tag{40.3}$$

The directional symmetry vector (*DS*) is:

$$DS_i = \frac{W_{\mathrm{Li}}}{W_{\mathrm{Ri}}}, \quad i = 1,\ldots,6 \tag{40.4}$$

where each component represents the adirectional symmetry proportion between the widths of the parts of the block B_i to the right and left of the symmetry axis.

Swinging feet coefficient (S_f):

$$S_f = \frac{\max(W_{L5}, W_{L6}) + \max(W_{R5}, W_{R6})}{H_T} \tag{40.5}$$

This parameter can be used for detecting activities related to gait and is a measure of the legs aperture.

Swinging-hands coefficient (S_h) is:

$$S_h = \frac{\max(W_{L3}, W_{L4}) = \max(W_{R3}, W_{R4})}{H_r} \tag{40.6}$$

This is an important parameter when studying periodicity, type of movement, and different actions associated with the movement of the arms.

In addition, we use spatial pyramid method to extract the local features in each block. As shown in Fig. 40.1a, the B_5 is subdivided into four sub-blocks and the raw pixel values of the silhouette are integrated over the domain of each sub-block. The Block-Based Model is composed of six blocks, yielding a 24(4 × 6)—dimensional histogram vector and normalize it as P. Therefore, the contour descriptor is obtained from the following feature vector $(CW^T, \Delta CM^T, DS, S, S_f, S_h, P)$, thus, feature vector of this descriptor has 46 dimensions.

40.2.2 Optical Flow Feature Extraction and Representation

We extract the local optical flow in motion region and then propose a new stable and accurate motion descriptor to denote motion information. We use Lucas-Kanade algorithm to calculate the optical flow inside the normalized bounding box [8]. The optical flow field is splitted into vertical and horizontal channels which gives us two real-valued channels F_x and F_y as illustrated by Fig. 40.2.

Given an optical flow field OFF, we will treat it as complex numbers or vectors, i.e., $\text{OFF} = F_x + F_y i$ The corresponding polar coordinate representation is $\text{OFF\_PC} = (\rho, \theta)$, where ρ is the intensity and θ is the phase angle. In this manner, the optical flow intensity and phase angle are treated as two separated channels of the OFF.

The centroid of intensity channel is computed as

$$C_\rho = \frac{1}{N}\sum_{i=1}^{N}\left(1 - \exp\left(-\frac{\rho_i^2}{2\sigma^2}\right)\right) \tag{40.7}$$

where N is the number of pixels, ρ are all normalized, and σ is the standard deviation of Gaussian kernel used in normalization of ρ.

Fig. 40.2 Horizontal and vertical optical flow

The centroid of phase angle channel is computed as

$$C_\theta = \mathrm{ANGLE}\left(\sum_{j=1}^{N} \frac{x_j + y_j i}{\rho_j}\right) \tag{40.8}$$

where $x_j + y_j i$ is the element on OFF, and the function ANGLE is to compute the phase angle of a vector.

Assuming that $\xi_{i,j}$ is the jth element of channel i, $i \in \{\rho, \theta\} \cdot \xi_{\rho,j}$ should be normalized as ρ in (40.7). The relation between the centroid of channel and $\xi_{i,j}$ is solved by:

$$R_{\xi_{i,j}} = \exp\left(-\frac{\left(\xi_{i,j} - C_i\right)^2}{2\sigma_i^2}\right) \tag{40.9}$$

where σ_i is standard deviation of channel i, $i \in \{\rho, \theta\}$, and C_i is the centroid of channel. The motion area is represented as relations of two independent channels.

We adopt the method of subregional radial histogram to represent the optical flow features [8]. As is shown in Fig. 40.3, the normalized bounding box is divided into 2 × 2 subwindow. We extract the two kinds of channel information R in the

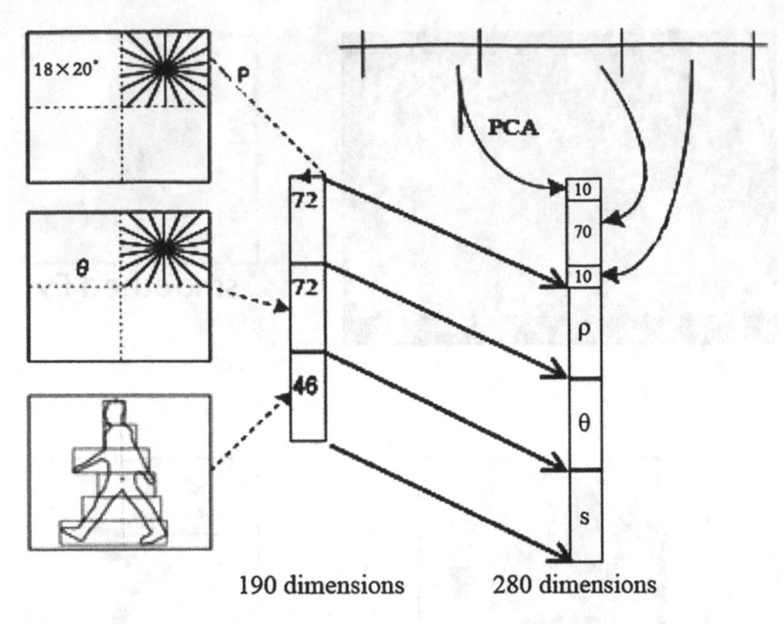

Fig. 40.3 Feature extraction flow chart

subwindow. Each subwindow is then divided into 18 pie slices, each of them covers 20° and the slices do not overlap. We treat the center of the subwindow as the center of the pie. Therefore, the bounding box is divided into 72 subregions. Let $S_{k,i}$ represents the kth subregion for relations of channel i, thus the value of channel i over the domain of the kth interval is:

$$M_{k,i} = \frac{\sum_{j,\, R_{\xi i,j} \in S_{k,i}}^{N} R_{\xi i,j}}{\sum_{j}^{N} R_{\xi i,j}} \tag{40.10}$$

where M is a histogram which denotes the ratio of the summation of relations in $S_{k,i}$ to the total relations of channel i. As a result, we can get a $72(2 \times 2 \times 18)$—dimensional histogram for channel ρ and θ.

Assuming OFF is the optical flow field with some level of noise and *OFF is the one computed precisely on each frame. Due to the effect of noise, C_ρ and C_θ may be not equal to them of *OFF. However, since M describes the relation between $\xi_{i,j}$ and the centroid of channel i, the R of OFF have almost the same errors with *OFF respectively. In Sect. 40.4.1, we have verified our motion descriptor composed of the resulting statistics performs reliably under the noisy data and is a discriminative feature for human action recognition.

40.2.3 Motion Context Descriptor

The optical flow field gives us two real-valued channels ρ and θ. The Block-based model gives us the third (binary) channel S. By concatenating the histograms of all three channels we obtain a 190-dimensional frame descriptor, which reflects the appearance of human action and motion information. Our action descriptor also borrows the idea of [8] to obtain the motion context descriptor. We use 15 frames around the current frame and split them into three blocks of five frames: past, current, and future. For a five-frame window, the frame descriptor of each block is then projected onto the first K principal components using PCA. We pick the first 70, 10, and 10 dimensions for the current, past, and future blocks, respectively. Therefore, our descriptor is formed by the 90-dimensional context descriptor appended to the current 190-dimensional frame descriptor. The extraction process of the final 280-dimensional motion context descriptor is showed in Fig. 40.3.

After extracting the motion descriptor, we adopt the well-known BoW model to obtain the video level representations. In its basic form, by utilizing the *K*-means clustering, we could acquire the target action codebook $CBs = \{cs1, cs2, \ldots csV\}$, in which cj represents a visual word, namely the cluster center. The center of each resulting cluster is defined to be a visual word.

Note that the extracted codebook features are very different from the spatio-temporal interest points in [9]. In this paper, each frame corresponds to a visual words after clustering. Although the features are denoted in different ways, all of them are based on BoW model.

40.3 Recognizing Actions

After the feature extraction step, we use multiclass SVM as our classifier [10]. Note that the action recognition models represented by histogram generally need specific nonlinear kernel function, which leads to a high complexity during training. By testing on different kernel functions, we found that the intersection kernel is superior to other kernel functions for our action recognition model:

$$K(X, y) = \sum_{i=1}^{m} \min(x_i, y_i) \tag{40.11}$$

where x_i and y_i are feature vectors extracted from the video set. The kernel has no free parameters so it works well for models that require real-time computation.

In addition to our local motion information (optical flow histogram), we also utilize an additional global velocity information to improve the overall recognition performance. Here, we propose a simple feature, which is the overall speed of the target object in motion. If we want to discriminate two actions: "bend" versus "running." The probability of the person is running would be quite low only if the velocity of him is equal to zero.

Based on this idea, we use a two- level classification system [11]. In the first level, we use the global velocity information for rough classification. In the second level, we evaluate the sequences using our contour and flow descriptor. We take the maximum response of the SVM for the testing sequences as our classification decision.

40.4 Experimental Methods and Results

Our system has been fully implemented by MATLAB/C ++ and all the experiments were conducted on a 2 GHz Intel Core2 Duo under Windows 7. First, we use the above motion context descriptor to obtain a high-dimensional feature vector, then K-means clustering is applied to generate a BoW representation. The optimal value of K can be obtained by repeating experiments, each sequence can be expressed as a K-dimensional vector. In order to obtain an unbiased accuracy estimation, we also employ leave-one-out manner to verify the experimental results, which tests each original video clips successively while training all of the remaining clips. The flipped version of test video sequence is excluded from the training set. We report the average accuracy at the optimal value of K.

40.4.1 Performance on Weizmann Dataset

The Weizmann dataset is a standard dataset to evaluate various human action recognition algorithms. It consists of ten types of human actions belonging to 10 action classes including bend, jack, jump, pjump, run, side, skip, walk, wave1, and wave2. Although the recognition accuracy of this dataset has reached 100 % in several papers, they have used fully 3D model which are computation intensive. We also achieve 100 % on this state-of-art dataset. Especially, our histogram features are simple to extract so as to significantly reduce the computational cost, while the classification performance is also maintained.

The respective confusion matrix is shown in Fig. 40.4a, which reveals that the average accuracy is close to 100 %. In Fig. 40.4b, comparisons are conducted using our motion descriptor and the HOOF [2]. We show that when K is changing, our motion descriptor is more effective in extracting accurate local motion from the noisy optical flow. We demonstrate that our approach achieves state of the art discriminative performance.

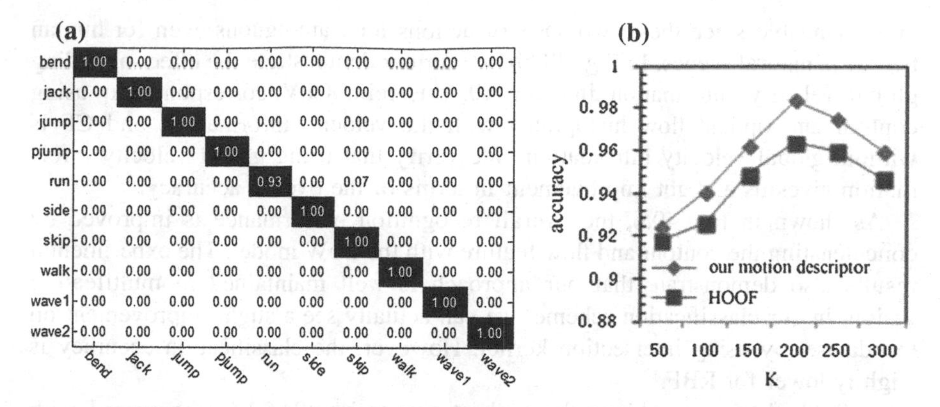

Fig. 40.4 **a** Confusion matrix for the Weizmann dataset; *rows* are ground truth, and columns are model results. The action models using 200 codewords show an average performance of 99.3 %. **b** Classification accuracy obtained using our motion descriptor versus the HOOF [2]

40.4.2 Performance on KTH Dataset

KTH database is a realistic and challenging database, including inter-person variations due to different physical bodies and different motion styles. It contains six kinds of human actions, including walking, jogging, running, boxing, waving, and clapping. Such that a video can be represented as a collection of visual words from the codebook. The effect of the codebook size K is explored in our experiments. In Fig. 40.5b, our results show that $K = 200$ gives the best classification accuracy. The confusion matrix is shown in Fig. 40.5a, which shows that all the classes are classified with satisfying precision except the pair of jogging and running. The confusion occurs between jogging and running actions is quite

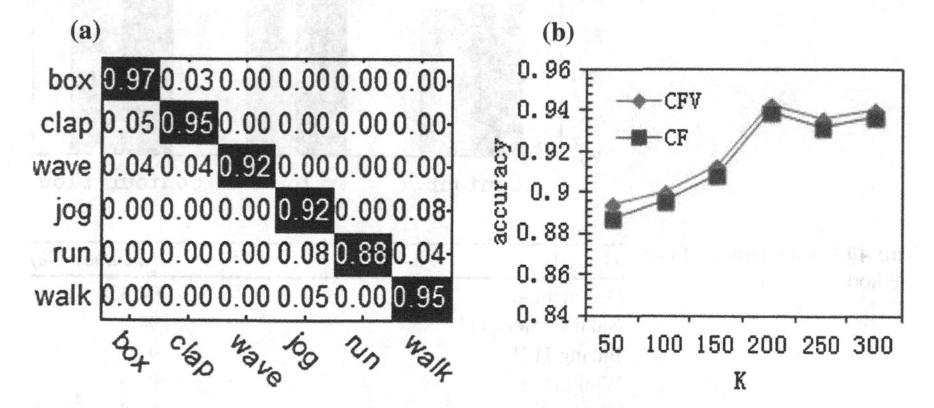

Fig. 40.5 **a** Confusion matrix for the KTH dataset using 200 codewords (performance average = 94.32 %); **b** Classification accuracy versus codebook and including global velocity information

understandable since these two sorts of actions look ambiguous even for human beings in natural scenes. In Fig. 40.5b, we further demonstrate the effect of adding global velocity information in Sect. 40.3. Herein, CFV corresponds to using contour and optical flow histograms with the velocity information, and CF is without global velocity information. We verify that using global velocity information gives us a slight improvement in terms of the overall accuracy.

As shown in Fig. 40.6, the overall recognition performance is improved by concatenating the contour and flow feature with the BoW model. The experimental results also demonstrate that our approach is well maintained in multifeature fusion. In our classification scheme, we can actually see a slight improvement on the dataset by using intersection kernel. However, the classification accuracy is slightly lower for RBF.

In Table 40.1, we achieve the highest accuracies (94.32 %) compared with other state-of-the-art results which are reported on this dataset. It demonstrates that the performance of our approach discriminates action instances successfully. In addition, compare to other encouraging results, our feature can be extracted and characterizated easily, without losing the high reliability. It can avoid complicated operation on feature selection based on 3D body posture model. Consequently, it greatly reduces the computational complexity for training the classifier.

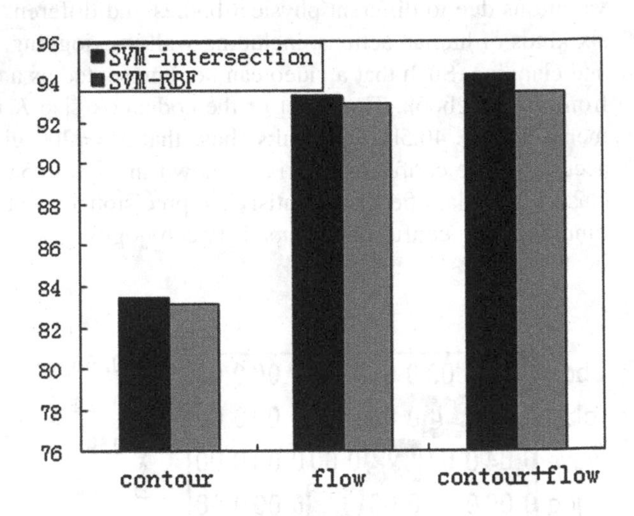

Fig. 40.6 Comparison of different feature and SVM kernel

Tab 40.1 Comparison of our method to others

Method	Accuracy (%)
Our method	94.32
Nazli Ikizler [11]	94.0
Jhuang [12]	91.7
Wong [13]	91.6
Niebles [6]	81.5
Doll'ar [9]	81.2

40.5 Conclusion

This paper presents a compact and novel representation for human action recognition using Block-based model and optical flow histograms. By adopting this novel representation, we reduce computational complexity substantially. Within this framework, the performance of the method depends on a reasonable background subtraction technique, and is shown to be comparable with other state-of-the-art methods on KTH and Weizmann datasets. Experimental results prove that our novel contour and motion descriptor is quite successful in action recognition. In future work, we will investigate more effective features and utilize more complicated classification schemes. It is also necessary to further investigate and justify the mechanism of feature fusion, so that our method can be extended to detect more complex human actions.

Acknowledgments This work is funded by National Science Foundation of China (61175009).

References

1. Folgado E, Rincón M, Carmona EJ et al (2011) A block-based model for monitoring of human activity. Neurocomputing 74(8):1283–1289
2. Chaudhry R et al (2009) Histograms of oriented optical flow and binet-cauchy kernels on nonlinear dynamical systems for the recognition of human actions. In: IEEE conference on computer vision and pattern recognition CVPR, June 2009
3. Laptev I (2005) On space-time interest points. Int J Comput Vis 64(2-3):107–123
4. Blank M et al (2005) Actions as space-time shapes. In: 10th IEEE international conference on computer vision ICCV, vol 2
5. Dollár P et al (2005) Behavior recognition via sparse spatio-temporal features. In: 2nd Joint IEEE international workshop on visual surveillance and performance evaluation of tracking and surveillance, October 2005, pp 65–72
6. Niebles JC, Wang H, Fei–Fei L et al (2006) Unsupervised learning of human action categories using spatial-temporal words. In: British machine vision conference BMVC, 2006
7. Maji SB, Berg AC, Malik J (2008) Classification using intersection kernel support vector machines is efficient. In: IEEE conference on computer vision and pattern recognition, 2008, pp 1–8
8. Du T, Sorokin A (2008) Human activity recognition with metric learning. In: European conference on computer vision, 2008, pp 548–561
9. Dollar P, Rabaud V, Cottrell G, Belongie S (2005) Behavior recognition via sparse spatio-temporal features. In: IEEE international workshop on visual surveillance and performance evaluation of tracking and surveillance, 2005, pp 1–8
10. Yang J, Yu K, Gong Y, Huang T (2009) Linear spatial pyramid matching using sparse coding for image classification. In: IEEE conference on computer vision and pattern recognition, 2009
11. Ikizler N, Cinbis RG, Duygulu P (2008) Human action recognition with line and flow histograms. In: 19th IEEE international conference on pattern recognition, ICPR, 2008, pp 1–4

12. Jhuang H, Serre T, Wolf L, Poggio T (2007) A biologically inspired system for action recognition. In: 11th IEEE international conference on computer vision ICCV, 2007
13. Wong S-F, Kim T-K, Cipolla R (2007) Learning motion categories using both semantic and structural information. In: IEEE conference on computer vision and pattern recognition (CVPR 2007), Minneapolis, USA, June 2007

Printed in the United States
By Bookmasters